SEASHELLS OF EASTERN ARABIA

SEASHELLS
OF EASTERN ARABIA

DONALD T BOSCH
S PETER DANCE
ROBERT G MOOLENBEEK
P GRAHAM OLIVER

EDITED BY S PETER DANCE

WITH COLOUR PHOTOGRAPHS OF THE
SPECIES BY NEIL FLETCHER

MOTIVATE
PUBLISHING

Published by Motivate Publishing

Dubai: PO Box 2331, Dubai, UAE

Abu Dhabi: PO Box 43072, Abu Dhabi, UAE

London: London House, 19 Old Court Place, Kensington High Street
London W8 4PL

First published 1995

© 1995 Donald T Bosch, S Peter Dance,
Robert G Moolenbeek, P Graham Oliver,
and Motivate Publishing

All rights reserved. No part of this publication may be reproduced in any material form (including photocopying or storing in any medium by electronic means) without the written permission of the copyright holder. Applications for the copyright holder's written permission to reproduce any part of this publication should be addressed to the publishers. In accordance with the International Copyright Act 1956 or the UAE Federal Copyright Law No 40 of 1992, any person acting in contravention of this copyright will be liable to criminal prosecution and civil claims for damages.

ISBN 1 873544 64 2

British Library Cataloguing-in-Publication Data. A catalogue record for this book is available from the British Library.

Printed by Emirates Printing Press, Dubai, UAE

Dedication

To the memory of
the late Kathie Smythe,
enthusiastic pioneer of
eastern Arabian malacology.

Contents

INTRODUCTION

    FOREWORD 9

    OUR PATRONS 11

    THE GENESIS OF THIS BOOK AND ITS PURPOSE 12

    HISTORY OF SHELL COLLECTING AND MALACOLOGY IN EASTERN ARABIA 13

    EASTERN ARABIA AND FACTORS INFLUENCING ITS MOLLUSCAN FAUNA 18

    ARABIC PLACE NAMES USED IN EASTERN ARABIA 22

    MAP OF EASTERN ARABIA 23

SYSTEMATIC DESCRIPTIONS

    GASTROPODS (GASTROPODA) 24

    TUSK SHELLS (SCAPHOPODA) 186

    CHITONS (POLYPLACOPHORA) 188

    CEPHALOPODS (CEPHALOPODA) 191

    BIVALVES (BIVALVIA) BY P.G. OLIVER* 194

GLOSSARY 282

REFERENCES 286

PHOTOGRAPHIC CREDITS 289

INDEX 290

ACKNOWLEDGEMENTS 296

* This section may be referred to as: P.G. Oliver, Bivalvia *in* S.P. Dance (Ed.) *Seashells of Eastern Arabia*.

Foreword

In recent years, mankind has become increasingly aware of the biodiversity in nature and of the extinction of many plants and animal species. It follows that current studies of molluscs could be very important. Some of the rarer living species of molluscs inhabit the relatively unknown coastal waters covered by this book, namely, those of Oman and the Arabian Gulf. Following the exploitation of oil, interested scientists from ARAMCO (Arabian American Oil Co) and others, included the investigation of molluscan life in their studies and publications.

In 1982, Kathleen Smythe, working with Michael Gallagher and the Oman Natural History Museum, brought out a small handbook, *Seashells of the Arabian Gulf* (George Allen and Unwin Ltd). Meanwhile, in Oman, Donald T Bosch and his wife Eloise had been studying and collecting molluscs, aided by the American Museum of Natural History in New York and Zoological Museum of Amsterdam in the Netherlands. These efforts resulted in two publications, *Seashells of Oman* (Longmans) in 1982 and *Seashells of Southern Arabia* (Motivate) in 1989.

The ground had now been laid for a much broader study and one which required the knowledge and expertise of professional malacologists. Dr Bosch, supported by many Omani friends, was pleased to discover another interested individual, Mr S Peter Dance of Carlisle, UK, who had long been interested in the molluscs of the region. Later, Dr P Graham Oliver of the National Museum of Wales, author of *Bivalved Seashells of the Red Sea* (Hemmen, 1992) and Robert G. Moolenbeek, a micromollusc specialist of the Zoological Museum in Amsterdam, were recruited. A substantial book was now a possibility.

Numerous expeditions, involving some or all of the four authors, have been conducted, aided in part by courtesy travel from British Airways and KLM. Many amateur collectors have assisted the authors in their exploration of hundreds of sites along the shoreline from Kuwait to the Yemeni border.

With the publication of this well-illustrated book, our understanding of the marine malacofauna of eastern Arabia will be greatly enhanced. The authors and their many sponsors and helpers are to be congratulated for having brought a difficult project to so successful a conclusion.

Dr Henry E Coomans
Chairman
Department of Malacology
Zoological Museum of the
University of Amsterdam

Our Patrons

This book, the realisation of an ambitious idea, could not have been prepared, much less published, without the financial support of numerous friends and organisations. That support, given unconditionally and in advance, has been munificent. At an early stage we were honoured to be given substantial help by His Highness Sayyid Thuwainy bin Shehab, His Highness Sayyid Fahar bin Taimur, His Highness Sayyid Faisal bin Ali, His Highness Sayyid Shabib bin Taimur, His Highness Sayyid Kais bin Tariq and His Highness Sayyid Shahab bin Tariq.

At the same time we were delighted to receive donations from His Excellency Dr Omar bin Abdul Muneim Al Zawawi (and the companies with which he is associated), His Excellency Qais bin Abdul Muneim Al Zawawi, His Excellency Sayyid Hamid bin Hamoud and His Excellency Abdul Hafith bin Salim bin Rajab.

Welcome too was an early and significant contribution from Petroleum Development Oman (PDO).

Later, other friends and organisations backed the project generously. We are pleased to acknowledge them here, in alphabetical order:

Mr Abdulla Ali Al Araimi, Mr Abdulla Moosa Abdulrahman, Al Harthy Complex, British Airways (for courtesy travel), British Bank of the Middle East, Comprimo Oman, Dowell Schlumberger (Western) SA, Halliburton Worldwide Limited, International Petroleum Corporation, Japex Oman Limited, Mr and Mrs Kamal Sultan, Khimji Ramdas Company, KLM Airlines (for courtesy travel), The Linnean Society of London (administrators of the Percy Sladen Trust Fund), for a grant toward travel expenses, Mohammed Al Barwani Petroleum Services LLC, Dr and Mrs Mohammed Sultan, Mr Mohsin Haidar Darwish, Muna Noor Incorporated, Occidental of Oman Limited, Oman Deutag Drilling Company LLC, Oman Refinery Company LLC, Oman United Agencies LLC, Schlumberger Overseas SA, Sea and Land Drilling Contractors Inc, Mr and Mrs Shawqi Sultan, Tawoos LLC, WJ Towell and Company LLC and Mr Yahya Mohammed Nasib.

We cannot thank all these patrons enough for their generosity and encouragement, but we can assure them that their influence is manifest on every page of the book.

Donald T Bosch
S Peter Dance
Robert G Moolenbeek
P Graham Oliver

The Genesis of This Book and Its Purpose

When not fulfilling the often demanding duties of a surgeon in Muscat Donald Bosch would go down to the sea and collect shells. His wife Eloise and their children would often accompany him and, in time, their collecting trips would take them farther afield, even unto the then mysterious shores of Masirah Island. It was as if they were explorers in an uncharted territory where rare treasures lurked under rocks, buried in the sand or cast up on the beach. Soon these family activities began to have an impact in scientific circles. Some of the shells they collected were new to science, others long-lost species rediscovered. The name of Bosch, often incorporated into the scientific name of a new species, became familiar to shell enthusiasts the world over. Don and Eloise wrote an illustrated book, published in 1982 as *Seashells of Oman*. Seven years later another illustrated book appeared under their names, *Seashells of Southern Arabia*. Since arriving in Oman in 1954 they had made their mark, first in medicine and education, then in malacology. But it was not enough to have brought out two books. Don knew there was room for something much more substantial.

He first proposed the idea of a comprehensive treatment of the seashells of eastern Arabia to the late Kathie Smythe, the scientific editor of *Seashells of Oman* and leading authority on the molluscs of the region. Kathie took up the challenge enthusiastically, but the poor state of her health hindered — and ultimately prevented — her from making real progress. So Don approached three other students of molluscs: Peter Dance, Robert Moolenbeek and Graham Oliver, the first already known for several books about shells, the second for his work on microshells and many published articles, the third for his work on bivalves and authorship of the authoritative *Bivalved Seashells of the Red Sea* (1992). They agreed to collaborate with him on a book which would describe and illustrate a significant part of the shelled molluscan fauna of Oman and the Arabian Gulf. He agreed to take care of the financial side of the project and help to arrange trips to Oman and the Arabian Gulf.

Now, after four or five years of intermittent and sometimes exhausting work, the book is finished. If it looks good it is largely because of the many excellent colour photos taken by Neil Fletcher which illustrate it. If it does what it sets out to do it does so because the authors have been able to work together harmoniously to achieve a commonly agreed objective over a long period. That it exists at all is, of course, due to the generosity and goodwill of our patrons.

So what was the objective? Briefly, it was to provide a well-illustrated guide to most of the species of shell-bearing Mollusca living around the coasts of Oman and the Arabian Gulf, including

Don Bosch and Peter Dance collecting among loose rocks, Raysut, Oman.

Robert Moolenbeek sieving for shells in a *khor* (lagoon) at Juzor Al Halaaniyaat (Kuria Muria Islands).

some occurring only in deeper water. Many of the shells found in this region are tiny or minute and belong to species about which we know very little. So it has not been possible to describe and illustrate more than a fraction of the microshell fauna. The status of many of the larger species, too, is still unclear which is why the symbol "cf" (*confer* or compare) is often associated with a specific name. As far as possible we have not used technical expressions but some are unavoidable in this kind of book. The substantial Glossary defines most of those we use.

Most of all we have tried to produce a book which shall satisfy the amateur shell collector and the professional biologist alike, aware that we may blind the first with science and annoy the second with over-simplification. Adopting this middle course has often been difficult because we have had to rely for much of our information on the work of specialists who tend to develop a specialised language, but we believe the attempt has been worth while.

It may be worth saying something about the role played by each of the authors. The collections and notes made by Donald Bosch over the years, principally in Oman and the island of Masirah, provided the basic material for research. This was supplemented by material from many other collections, including those of several major museums. Robert Moolenbeek did the original outline research for all gastropod groups, later providing detailed information about most of them, instigated the publication of a series of articles intended primarily to give descriptions of many of the new or little-known species recognised during his researches, and supplied most of the scanning-electron micrographs of minute species which appear in this book. Graham Oliver was responsible for the entire Bivalvia section, including the preparation of shells for photography, and this section rightly bears his name as sole author. He also contributed many items to the Glossary and provided useful notes about the biogeography of eastern Arabia. Apart from the place names appearing on pages 22 and 23, which were provided by Don Bosch, Peter Dance was responsible for writing the rest of the book and for editing the whole. He wishes to place on record his indebtedness to John Baxter, however, who provided the text for the chiton section which is printed here virtually unaltered, and Phil Palmer who provided most of the information for the scaphopod text. Peter also prepared all the non-bivalve shells for photography and made all the drawings of scaphopods. The other line illustrations, except those in the introduction to the Bivalvia, which are reproduced here by permission of Graham Oliver, and those illustrating parts of gastropod shells on pages 26 and 27, which are by Mathilde Duffy, are virtually all reproductions of those accompanying the many publications of Melvill and Standen. Whatever errors may occur in the book are also likely to have been his work and for these he offers his apologies in advance. And this is as good a place as any to mention that he has followed Kay Vaught's *A classification of living Mollusca* (1989) in the arrangement of the systematic part of the book.

N.B. For all shells of Bivalvia maximum dimensions are given (in millimetres). For shells of all other groups average dimensions are given.

History of Shell Collecting and Malacology in Eastern Arabia

Molluscs and their shells have played an important part in the economies of eastern Arabian territories for thousands of years. Oysters and abalones, for instance, have been and still are collected for human consumption; and the pearling industry was once significant in the Arabian Gulf. But the collecting of molluscs and their shells for the sake of admiring and studying them may have begun not much earlier than 1840. Some eastern Arabian seashells would have found their way into European cabinets of curiosities before then, through the casual collecting activities of travellers and seamen, but few authentically documented examples have survived. One species, for long among the great rarities of the shell world, may have been first collected from Masirah Island or the neighbouring Oman coast during the first half of the nineteenth century. For this island is now known to be the principal haunt of *Cypraea teulerei*, described in 1845 from the Red Sea port of Mocha (where it does not occur). The original specimens of this distinctive cowry may have originated from Masirah.

The earliest published evidence of shell collecting in eastern Arabia dates from 1865 in an article by the Italian malacologist Arturo Issel. This gives an account of 17 species collected at Bundar Abbas and Hormuz in the Arabian Gulf by G Doria and R A Philippi. Then, in 1872, Edgar A Smith, zoologist at the British Museum in London, published descriptions of some small opisthobranch shells, new to science, collected by Lewis Pelly, Political Resident at Bushehr, towards the north-eastern end of the Arabian Gulf. Subsequently Smith described other supposedly new species, particularly Terebridae, collected by Pelly at or near Bushehr. In 1874 Eduard von Martens, malacologist at the Berlin Zoological Museum, listed 119 species collected at Bushehr by Dr Haussknecht, including some new to science. In 1891 the French malacologist Paul Fischer produced a list of 33 species collected in the Arabian Gulf by F. Houssay, of which 14 were then newly reported. Up to that time about two hundred species had been listed, but the waters of eastern Arabia had been scarcely ruffled in the search for molluscs, the Gulf of Oman being virtually a *terra incognita* for them then. Soon that situation was to change dramatically.

In 1895, in a brief article, G B Sowerby (3rd of the name) described and illustrated two new species from the 'Persian Gulf'. He christened one *Donax townsendi*. Later that year he published two similar articles, the title of the second one mentioning F W Townsend as the collector. Thus did Frederick W Townsend walk unostentatiously onto the stage of eastern Arabian malacology; he was to become an important actor on that particular stage. Shortly afterwards another actor, scarcely less important, joined him. In 1897 James Cosmo Melvill, an influential and wealthy English malacologist, produced the first of a long series of articles describing a host of molluscan novelties collected by Townsend. In it he described 34 species as new to science collected at various localities in the Arabian Gulf and along the Mekran coast as far as Karachi. At least one came from debris brought up on a submarine telegraph cable.

This cable (actually there was more than one) was the source of hundreds of remarkable shells obtained by Townsend and deserves special mention here. From 1890 to 1914 Townsend worked for the Indo-European Telegraph Department, the organisation responsible for the submarine cable laid down from Karachi, via Hormuz and Bushehr, to the entrance of the Shatt al Arab at the north-western corner of the Arabian Gulf. Among his duties on board the steamer "Patrick Stewart", of which vessel he became the captain, was the periodic taking up and cleaning of such cables. He

F W Townsend, 1897. Below, a sample of his handwriting and his signature, from a letter written by him to J C Melvill in 1898.

obtained thousands of shells from them, including many new to science. Together with those he obtained by dredging they represented a collection of such outstanding scientific importance that it transformed our knowledge of the eastern Arabian molluscan fauna. An article he published in 1928 shows the time and effort he put into his own collecting activities and makes fascinating reading.

Over several years Townsend sent packets of fully documented shells to Melvill who, on his own or in collaboration with Robert Standen, curator of zoology at the Manchester Museum, studied and wrote reports upon them, most notably in two summarising articles published in 1901 and 1907. The last of a seemingly endless succession of these well-illustrated and carefully written reports, correcting and amplifying the previous ones, appeared in 1928. Perhaps not less than six hundred supposed new species and varieties were described altogether, a remarkable achievement for that period. Fortunately most of Townsend's shells are preserved in three institutions in the United Kingdom: The Natural History Museum in London, the National Museum of Wales at Cardiff, and the Manchester Museum at Manchester University. The Townsend material in these three institutions has provided the bedrock for the present study.

Before discussing the contributions of others it is worth highlighting a seemingly insignificant event which was to have a considerable impact on the outcome of Townsend's collecting activities. On 7 April 1903 his ship was anchored well out at sea in the Gulf of Oman, about 24° 58 N, 56° 54 E. During the night the anchor dragged for some distance and when drawn up, from the considerable depth of 156 fathoms, Townsend saw that a sizeable ball of muddy sand was stuck to it. He retrieved the ball, dried it out and eventually handed it over to his scientific colleagues in England. That fortuitously obtained piece of sea bed kept specialists busy for many months. Trapped in it were countless small and minute shells, among them representatives of two hundred or more species new to science, many of them not found since. The locality "156 fathoms off Muscat" is one of the most productive in the annals of systematic malacology and all because someone had had the foresight to scrape a ball of muddy sand off a ship's anchor.

Townsend did little shore work and so his collections lacked many common intertidal species. That left some easy pickings for later collectors, but these took a long time to make their impact.

The Rev H E J Biggs, an English missionary, was among the first of them. He collected sporadically at various places in the Arabian Gulf from about 1922 to 1935 but did not publish any of his findings until 1958. Other articles followed, the most comprehensive appearing in 1973 and dealing with the molluscs collected by a team of geomorphologists along the coast of what is now the United Arab Emirates. Among other articles by Biggs was one published in 1969 on shells from Masirah Island, the precursor of many articles which other workers would publish highlighting that island's remarkable molluscan fauna. Biggs listed many species, some of them described as new to science, additional to those in the already impressively long list recorded by Melvill and Standen.

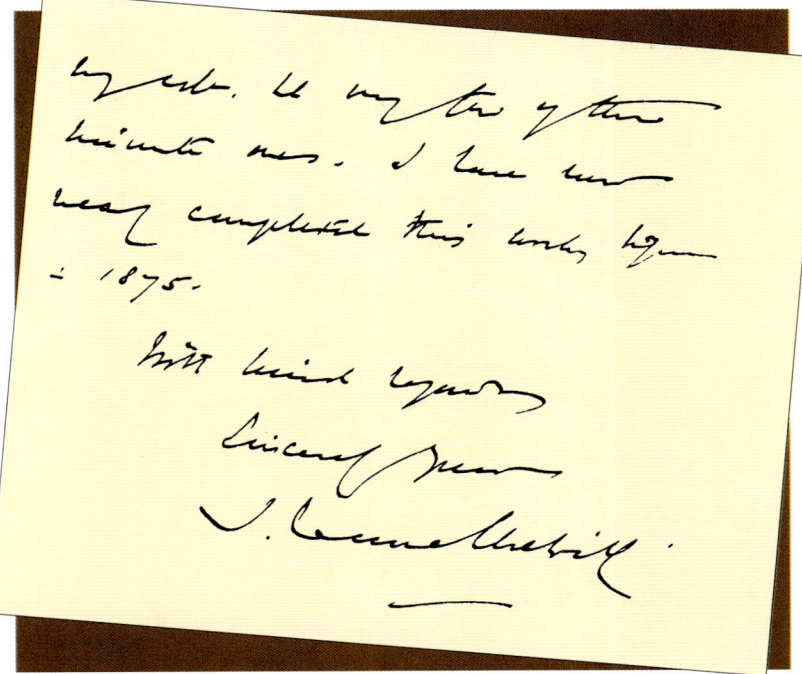

J C Melvill, about 1915, and (below) a sample of his handwriting and his signature, from a letter to R Winckworth.

In 1950 the Peabody Museum Harvard Expedition to the Near East collected some molluscs from western coasts of the Arabian Gulf and these were the subjects of a short and uncritical list published in 1952 by the German malacologist Fritz Haas, then at the Field Museum of Natural History at Chicago. Two years later, in an article listing some shells from Dubai, he made the significant comment, "It is strange that we should know comparatively more about the deepwater shells, obtainable only through dredging operations, than about the shore fauna, which can be collected without the use of instruments; nevertheless, this is the case".

The next contribution to our knowledge of the molluscan fauna of eastern Arabia came from test borings put down by the Iraq Petroleum Company near Basrah in Iraq. These borings, none more than 18 metres deep, produced an abundance of mainly small gastropod and bivalve shells. Some of these were described as new to science, including the aptly named *Abra cadabra* — the specimens named, after all, had been a long time dead — which also occurs in the Arabian Gulf and appears in this book, regrettably perhaps, under the less arresting name *Theora cadabra*.

The results of further borings in south-eastern Iraq were published ten years later when some more new species were described, several having occurred alive since in eastern Arabian waters.

About 1971 Major M D Gallagher, then with the British army and later well known for his collecting activities in Oman and elsewhere in eastern Arabia, began collecting molluscs along the coast of Bahrain, helped by friends. The resulting collections were written up in 1972 by Kathleen R Smythe in the first of several articles she would devote to eastern Arabian molluscs. Later this energetic lady involved herself closely with the molluscan fauna of eastern Arabia, producing a series of articles and two small books. Her published surveys of the molluscs of the United Arab Emirates and of Kuwait, the latter in collaboration with B A and D T Glayzer, have been particularly useful to the authors of this book. In her other published articles she described several new species, most of them from Masirah or the coast of Oman. Some of the coloured photos of shells reproduced in this book are of shells from her collection which is now housed in the Natural History Museum, London.

Studies on the molluscs of the Iranian side of the Arabian Gulf had languished since Townsend's day, but M Tadjalli-Pour, an Iranian student, chose to make a prolonged survey of the distribution of molluscs along the Iranian coast in the 1960s and early 1970s. Presented as a doctoral thesis in 1974, his survey involved the identification of 17,000 specimens weighing 80 kilos — a late holiday added more specimens — and a comparison with the work of his predecessors. Mention should also be made here of the work of Mahmoud M Ahmed, a biologist at the University of Basrah, who studied two collections of shells from the north-western coastal region of the Arabian Gulf and published a fully illustrated account of his findings in 1975. It is useful as an indicator of which molluscs occur off the Iraqi coast.

Meanwhile, Donald Bosch, helped substantially by members of his family, had been collecting shells along the coasts of Oman and around Masirah Island since the 1960s. His efforts were rewarded at an early stage by the discovery, at Masirah, of some striking new species. *Cymatium boschi* was described in print in 1970. This was followed, in 1973, by an article describing *Acteon* (now *Punctacteon*) *eloiseae*, perhaps the best known of all eastern Arabian shells by virtue of its exotic colour pattern. The list of species added to the eastern Arabian fauna by him and his family is impressive and includes several outstanding for their beauty and rarity, nearly all picked up in shallow water without the aid of expensive collecting equipment. With his wife, Eloise, he is also the author of two well-illustrated books about the seashells of Oman, published in 1982 and 1989 respectively. In recent years Harry Henseler and Martyn Day have made valuable collections at Masirah; Steve Green has collected extensively at Bahrain; and Horst Kauch of Dubai has collected many shells, often in hazardous circumstances, from around the coasts of the United Arab Emirates. Many of the colour photos in this book portray shells acquired over the years by these industrious collectors. While working on this book the authors themselves have collected assiduously

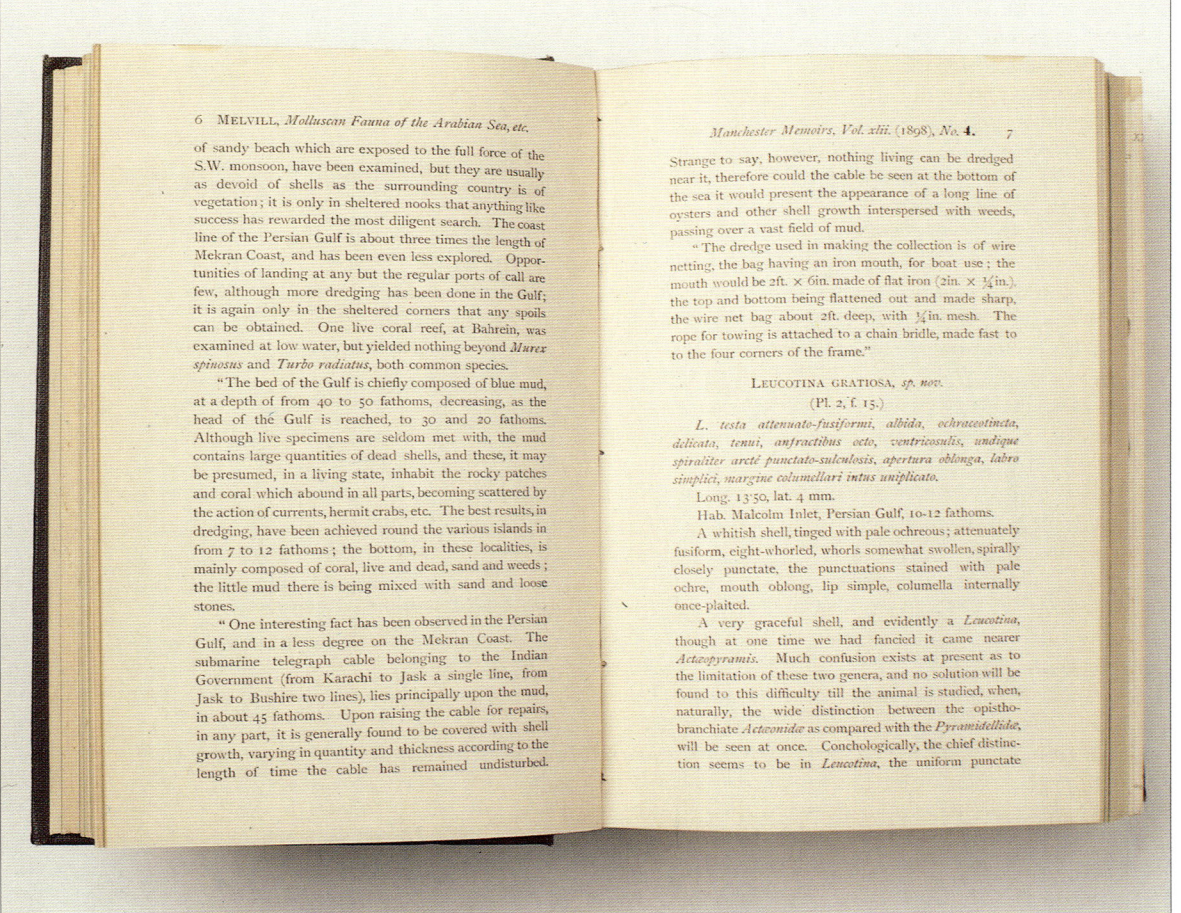

Pages from *Memoirs and Proceedings of the Manchester Literary and Philosophical Society* vol. 42. 1898, in which J C Melvill describes shells collected by F W Townsend in eastern Arabia.

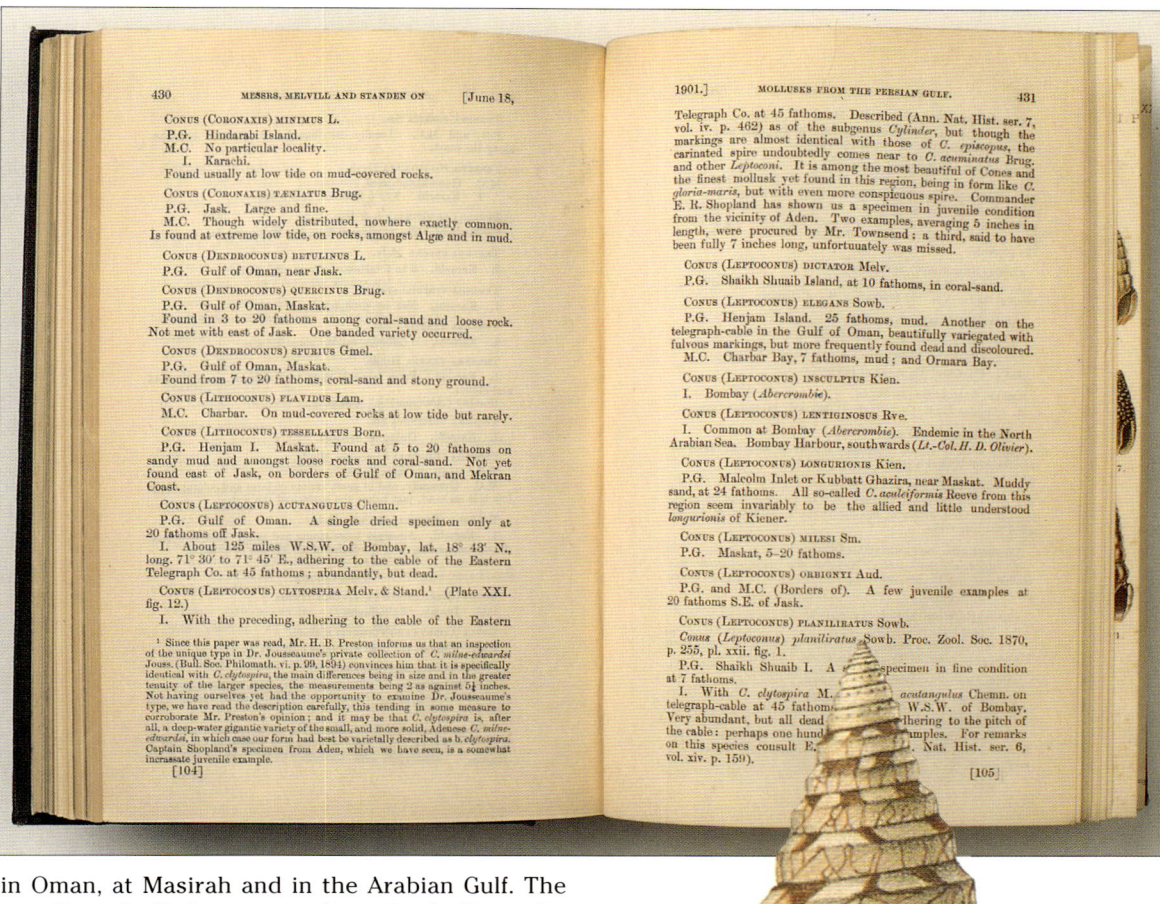

Pages from *Proceedings of the Zoological Society of London* for 1901, in which J C Melvill and R Standen refer to a cone shell they had described in 1899 under the name *Conus clytospira* (now known as *Conus milneedwardsi*). The colour picture of one of the two specimens collected by F W Townsend accompanied their 1901 article.

in Oman, at Masirah and in the Arabian Gulf. The results of their researches, including the descriptions of many new species and genera, have already begun to appear in scientific journals.

Many more species are known from eastern Arabia now than were known to Melvill and his collaborators, but it is certain that many others remain unknown to us still. In 1904, when describing some of the small shells brought up on the ship's anchor from 156 fathoms, Melvill said it was his belief that Townsend's explorations had thus far "only touched the threshold of benthal life, and that these scarcely known seas possess within their profounder recesses many wonderful forms, waiting to be revealed some day". There is little doubt that Melvill's statement will prove to be prophetic. But, as the Bosch family and others have shown, even the casual shore-bound collector may expect to make discoveries, exciting discoveries which may become part of the history of shell collecting in eastern Arabia.

References for this section only: Abbott, 1973; Abbott & Lewis, 1970; Ahmed, 1975; Biggs, 1958b, 1969, 1973; Bosch & Bosch 1982, 1989; Dance & Eames, 1966; Eames & Wilkins, 1957; Fischer, 1891; Glayzer, Glayzer & Smythe, 1984; Haas, 1952, 1954; Harris, 1969; Hudson, Eames & Wilkins, 1957; Issel, 1865; Martens, 1874; Melvill, 1897b, 1904c, 1928; Melvill & Standen, 1901, 1907; Smith, 1872, 1877; Sowerby, 1895a, 1895b, 1895c; Smythe, 1972, 1979a, 1982, 1983; Tadjalli-Pour, 1974; Townsend, 1928.

Eastern Arabia and Factors Influencing its Molluscan Fauna

Two distinct marine areas are covered by this book:
 a) the entire Arabian Gulf and
 b) the Gulf of Oman and the Arabian Sea, bordered by the coasts of Oman and the eastern side of the United Arab Emirates, from Hormuz to the Yemen border.

The two areas, labelled here for convenience the eastern Arabian region, form part of the Arabian sub-unit of the Indian Ocean faunal province. They differ from each other in several ways and the differences are reflected in the composition of their respective molluscan faunas, although many species are common to both.

Before discussing these two areas as they are now it is worth examining some aspects of the geological history of the Arabian peninsula and the surrounding area. At one time Africa and Arabia were joined but over 100 million years ago the earth's crust began to break. As this rift widened it separated Arabia from Africa. Simultaneously this shifting of the earth's crust pushed up the nearby ocean bed in a westerly direction and over what became the land mass of Arabia. This is why it is now possible to find fossil seashells in the mountains of Oman (and on the highest part of Masirah). During the next 50 million years, as Arabia continued to separate from Africa, it moved in a north-easterly direction towards the land mass of Asia. During this process the large body of water separating Arabia from Asia and known as the Tethys Sea became narrower.

For about 20 million years the Tethys Sea, though narrower, still connected what is now the Mediterranean Sea to what is now the Gulf of Oman. The rifting which created the Red Sea continued the north-eastern movement of Arabia and so completed the separation of Arabia from Africa. The Tethys Sea closed to the north during the next 10 million years and the Mediterranean became separated from the Arabian Gulf. Meanwhile the Red Sea continued to widen and became deeper. Thus was created the situation as we see it today: a deep Red Sea, a deep Gulf of Oman, and a shallow Arabian Gulf which is closing up gradually.

It seems likely from this sequence of events that the molluscan faunas of the Mediterranean and the Gulf of Oman could be similar. This is not the place to look for proof of this; suffice it to say that there are shells which seem to provide it. Conversely, it may be expected that the molluscan fauna of the deep Red Sea would not correspond closely to that of the shallow Arabian Gulf. Today a collection made in the first would differ considerably from one made in the second. These geological observations and speculations provide a backdrop for our examination of the two marine areas of eastern Arabia.

The Arabian Gulf is about 800 kms long and shallow, with an average depth of only 35m and

Fossil Eocene bivalves on Jebel Hamrah, Masirah Island.

The broad beaches of Masirah Island.

nowhere exceeding 100m. Because it is shallow its water temperature is higher, sometimes much higher, than that of the Gulf of Oman, in places averaging about 38°C (100°F) in summer. In some Arabian Gulf biotopes, indeed, temperatures fluctuate far more than anywhere else in the world. The water also has a high salinity content, compounded by the warm temperature, high evaporation and the paucity of fresh water flowing into it. This has a marked effect on the Gulf fauna and flora. Some molluscan species, for instance, may not grow as large there as they do elsewhere in their range where salinity and temperature are generally lower, but more normal growth seems to be the rule towards the exit of the Gulf and in the waters of Kuwait which benefit from the ingress of fresh water through the Shatt al Arab. Broadly speaking, although there is a wealth of suitable biotopes in the Arabian Gulf, life is hard for many organisms. Molluscs may be plentiful but they are less diverse there than elsewhere in eastern Arabia. For instance, few species of cowries and cones live there but along the Oman coast these two groups are well represented.

Water currents flow inwards from the Gulf of Oman through the Strait of Hormuz bringing free-swimming molluscan larvae from the eastern Indian Ocean. The Musandam peninsula, with its deep, fjord-like inlets and unusual coral communities, provides a congenial biotope for them and for other molluscs which lack a free-swimming, developmental stage. Shell collecting around that peninsula is a rewarding experience. Nowhere else in the region, for instance, is the large *Chicoreus ramosus* so plentiful and so easily collected. Rewarding, too, is the rocky coast of Ras al Khaimah, inside the Gulf north of Dubai. Having entered the Strait of Hormuz the water current flows along the coast of Iran, bends westwards past the Shatt al Arab and turns south towards Bahrain. Changing salinities experienced along this route by incoming larvae may explain why many molluscan species are absent from western shores in the Gulf. Neither are such conditions favourable for molluscs. It is possible to make a large, representative collection of shells in the Arabian Gulf but it is essential to find the right conditions first.

The deeper and cooler waters of the Gulf of Oman and the Arabian Sea provide much better conditions for the development of molluscan life along the coast of Oman and its few islands, most notably the island of Masirah. In some places, especially along the Dhofar coast of Oman, those conditions are influenced dramatically by what is known as the south Arabian upwelling. This water movement is caused by the onset of the south-west monsoon which drives surface water away from the Arabian peninsula. The surface water is then replaced by cold water upwelling from considerable depths. Upwelling water is only 16 – 17°C, bringing temperate conditions during the summer months. Paddling or swimming in water that cold is an uncomfortable experience, especially when the air temperature is so high. Molluscs and other organisms benefit from the rise in nutrients and food productivity created by this phenomenon, especially where subsequent rapid warming helps maintain coral communities.

The most conspicuous result of the cold upwelling is the existence of endemic species where its influence is felt. The presence of endemic molluscs and other organisms along the coast of Dhofar and around Masirah Island has attracted the attention of many biologists. It played a large part in attracting the authors of this book to take up the study of eastern Arabian molluscs and will continue to be a focus of scientific attention in the foreseeable future.

HABITATS FOR MOLLUSCS IN EASTERN ARABIA

From the air the coastal fringes of eastern Arabia look mostly barren and forbidding. They seem to offer few opportunities for molluscs to gain a foothold and flourish. But tell-tale signs here and there show that conditions may be propitious for the development of molluscan life. Dark patches show the presence of grass beds offshore. Patches of coral reef may be seen occasionally but seldom are they more than a few hundred metres across. Green areas hugging the shore, clearly denoting vegetation which is at least partially above water, show the presence of mangroves and tidal flats. Coastal rocks, boulders and cliffs also often break the monotony of the sandy fringes. These different features show that what at first seemed a monotonous and unpromising environment for molluscs and other forms of life may, on a closer inspection, be varied and bountiful for them.

The exposed sandy shore, so characteristic of eastern Arabian coasts, is the haunt of burrowing bivalves and gastropods. It is often littered with the shells of sand-burrowing naticids, or moon snails, and the dismembered valves of tellins and other sand-burrowing bivalves. Unless varied with mats of eel grass and other vegetation it is seldom rich in species but it may abound with specimens.

The rocky shore is very different. To survive among wave-battered rocks, large boulders and cliff faces a bivalve must be cemented to them, attached to them by a bunch of byssus threads, wedged into a cranny or a hole, or ensconced in a tunnel it has bored into them. Its empty shell does not commonly survive intact. The rocky shore is a place for oysters and mussels, which cling to boulders and rocks, and for rock-boring date mussels. It is not a place for tellins and other sand burrowers. In eastern Arabia it is pre-eminently a place for chitons, limpets, nerites, top shells, muricids and oysters.

Intertidal flats, muddy and sandy by turns, often variegated with rocks, stones and seaweeds and protected from violent wave action, provide some of the best conditions for molluscs to flourish. Here you may find cowries, cones, mitres, turrids, bubble shells, pen shells, cockles and scallops all living in close proximity; even the most delicate shells may survive intact. Many rare and attractive eastern Arabian species have been found in such places.

Mangroves may develop on intertidal flats and form a definite zone having its own distinctive fauna. Fine-grained mud is trapped by the extensive mangrove-root system which helps to extend the flat seawards. Here is the place to look for bivalves,

Red rocks on the coastline of Juzor Al Halaaniyaat (formerly known as the Kuria Muria Islands).

Low tide on the coral-strewn beach at Sur Masirah, Masirah.

such as certain oysters which may occur in clumps among the mangrove roots, winkles and the heavy shells of *Terebralia palustris*.

Coral reefs are poorly developed in eastern Arabian waters by comparison with those in the Red Sea but they are common off the Saudi Arabian coast. They also occur off rocky coasts in the Gulf of Oman but are not very extensive. The corals may be of the hard or soft variety and molluscs are often associated with them. The gastropod *Magilus antiquus* actually grows within certain corals. Most of the corals found around the Omani coast, however, are solitary and do not form reefs. But corals die, break up and form coral sand into which many molluscs burrow and live out most of their lives — and there is plenty of coral sand around eastern Arabian coasts.

Each of these different kinds of habitat is an entity with its own characteristic topography, fauna and flora (known collectively as a biotope), and each may be subdivided further into other more precisely defined biotopes. There are other kinds of biotopes, too many to describe here, in which molluscs occur. They include some found only in deeper water, such as exists in the Gulf of Oman, but our knowledge of them, for obvious reasons, is sketchy. We do know that in certain places, at considerable depths, molluscs occur in astonishing numbers. The molluscan fauna of those lightless — and for most of us unattainable — regions is given due consideration in this book but who knows what may still be down there awaiting discovery? The best may be yet to come.

Top: Coral exposed at the Damaniyat Islands. Above: Mangroves at Quriyat, Oman.

References for this section only: Basson, Burchard, Hardy & Price, 1977; Currie, Fisher & Hargreaves, 1973, Sheppard, Price and Roberts, 1992.

INTRODUCTION

ARABIC PLACE NAMES USED IN EASTERN ARABIA

The following is an alphabetical list of Arabic names used in eastern Arabia some of which occur on the accompanying map and in the text. It should be noted that transliterated names are subject to different spellings depending upon varying auditory perceptions. As far as possible we have utilised the official spellings adopted by the various countries. No abbreviations are used except UAE for United Arab Emirates and N, S, E and W to indicate North, South, East and West.

Abadan – river port at northern end of Arabian Gulf
Abu Dhabi – UAE, Arabian Gulf
Abu Hayl – UAE, half way between Dubai and Abu Dhabi
Ad Dawhah – capital city of Qatar, also spelled Doha
Ad Duqm – Oman, 100 km. SW of Barr Al Hikman
Aden – South Yemen at southern end of Red Sea
Ajman – UAE, 20 km NE of Sharjah
Al Ayjah – Oman, Masirah Island
Al Bustan – Oman, at Muscat
Al Fujairah – UAE, on Gulf of Oman
Al Jubayl – Saudi Arabia, 100 km NW of Dhahran
Al Khaburah – Oman, between As Sib and Sohar
Al Khobar – Saudi Arabia, port city of Dhahran
An Nuqdah – Oman, on mainland across from Masirah Island
As Sib – Oman, 30 km W of Muscat
As Sifah – Oman, 30 km SE of Muscat
Azaibah – Oman, 15 km W of Muscat

BAHRAIN – island country on west coast of Arabian Gulf
Bandar Abbas – Iran, at Strait of Hormuz
Bandar-e Lengeh – Iran, north coast of Arabian Gulf
Bandar Jissah – Oman, 10 km SE of Muscat
Barr Al Hikman – Oman, mainland across from Masirah Island
Batina – Oman, the long beach from Muscat to UAE border
Bushehr – Iran, north coast of Arabian Gulf

Damaniyat Islands – Oman, small island chain about 20 km N of As Sib
Dammam – Saudi Arabia, 10 km N of Dhahran
Das Island – island in Arabian Gulf N of Abu Dhabi
Dawwah – Oman, Masirah island
Dhahran – Saudi Arabia, across from Bahrain
Doha – capital city of Qatar, also spelled Ad Dawhah
Dubai – UAE

Fahal Island – Oman, at Muscat
Failaka Island – Kuwait
Filim – Oman, next to Barr Al Hikman
Fujairah – UAE, on Gulf of Oman

Green House – Kuwait

Hamriya – UAE, 25 km NE of Sharjah
Haql – Oman, Masirah Island
Hasik – Oman, about 150 km E of Salalah
Hilf – Oman, Masirah Island
Hormuz – Oman, Straits at entrance to Arabian Gulf
Hulaylah – UAE, island near Ras Al Khaymah

Jabal Ali – UAE, 25 km SW of Dubai
Jabal Dhanna – UAE, 200 km W of Abu Dhabi
Jazirat A'Shaghpah – Oman, Masirah Island
Juzor al Halaaniyaat – Oman, islands off southern coast (formerly Kuria Muria Islands)

Kalban – Oman, Masirah Island
Kangan – Iran, north coast of Arabian Gulf
Khan Creek – UAE, at Sharjah
Khasab – Oman, in Musandam
Khatmat Al Malaha – Oman, about 80 km NW of Sohar
Khor Fakkan – UAE on the Gulf of Oman
Kuria Muria Islands – Oman, off Southern Coast. Now called Juzor Al Halaaniyaat
KUWAIT – at north-west corner of Arabian Gulf

Madrakah – Oman, about 450 km NW of Salalah
Mahowt – Oman, next to Barr Al Hikman
Manifah – Saudi Arabia, 200 km NW of Dhahran
Marbat – Oman, 60 km E of Salalah
Masirah Island – Oman, off eastern coast
Mina Al Ahmedi – Kuwait
Mina Al Fahal – Oman, at Muscat
Muharek – Bahrain
Muscat – capital city of Oman
Musandam – Oman, at Strait of Hormuz

Neutral Zone – immediately S of Kuwait

Qalhat – Oman, 20 km NW of Sur
Qantab – Oman, 7 km SE of Muscat
QATAR – country on west coast of Arabian Gulf
Qeshm Island – Iran, in Strait of Hormuz
Quriyat – Oman, 50 km SE of Muscat
Qurm – Oman, Batina Coast beach near Muscat

Rakhyut – Oman, W of Salalah
Ras Abu Rasas – Oman, Masirah Island
Ras Al Barr – Bahrain
Ras Al Hadd – Oman, 30 km E of Sur
Ras Al Hamra – Oman, at Muscat
Ras Al Jazirah – Oman, Masirah Island
Ras Al Khaymah – UAE, Arabian Gulf
Ras Al Khayran – Oman, about 25 km SE of Muscat
Ras Al Ya – Oman, Masirah Island
Ras Hilf – Oman, Masirah Island
Ras Himari – Oman, Masirah Island
Ras Jazair – Bahrain
Ras Kaydah – Oman, Masirah Island
Ras Musharib – UAE, close to Qatar
Ras Qumayiah – Oman, 50 km S of Ras Al Hadd
Ras Sadr – UAE, near Abu Dhabi
Ras Suwaidi – Oman, about 100 km W of Muscat
Ras Tannurah – Saudi Arabia, near Dhahran
Raysut – Oman, W of Salalah
Ruwais – UAE at Qatar

Sadiyat Island – Arabian Gulf at Abu Dhabi
Salalah – Southern Oman
Salmiah – Kuwait
SAUDI ARABIA – large country occupying much of the Arabian Peninsula
Sawquirah – Oman, about 300 km NE of Salalah
Shaghaf – Oman, Masirah Island
Sharjah – UAE, 10 km NE of Dubai
Sheyk Shoeyb – Iran, north coast of Arabian Gulf
Shinas – Oman, 60 km NW of Sohar
Shinzi – Oman, Masirah Island
Sitrah – Bahrain
Sohar – Oman, 200 km NW of Muscat
SULTANATE OF OMAN – south-eastern corner of Arabian Peninsula
Sur – Oman, 30 km W of Ras Al Hadd
Sur Masirah – Oman, Masirah Island

Tarut Bay – Saudi Arabia at Ras Tannurah
Tiwi – Oman, between Sur and Quriyat

Umm Ar Rusays – Oman, Masirah Island
Umm Al Qaywayn – UAE between Sharjah and Ras Al Khaymah, also spelled Umm Al Quwain
Umq – Oman, Masirah Island
UNITED ARAB EMIRATES – Emirates on south-eastern coast of Arabian Gulf
Urf – Oman, Masirah Island

Yiti – Oman, 15 km SE of Muscat

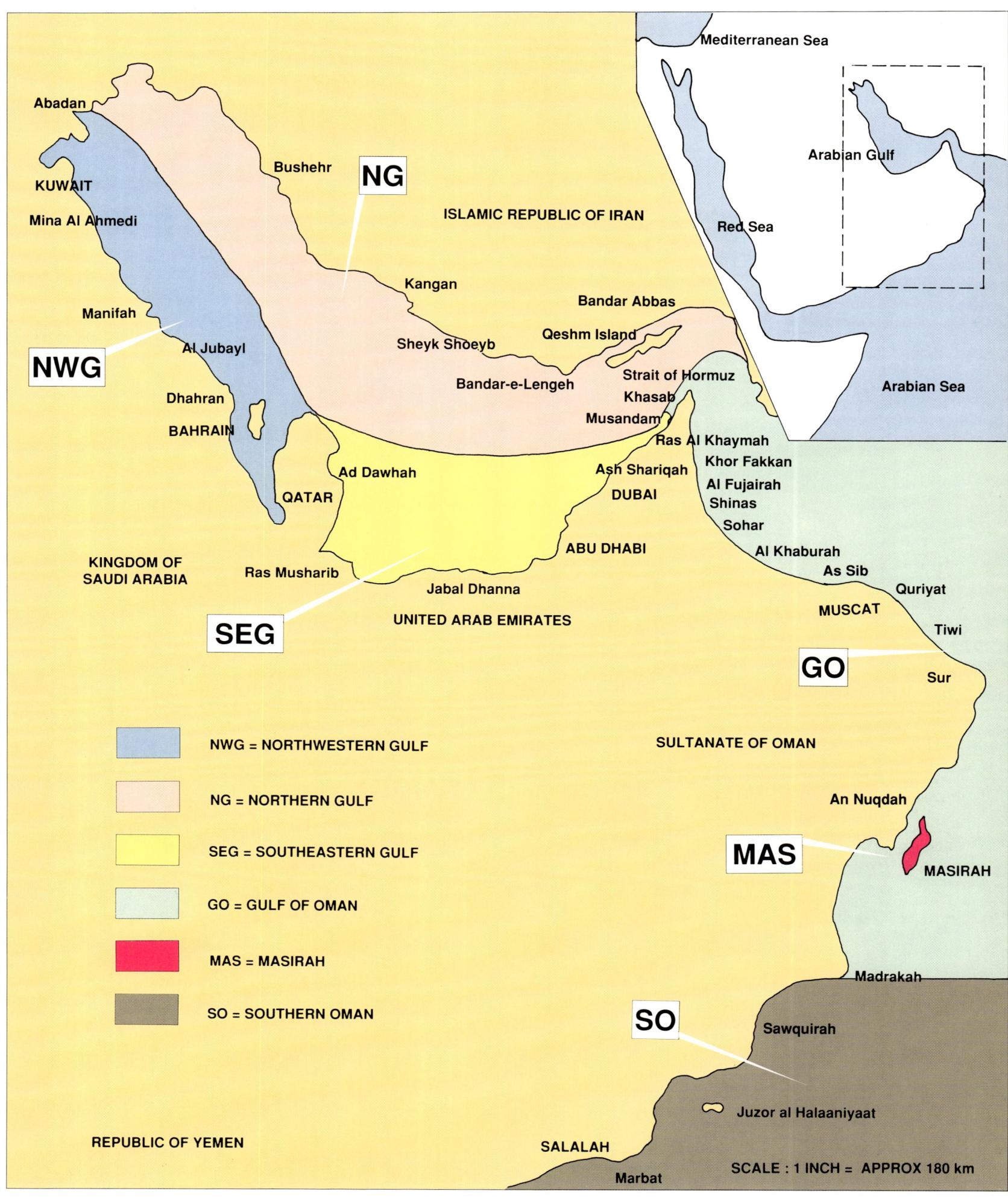

Gastropods
(Gastropoda)

with

Tusk Shells
(Scaphopoda)

Chitons
(Polyplacophora)

and

Cephalopods
(Cephalopoda)

Lambis truncata sebae
(see page 62)

GASTROPODS

THE GASTROPOD SHELL

The names of the shell features shown here are those which occur most often in the species descriptions which follow. Each shell drawing is accompanied by a drawing of the operculum relevant to it, except for the drawing of *Harpa* as no species of that genus has an operculum. The names and definitions of other features not illustrated here occur in the Glossary. In every instance the top end of the shell is posterior when the animal is moving (although the shells of the trochid, the turbinid and the naticid are generally carried askew). The figures are not reproduced to scale.

NOTES ON ILLUSTRATIONS

1. *Ptychobela*. Note the conspicuous posterior sinus. The chitinous operculum has a basal nucleus.

2. *Natica*. Peculiarly a feature of naticids, the funicle sometimes blocks the umbilicus. The operculum illustrated is calcareous and radially ribbed; some species have a chitinous operculum.

3. *Harpa*. Regular axial ribs are the edges of former apertural lips. The animal lacks an operculum.

4. *Murex*. Spines are arranged in rows on a varix. The operculum is chitinous and has a basal nucleus.

5. *Trochus*. Shell illustrated has a false umbilicus. The operculum is chitinous and spirally coiled.

6. *Turbo*. Sculpture is coarse. The operculum is calcareous, flat on the inside, dome shaped and often coloured in blotches on the outside.

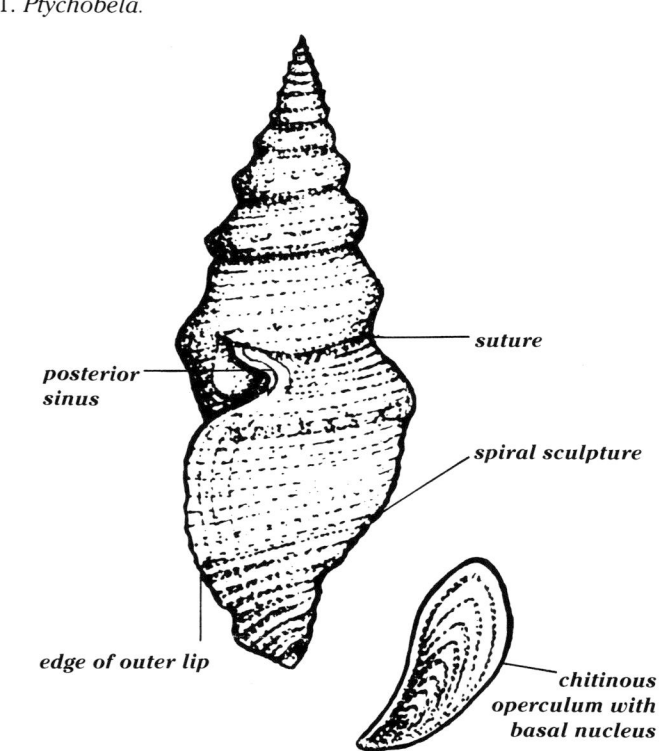

1. *Ptychobela*.

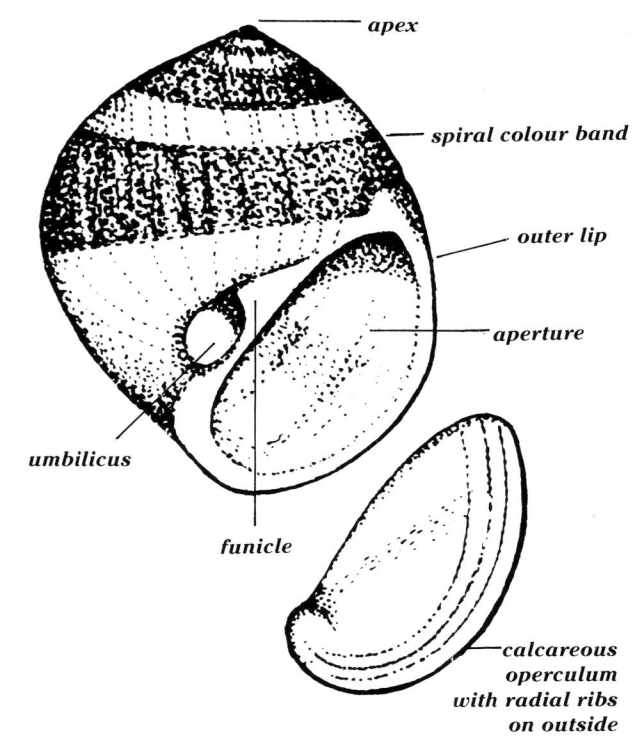

2. *Natica*.

3. *Harpa.*

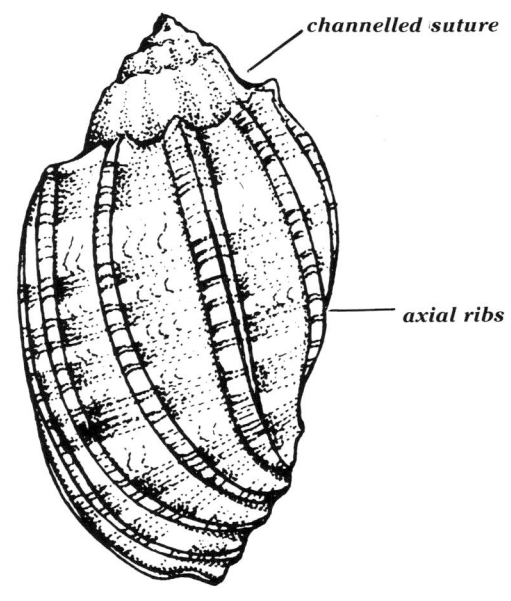

5. *Trochus.*

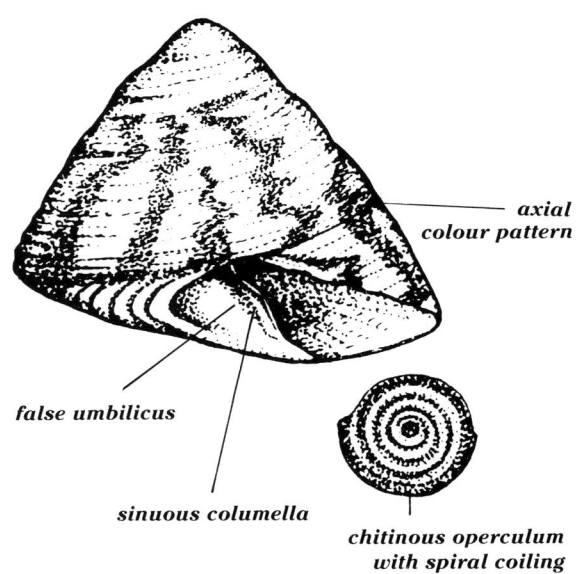

4. *Murex.*

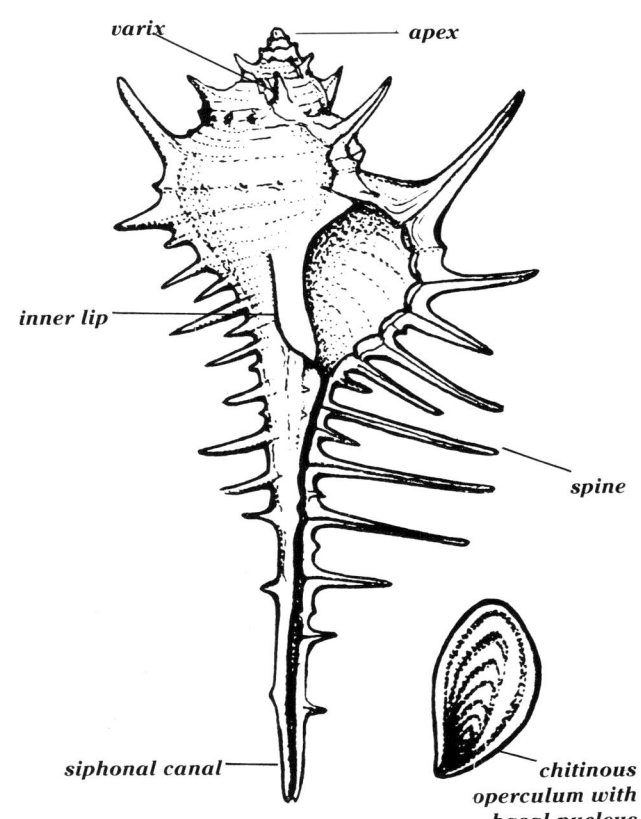

6. *Turbo.*

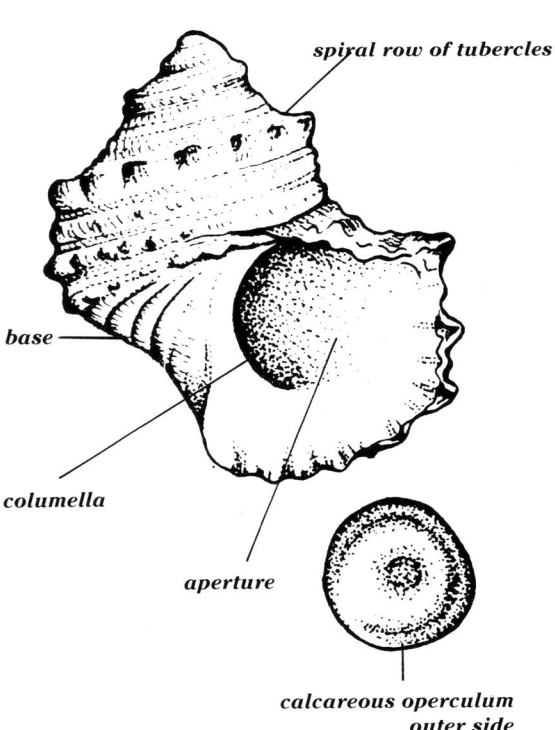

Class GASTROPODA

The gastropods are molluscs with a shell formed of one piece (but many species are shell-less) which may be spirally coiled. The name given to this class, however, implies that the stomach and associated soft parts sit on top of the animal's muscular foot where they are normally contained within the shell. Usually there is a recognisable head with a pair of eyes, stalked or unstalked. The subclasses are classified according to the different ways they cope with breathing. Most of them have a radula, a narrow ribbon beset with rows of hooked and pointed teeth, which enables them to rasp vegetation or other food in a vaguely tongue-like manner from surfaces. Outnumbering all the other classes of Mollusca combined, they are widely distributed in the sea, in freshwater and on land. Some air-breathing, non-marine species are included here because they occupy an essentially marine environment.

Subclass PROSOBRANCHIA

Prosobranch gastropods have their gills situated within the mantle cavity towards the front of the animal, an arrangement which has enabled them to occupy many different kinds of habitats. Prosobranchs have developed several types of radula which are so distinctive that they have been used as a basis for classification. Most radulae consist of transverse rows of teeth, variously arranged, but members of the superfamily Conoidea have harpoon-like, detachable barbs associated with a poison apparatus. Attached to the foot of many species is a chitinous or calcareous plate, the operculum. This structure, which assumes many different forms according to the species and is an important aid to identification, may enable the animal to seal the aperture of its shell but, in some groups, such as the cones, it is so reduced in size that it seems to serve no useful function.

Order ARCHAEOGASTROPODA

Archaeogastropods are herbivorous molluscs, many of them familiar and abundant intertidal species. Some have limpet-like shells and lack an operculum, others have top-like or spindle-shaped shells with an operculum. Often the shell has a nacreous lining. Many species have a free-swimming larval stage.

Anatoma aetheria 1 *Anatoma jacksoni* 2 *Sukasitrochus dorbignyi* 3

Superfamily PLEUROTOMARIOIDEA

Members of this superfamily, the most primitive of all living gastropods, have a slit, a hole or a series of holes in the shell. Except for the Haliotidae all eastern Arabian species are small or tiny.

Family SCISSURELLIDAE

Members of this family may be considered diminutive cousins of the glamorous and rare *Pleurotomaria* shells. Their shells are usually fragile and colourless and have few whorls. A slit (or hole) in the expanded outer lip continues as a narrow channel (the selenizone) encircling the whorls. The operculum is multispiral. Some eastern Arabian species occur intertidally, others in deep water. Little is known about their biology.

Anatoma aetheria (Melvill & Standen, 1903): **1**
1.3mm. Fragile, globose-trochoidal, slit relatively long and narrow. Umbilicus narrow; protoconch planorboid. Fine, reticulate sculpture overall. Translucent white. **Habitat:** deep water. **Distribution:** GO.

Anatoma jacksoni (Melvill, 1904): **2**
1.75mm. Fragile, oblong-ovate, selenizone well above periphery. Umbilicus broad and deep. Protoconch planorboid. Fine reticulate sculpture. Translucent white. **Habitat:** offshore and deep water. **Distribution:** NWG, SEG, GO.

Sukasitrochus dorbignyi (Audouin, 1826): **3**
1.4mm. A depressed species with strong spiral keels. Known from GO.

Scissurella rota Yaron, 1983: **4**
1.5mm. A flat-topped species with prominent, curved axial ribs and spiral threads between them. Recorded by Yaron, 1983:269, from Ain Sukhna, Arabian Gulf (NG); also known from GO.

Scissurella rota 4

Family HALIOTIDAE

The abalones (also known as ear shells or ormers) have flattened, loosely-coiled shells with a pearly interior. A line of small respiratory holes, arranged serially, perforates the expanded body whorl. The broad foot allows the animal to cling tenaciously to rocky surfaces and is sometimes greatly valued as a nutritious food. There is no operculum. Widely distributed in warm waters where they browse algae, the many species probably belong to one genus, *Haliotis*, although many other generic names have been proposed for them.

Haliotis mariae Wood, 1828: **5**
100mm. Roundly ovate, very compressed; spire no higher than body whorl; 6 or 7 open holes; broad, flat and raised spiral cords; deep furrow between rim and holes. Chocolate, beige, red and white, often with zigzag pattern. **Habitat:** among rocks offshore. **Distribution:** Mas, SO.

Haliotis pustulata Reeve, 1846: **6**
35mm. Elongate-ovate; spire raised above body whorl, suture distinct; 6 open holes or fewer; broken spiral cords form elongated pimples; shallow furrow between rim and holes. Green, brown, orange and white, variously mottled. Colour and ornament vary greatly. **Habitat:** under intertidal rocks. **Distribution:** GO.

Above: *Haliotis pustulata* 6. Below: Empty shells of *Haliotis mariae* on a beach near Salalah, Dhofar.

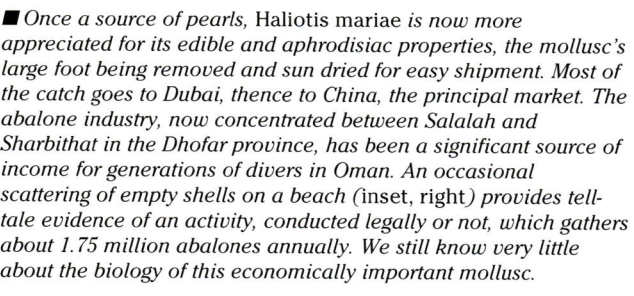

■ Once a source of pearls, Haliotis mariae is now more appreciated for its edible and aphrodisiac properties, the mollusc's large foot being removed and sun dried for easy shipment. Most of the catch goes to Dubai, thence to China, the principal market. The abalone industry, now concentrated between Salalah and Sharbithat in the Dhofar province, has been a significant source of income for generations of divers in Oman. An occasional scattering of empty shells on a beach (inset, right) provides tell-tale evidence of an activity, conducted legally or not, which gathers about 1.75 million abalones annually. We still know very little about the biology of this economically important mollusc.

Haliotis mariae 5

Superfamily: FISSURELLOIDEA

Superfamily FISSURELLOIDEA
Contains only one family. For details see family description below.

Family FISSURELLIDAE
The keyhole limpets and their allies are small to medium-sized gastropods with conical to almost flat, cap-like shells. They differ from true limpets by the presence of a long or short slit anteriorly, or an apical or almost marginal orifice (lacking in *Scutus*) through which waste matter is expelled. Inside there is a horseshoe-shaped muscle scar, open anteriorly. The foot may be larger than the shell; and in some genera (including *Scutus*) the mantle covers the shell completely. Some display a homing instinct, returning from feeding forays to the spot whence they departed. Many species worldwide, all found on or under flat surfaces where they feed on algae. The family is well represented in rocky places throughout eastern Arabia.

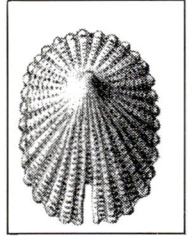

Emarginula camilla 8

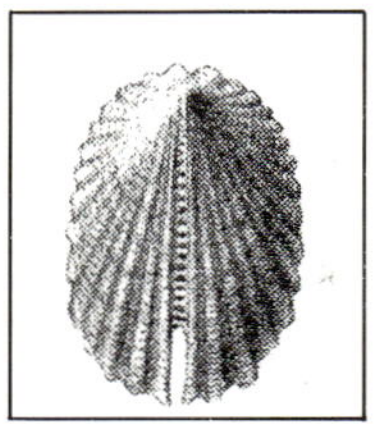

Emarginula undulata 10

Subemarginula panhi 11

Subfamily EMARGINULINAE

Emarginella incisura (A. Adams, 1852): **7** NOT ILLUSTRATED
Listed by Glayzer et al, 1984:318, from Kuwait (NG).

Emarginula camilla Melvill & Standen, 1903: **8**
6.5mm. Described from shells obtained at 156 fathoms off Muscat (GO).

Emarginula peasei Thiele, 1915: **9**
7mm. Moderately thick, flattened, oval in outline, apex points backwards; slit 4 times as long as wide; well differentiated radial ribs and concentric ridges resulting in cancellate sculpture; margin thickened and corrugated. All white. Animal milky white. Syn: *E. clathrata* Pease, 1863, (sensu Glayzer et al, 1984:318.) **Habitat:** among intertidal rocks. **Distribution:** NWG, GO, Mas SEO.

Emarginula undulata Melvill & Standen, 1903: **10**
5.5mm. Described from shells obtained at 156 fathoms off Muscat (GO).

Subemarginula panhi (Quoy & Gaimard, 1834): **11**
18mm. Solid, oval, high- or low-conical, apex pointed anteriorly, rounded anterior has concave slope, angular posterior has convex slope. Central ridge has a broader radially ribbed ridge either side; alternating broad and narrow radial ridges on anterior half; margin smooth inside. Yellowish white. **Habitat:** beached. **Distribution:** Mas.

Emarginula peasei 9

Subemarginula subrugosa (Thiele, 1916): **12**
12mm. Elongate, depressed, apex pointed anteriorly; rounded anterior has concave slope, angular posterior has convex slope. Central ridge has a broader radially ribbed ridge either side; alternating broad and narrow radial ridges on anterior half; margin smooth inside. Yellowish olive green. **Habitat:** beached. **Distribution:** Mas.

Rimula* cf *cumingii A. Adams, 1853: **13**
4mm. Thin, oval, highest point above apex which is down-turned; concave slope below apex. Orifice pointed posteriorly, rounded anteriorly. Fine radial ribs crossed by equally strong concentric ribs; central ridge well raised. White. **Habitat:** deep water. **Distribution:** GO.

Scutus unguis (Linnaeus, 1758): **14**
40mm. Elongate, greatly flattened, apex nearer posterior margin, anterior margin slightly indented; weak concentric growth lines; shiny inside. White or cream. Animal dark brown to black. **Habitat:** among rocks and coral. **Distribution:** GO, Mas.

Subfamily DIODORINAE

Diodora funiculata (Reeve, 1850): **15**
30mm. Conical with convex or concave sides, orifice nearer anterior end, base not arched. Widely spaced, strong radial ridges with finer intermediate ones crossed by fine concentric ridges; margin corrugated. Creamy white usually with reddish brown or blue-black rays or concentric rings, rarely dashes; margin edge tinged same colour; white rim around orifice inside. **Habitat:** among rocks and stones. **Distribution:** all.

Subemarginula subrugosa 12

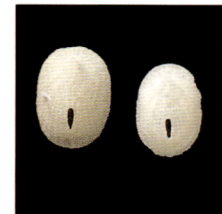

Rimula cf *cumingii* 13

Diodora funiculata 15

Scutus unguis 14

Superfamily: **FISSURELLORDEA, PATELLOIDEA**

Diodora rueppellii (Sowerby, 1834): **16**
15mm. Steeply conical, robust, orifice much nearer anterior than posterior end, base conspicuously arched; radial ridges stronger than concentric ridges, noduled at intersections; margin corrugated; internal rim of orifice straight-edged posteriorly. White with reddish or brown rays. **Habitat:** intertidal among rocks and stones. **Distribution:** all.

Diodora singaporensis (Reeve, 1850): **17**
20mm. Elongate, compressed-conical, each end rounded; orifice nearer anterior end, base gently arched; radial ridges crossed by thinner, more crowded concentric ridges. Thick callus around orifice inside. White often with brownish rays. Syn: *D. bombayana* (Sowerby, 1862). **Habitat:** intertidal rocks and stones. **Distribution:** all.

Subfamily FISSURELLINAE
Macroschisma megatrema A. Adams, 1851: **18**
6mm. Delicate, elongate-oval in outline with straight, compressed sides, rounded posteriorly, anterior edge straight; orifice large, wide anteriorly, V-shaped posteriorly. Coarse radial ribs crossed by concentric threads; glossy inside. White with 2 broad rays anteriorly, mottled and spotted posteriorly. Syn: probably the species listed by Glayzer et al, 1984:318, (as *M. elegans* Preston, 1908), from Kuwait. **Habitat:** beached. **Distribution:** Mas, NWG.

Medusafissurella dubia (Reeve, 1849): **19**
30mm. Moderately thick, flattened, saddle-shaped cone, almost oval in outline, slightly raised anteriorly and posteriorly; keyhole-shaped orifice central, its sides evenly raised. Widely spaced radial ribs crossed by scaly, concentric ridges. White or yellowish with brown or red rays and occasional concentric rings; inside greyish white. **Habitat:** intertidal among stones and rocks. **Distribution:** Mas, SO.

Medusafissurella melvilli (Sowerby 1882): **20**
50mm. Steeply conical, oval in outline with almost central, circular or oval orifice; shell margin level or uneven; strong, irregularly spaced radial ribs, irregular concentric growth ridges; margin smooth. Red or pink, often faded, sometimes rayed; inside white. Larger shells may be thick and distorted. Syn: *M. gallagheri* Smythe, 1988. **Habitat:** among stones offshore and beached. **Distribution:** SO.

Medusafissurella salebrosa (Reeve, 1850): **21**
30mm. Thick, heavy, flattened, saddle-shaped cone, narrower and conspicuously raised anteriorly; orifice central, almost circular, sharp-edged, raised posteriorly. Strong, widely-spaced, radial ribs which are smooth or coarsely scaled; margin lightly corrugated or worn smooth. White, yellowish or pink, often brown-rayed; inside porcellaneous, white. **Habitat:** intertidal among stones and rocks. **Distribution:** GO, Mas, SO.

Superfamily PATELLOIDEA
Includes the true limpets (Patellidae) and other limpet-like groups with shells lacking an orifice or marginal slit. The animals browse on algae.

Family LOTTIIDAE
Cap-shaped or saucer-shaped shells usually with a circular outline and a glossy interior. The animal lacks an operculum. Worldwide and intertidal but sometimes well above high-water mark adhering to rocks or other objects. Only two small species recorded from the region.

Subfamily PATELLOIDINAE
Patelloida maraisi (Kilburn, 1977): **22**
4mm. Thin, silky, elevated, oval or nearly circular in outline with almost central apex; base even or slightly concave. Fine and coarse concentric growth lines. Fresh shells have a brown U-shaped mark around the apex surrounded by radially arranged golden brown dots; worn shells are translucent with white rays. **Habitat:** intertidal rocks. **Distribution:** Mas.

Diodora rueppellii **16**

Medusafissurella dubia **19**

Diodora singaporensis **17**

Medusafissurella salebrosa **21**

Medusafissurella melvilli **20**

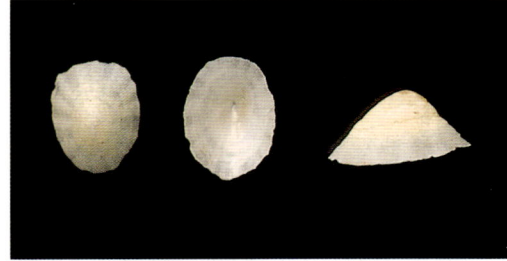
Patelloida maraisi **22**

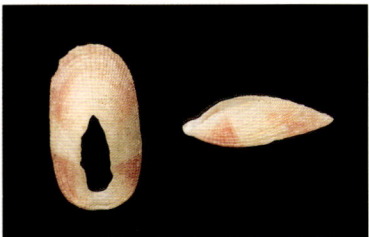
Macroschisma megatrema **18**

Superfamily: **PATELLOIDEA, TROCHOIDEA**

Patelloida profunda **23**

Patelloida profunda (Deshayes, 1863): **23**
20mm. Thick, elevated, oval or almost circular in outline with nearly central apex; base often concave. Fine radial riblets, usually eroded. White or creamy with irregular reddish brown rays; inside white with orange central area; juveniles rayed internally. **Habitat:** on rocks and other objects, often above high-tide level. **Distribution:** GO, Mas, SO.

Patella flexuosa **24**

Family PATELLIDAE
Saucer-shaped shells, often irregular in outline, elevated or depressed. Radiating ribs may be prominent and extend beyond margin. The horseshoe-shaped, internal muscle scar is often distinctively coloured. The animals lack an operculum and true gills. Shore dwellers, they adhere to smooth rocks, often tenaciously. They tend to return to the same location after grazing algae and may make shallow impressions in rocky surfaces. There are only two eastern Arabian species, both very variable in size, shape, ornament and colour pattern.

Subfamily PATELLINAE
Patella flexuosa Quoy & Gaimard, 1834: **24**
40mm. Thick, usually flattened, narrower posteriorly, with nearly central apex. Older shells often more circular and more elevated. Stronger radiating ribs intermittently fused and prolonged marginally. White or yellow with interrupted, reddish brown rays; inside white with central orange-brown blotch or ring. Syn: *P. pica* Reeve, 1854. **Habitat:** intertidal rocks. **Distribution:** GO, Mas, SO.

Subfamily NACELLINAE
Cellana rota (Gmelin, 1791): **25**
45mm. Thick or thin, elevated or low, oval in outline with apex towards narrower anterior end. Coarse or fine, sometimes nodular, radiating ribs and fine, irregular, concentric growth lines. Golden brown, to blackish brown, often with paler rays and thin, interrupted, radiating lines. External colour pattern may show internally; muscle scar truncated anteriorly and coloured white to orange. Syn: *Patella karachiensis* Winckworth, 1930. **Habitat:** intertidal rocks. **Distribution:** SEG, GO, Mas, SO.

Superfamily TROCHOIDEA
Most species in this extensive superfamily belong to one of two families: Trochidae or Turbinidae. Superficially their shells resemble each other, most of them being top-shaped or turbinate and often colourful. But the operculum, an important aid to identification, is either chitinous and thin (Trochidae) or calcareous and thick (Turbinidae).

Family TROCHIDAE
Extensive, worldwide family of small to large species with conical or globose shells having a flat or convex base and an oval aperture. Body whorl rounded or angular, columella smooth or with a single blunt tooth, base with or without an umbilicus; outer lip smooth or ridged internally. Often brightly coloured and boldly patterned, top shells are pearly within. Operculum chitinous, thin, flexible, circular and usually multi-spiralled (thus differing markedly from that of the Turbinidae). Top shells are predominantly rock dwellers browsing algae. Many so-called species have been reported from eastern Arabian waters, most of them small, and some of them are not included here.

Subfamily MARGARITINAE
Euchelus asper (Gmelin, 1791): **26**
20mm. Thick, dull, conical, whorls well rounded, almost circular aperture; poorly developed tooth at base of columella; outer lip finely ridged; umbilicus narrow but deep, closed or open. Strong spiral ridges crossed by fine grooves. Grey often mottled black; inside pearly. **Habitat:** intertidal under rocks or in sand. **Distribution:** all.

Granata sulcifera (Lamarck, 1822): **27**
20mm. Thin, dull, ear-like, much

Cellana rota on a rock face at low tide at Rakhyut, Dhofar **25**

Cellana rota **25**

Euchelus asper **26**

Superfamily: **TROCHOIDEA**

broader than tall, whorls rounded, sutures impressed; columella gently curved, no umbilicus. Alternately strong and weak spiral ribs, closely spaced axial ridges between. Brown fading to pink with darker spots on ribs, iridescent inside. Thin operculum about half length of aperture. **Habitat:** intertidal under rocks. **Distribution:** NWG, Mas, SO.

Perrinia stellata (A. Adams, 1864): **28**
10mm. Solid, dull, elongate-conical, of about 5 whorls, with pointed apex, deep sutures and flattened base; aperture lirate, 1 or 2 denticles on columella; no umbilicus. Whorls sharply keeled, almost frilled, with flattened lines above and below keel; base spirally ribbed with rows of pits between ribs. White. **Habitat:** offshore and beached. **Distribution:** NWG, SEG, GO, Mas, SO.

Vaceuchelus angulatus (Pease, 1868): **29**
6mm. Thick, ovate-conical, about 5 whorls, almost circular aperture and no umbilicus. Keeled spiral cords crossed by strong axial ribs produce series of deep, elongate pits; columella smooth. White sometimes spotted reddish brown; aperture white. Syn: *E. foveolatus* (A. Adams, 1851) (sensu Smythe, 1979a:64, and Glayzer et al, 1984:318). **Habitat:** under rocks. **Distribution:** NWG, SEG, GO, Mas.

Subfamily MONODONTINAE
Monodonta nebulosa (Forsskål, 1775): **30**
28mm. Thick, dull, low spired, well rounded whorls, apex usually eroded, shallow sutures, rounded base, no umbilicus; shallow spiral grooves, otherwise smooth; aperture lirate with thick rib at top; base of columella has strong, often pointed tooth. White overlaid violet, pink, brown or green in spiral lines and dashes. Bright green alga often covers shell. Syn: *M. vermiculata* Fischer, 1874. **Habitat:** on and under rocks. **Distribution:** all.

Subfamily GIBBULINAE
Agagus agagus Jousseaume, 1894: **31**
10mm. Solid, dull, steeply conical, straight sided, sutures shallow; base flat, umbilicus deep. Spiral cords with fine, obliquely axial threads between them; thick, nodulous cord at periphery and above sutures; strong spiral cords on base. White with axial, bluish green flames. Syn: *Gibbula (Enida) townsendi* Sowerby, 1895. **Habitat:** intertidal. **Distribution:** Mas.

Rubritrochus pulcherrimus (A.Adams, 1855): **32**
24mm. Thick, dull, as tall as broad, whorls swollen, sutures deep, apex blunt; base gently convex, umbilicus broad and deep. Coarse spiral ridges cross broad, widely spaced axial folds, base lacks folds; beaded band above suture, strong keel around umbilicus. White with reddish blotches; spiral rows of darker spots on base. Syn: *Forskaelena declivis* (Forsskål, 1775) (sensu Smythe, 1979a:64, and Glayzer et al, 1984:318). **Habitat:** intertidal among rocks and silt. **Distribution:** NWG, SEG, GO, Mas, SO.

Subfamily CALLIOSTOMATINAE
Calliostoma fragum (Philippi, 1848): **33**
14mm. Solid, silky, slightly taller than broad, whorls slightly concave, base flat and keeled, apex blunt; no umbilicus. Columella straight; each spire whorl has 5 spiral rows of beads, lowermost projecting; base spirally corded. Amber with brown blotches, beads white and brown; thin, brown, spiral lines between beads and cords; columella and aperture nacreous. **Habitat:** intertidal and offshore in coarse sand. **Distribution:** NG, GO, Mas.

Calliostoma funiculare Melvill, 1897: **34**
15mm. Solid, silky, conspicuously longer than broad, whorls straight sided, base slightly convex; no umbilicus. Columella straight; each spire whorl has 4 or 5 spiral ridges, 2 at periphery of last whorl plain, others beaded; base spirally corded. Yellowish white with amber spots on ridges and cords. **Habitat:** offshore. **Distribution:** NG.

***Granata sulcifera* 27**

***Agagus agagus* 31**

***Vaceuchelus angulatus* 29**

***Perrinia stellata* 28**

***Monodonta nebulosa* 30**

***Rubritrochus pulcherrimus* 32**

***Calliostoma fragum* 33**

***Calliostoma funiculare* 34**

Superfamily: **TROCHOIDEA**

Trochus (Belangeria) scabrosus 37

Clanculus gennesi 35

Eloise Bosch, Peter Dance and Don Bosch sorting a large collection of *Clanculus pharaonius* at Masirah Island

Trochus (Infundibulops) erithreus 38

Trochus (Infundibulops) firmus 39

Clanculus pharaonius 36

Subfamily TROCHINAE

Clanculus gennesi Fischer, 1901: **35**
10mm. Thick, dull, almost straight-sided cone, about 6 whorls with almost flat base and pointed apex; sutures channelled, umbilicus narrow but deep. Beaded spiral cords strongest at periphery; fine axial growth lines. White mottled or streaked with dark or light brown; columella white. **Habitat:** intertidal under rocks. **Distribution:** NWG, SEG, GO, Mas, SO.

Clanculus pharaonius (Linnaeus, 1758): **36**
25mm. Robust, dull, much broader than tall, sutures impressed; base gently rounded, umbilicus deep; folds on columella, basal one strongest; inner lip ridged. Spiral rows of strong and weaker beads on whorls, spoke-like ridges around umbilicus. Rows of prominent, alternately black-and-white beads separated by rows of smaller, red beads; aperture white. **Habitat:** sand between rocks. **Distribution:** all.

Trochus (Belangeria) scabrosus Philippi, 1850: **37**
15mm. Thick, dull, slightly convex-sided cone, slightly broader than tall, sutures incised, last whorl keeled at periphery. Base rounded, false umbilicus narrow; columella toothed at base; outer lip lirate within, corrugated at base. Spiral rows of coarse beads. Cream or yellow-brown with zigzag, reddish brown or greyish brown flames. **Habitat:** intertidal among rocks. **Distribution:** NWG, GO, Mas.

Trochus (Infundibulops) erithreus Brocchi, 1823: **38**
35mm. Sturdy, straight-sided cone, about as tall as broad, sutures deep; base flat, umbilicus broad and deep; columella sinuous. Crowded, beaded, spiral cords; later whorls have broad, axially elongated knobs. Reddish brown to green with darker streaks and blotches; aperture white. **Habitat:** intertidal on rocks. **Distribution:** all.

Trochus (Infundibulops) firmus Philippi, 1849: **39**
30mm. Thick, dull, slightly broader than tall, slightly convex sided; columella sinuous; base slightly convex; false umbilicus deep and wide. Nodulose spiral ridges on all whorls; low, smooth ridges on base; thick spiral cord around false umbilicus. Greenish, pinkish or white with oblique, reddish brown streaks; base has spiral rows of dots and dashes encircling a green spiral cord. **Habitat:** intertidal under rocks. **Distribution:** GO, Mas.

Trochus (Infundibulops) fultoni Melvill, 1898: **40**
25mm. Thick, dull, about as broad as tall, whorls straight sided initially, then gently convex; columella gently curved; base almost flat; false

Superfamily: **TROCHOIDEA**

umbilicus deep and narrow. Spire whorls have 4 or 5 nodulose spiral ridges, last whorl 5 or 6. Greenish with obliquely axial, flame-like markings; base with spiral lines of red spots. **Habitat:** intertidal. **Distribution:** GO.

Trochus (Infundibulum) kochi Philippi, 1844: **41**
45mm. Solid, about 8 whorls, broader than tall, sutures shallow; base flat, periphery keeled, umbilicus funnel shaped; columella sharply angled towards upper end. Spire whorls have spiral rows of smooth beads, last whorl has widely spaced, interrupted, diagonal cords. Creamy with purple or greenish zigzag streaks, bright green around inside edges of aperture. Southern examples much larger. **Habitat:** intertidal among rocks. **Distribution:** Mas, SO.

Subfamily UMBONIINAE
Ethalia carneolata Melvill, 1897: **42**
12mm. Solid, glossy, twice as broad as tall, last whorl roundly keeled at periphery, sutures distinct, apex flat topped; umbilicus deep, mostly obscured by parietal callus. Columella thickened; all whorls with fine, spiral striae and finer growth lines. Pink or white blotched brown, spirally dotted white all over; callus white, umbilicus orange; aperture nacreous. The variety *rubrostrigata* Melvill, 1904, has a radially striped shell. **Habitat:** offshore. **Distribution:** NG, GO.

Ethalia minolina Melvill, 1897: **43**
10mm. Solid but lightweight, depressed conical, angled at periphery, base rounded, sutures shallow. Columella has a tongue-shaped callus which partly obscures the deep umbilicus. Surface finely, spirally striate, all but base being also finely, axially lined. Reddish pink or pale brown speckled and blotched white and darker brown, usually with interrupted, fine, spiral, white lines, columella usually whitish. **Habitat:** offshore. **Distribution:** NG.

Ethminolia degregorii (Caramagna, 1888): **44**
4mm. Solid, glossy, low spired, broader than tall, protoconch of 2.5 slightly raised whorls; columella straight; umbilicus very wide and deep. Whorls flattened below sutures then angled, last whorl gently keeled at periphery. Smooth except for growth ridges around edge of umbilicus. White with radiating, violet-brown streaks; encircled by spiral lines of white dashes, violet around umbilicus. **Habitat:** beached. **Distribution:** GO.

Ethminolia iridifulgens (Melvill, 1910): **45**
6mm. Thin, glossy-iridescent, slightly broader than tall, protoconch of 2 flattened whorls; columella sinuous; umbilicus wide and deep. Whorls well rounded, flattened below each suture, with low spiral threads, base smooth but threads around umbilicus. White heavily mottled, streaked and dotted violet, brownish, pink or greyish; umbilicus white; aperture nacreous. **Habitat:** deep water to intertidal. **Distribution:** GO, Mas.

Trochus (Infundibulops) fultoni **40**

Ethminolia iridifulgens **45**

Ethminolia degregorii **44**

Ethalia carneolata **42**

Ethalia carneolata variety *rubrostrigata* **42**

Ethalia minolina **43**

Trochus (Infundibulum) kochi **41**

SEASHELLS OF EASTERN ARABIA 35

Superfamily: **TROCHOIDEA**

Priotrochus aniesae **46**

Monilea chiliarches **51**

Pagodatrochus variabilis **52**

Pseudominolia climacota **53**

Osilinus kotschyi **47**

Priotrochus obscurus **48**

Priotrochus aniesae Moolenbeek & Dekker, 1993: **46**
15mm. Thick, dull, as tall as broad, whorls sloping below fine sutures, early whorls eroded; base very convex, umbilicus tiny or covered by spirally grooved parietal callus. Strong spiral ridges, heavily noduled at periphery; aperture lirate. White with reddish, wavy, axial, streaks and blotches; ridges minutely flecked red; callus and aperture white. **Habitat:** intertidal among muddy stones. **Distribution:** Mas.

Osilinus kotschyi (Philippi, 1849): **47**
20mm. Thick, dull, taller than broad when adult, whorls straight sided and stepped below incised sutures, apex pointed and often eroded; base slightly convex, umbilicus very small and deep. Coarse, nodular spiral ribs, nodules most prominent where stepped; ridges smooth around umbilicus. Greyish white with blackish axial streaks. **Habitat:** intertidal among muddy stones. **Distribution:** NWG, GO, Mas.

Priotrochus obscurus (Wood, 1828): **48**
20mm. Thick, dull, slightly taller than broad, whorls flat sided with angled shoulders, keeled at base; small, deep umbilicus. Columella has 1 tooth bearing denticles; coarse spiral ridges and large nodules below suture. White streaked and blotched green, brown or grey; aperture white. Varies in shape and colour, columella may have more than 1 tooth. Syn: *Trochus nabataeus* Issel, 1869; *Priotrochus sepulchralis* Melvill, 1899. **Habitat:** intertidal on rocks. **Distribution:** all.

"Umbonium" eloiseae Dance, Moolenbeek & Dekker, 1992: **49**
9mm. Solid, silky, compressed-discoidal, much broader than long, about 6.5 whorls, appressed sutures, last whorl roundly keeled, aperture half-moon shaped; umbilical callus surrounded by shallow groove. Smooth. White suffused rosy pink in varied patterns; apical whorls dark reddish brown or purple. **Habitat:** intertidal in sand. **Distribution:** Mas, SO.

Umbonium vestiarium (Linnaeus, 1758): **50**
12mm. Thick, glossy, compressed-trochoidal, much broader than long, about 7 whorls, appressed sutures. Periphery roundly keeled, umbilical area filled by broad callus pad extending to columella. Smooth. White, pink, grey, brown or purple, callus pad unicolorous, rest of shell variously coloured and patterned with chevrons, streaks and spiral bands; whole shell often unicolorous. **Habitat:** intertidal in sand. **Distribution:** NG, NWG, SEG, GO.

Subfamily SOLARIELLINAE
Monilea chiliarches Melvill, 1910: **51**
6mm. Solid, glossy, spire whorls concave sided, last whorl well

"Umbonium" eloiseae **49**

Umbonium vestiarium **50**

Superfamily: **TROCHOIDEA**

rounded, flat-topped protoconch of 1.5 whorls; umbilicus small. Spire whorls have 2 rounded, spiral cords with strong axial ridges between and above them; ridges become obsolete on top of last whorl (which has 6 strong, spiral cords) but encircle umbilicus. Amber, aperture nacreous. **Habitat:** offshore among rocks. **Distribution:** NG, GO.

Pagodatrochus variabilis (H. Adams, 1873): **52**
3mm. Thin, trochoidal, apex flat topped, base convex, spire whorls sloping above, their sides concave in profile and each bounded above and below by a thick spiral ridge, 4 spiral ridges at base; fine axial threads overall; umbilicus wide and deep. White with reddish brown blotches; aperture nacreous. Syn: *Minolia eutyches* Melvill, 1918. **Habitat:** intertidal. **Distribution:** NWG, GO.

Pseudominolia climacota (Melvill, 1897): **53**
14mm. Solid, lightweight, glossy, as tall as broad, whorls stepped at sutures, periphery of last whorl roundly keeled, apex blunt; umbilicus moderately large, deep. Columella sinuous and thickened; all whorls spirally ridged, channelled between ridges, base smooth, ridged around umbilicus. White with reddish brown triangles and zigzags becoming obsolete around umbilicus; aperture and columella nacreous. **Habitat:** offshore. **Distribution:** GO.

Pseudominolia gradata (Sowerby, 1895): **54**
8mm. Thin, as broad as tall, sutures impressed; blunt protoconch. Base gently rounded, columella slightly sinuous. Sharp, widely spaced, spiral ribs with crowded, oblique threads in channelled interspaces (threads often worn away); flat ribs on base continue inside broad, deep umbilicus. White with brown, red or pink, axial streaks; aperture and umbilicus white. **Habitat:** offshore and beached. **Distribution:** all.

Pseudominolia nedyma (Melvill, 1897): **55**
5mm. Thin, depressed-conic; blunt protoconch. Base gently rounded, columella almost straight. Spire whorls have strong keel at shoulder, almost smooth otherwise; last whorl with 2 strong, widely spaced keels and 5 or 6 flat-topped ribs around the large, deep umbilicus. Brown, orange, pink or off-white with spiral rows of white spots which are sometimes arrow-head-shaped. Syn: *P. biangulosa* (A. Adams, 1854) has been recorded from the Arabian Gulf but is easily confused with *P. nedyma*. **Habitat:** intertidal and offshore. **Distribution:** NWG, GO.

Pseudominolia nedyma **55**

Pseudominolia gradata **54**

Subfamily STOMATELLINAE
Small species with thin, few-whorled shells having a depressed spire and a large, oval or elongated aperture revealing the entire body-whorl interior; shell *Haliotis*-like but smaller and lacking holes. Shells in this subfamily lack an operculum. Mostly coral-reef dwellers, they are also found under intertidal rocks.

Broderipia iridescens (Broderip, 1834): **56**
8mm. Thin, fragile, limpet-like, much longer than broad, with very depressed, rounded apex close to posterior margin. Fine, regular axial striae crossed by irregular growth lines. White with reddish-brown flecks and radial lines which are most pronounced towards margin; iridescent inside. **Habitat:** intertidal under rocks. **Distribution:** GO.

Stomatella auricula Lamarck, 1816: **57**
12mm. Thin, fragile, last whorl elongated, occupying entire shell length; about 3.5 whorls, apex only slightly raised; sutures shallow but distinct. Columella continuous with outer lip which has slightly curved upper edge; no umbilicus. Fine spiral growth lines crossed by finer axial lines. Green, grey, pink or reddish brown with yellowish chevrons and triangles; iridescent inside. Syn: *S. varia* (A. Adams, 1850)? **Habitat:** intertidal under rocks. **Distribution:** NWG, SEG, Mas.

Stomatia phymotis **59**

Stomatella auricula **57**

Stomatella duplicata Sowerby, 1854: **58**
15mm. Thin, slightly broader than tall, spire whorls keeled, spire well elevated, apex prominent, sutures deep; columella smooth and gently curved, no umbilicus. Secondary keel above periphery; strong spiral ribs cover rest of shell; fine axial ridges between ribs. Yellowish with red or pale brown spots and blotches, similarly coloured inside. **Habitat:** intertidal under rocks. **Distribution:** NWG, SEG, GO, Mas.

Stomatia phymotis Helbling, 1779: **59**
30mm. Thin to thick, dull, spire whorls loosely coiled, last whorl greatly expanded and very wide open. Columella concave and continuous with outer lip; no umbilicus. 2 close-set, nodulous cords curve around periphery of last whorl. White to greyish-green with red blotches and streaks, aperture nacreous. **Habitat:** intertidal under rocks and beached. **Distribution:** NWG, SEG, GO, Mas, SO.

Broderipia iridescens **56**

Stomatella duplicata **58**

Superfamily: **TROCHOIDEA**

Stomatia* cf *rubra (Lamarck, 1822): **60**
6mm. Thin, dull, with trochoidal spire and very large last whorl, sutures impressed, umbilicus a mere chink. Each whorl has 2 widely spaced keels, base encircled by spiral ridges, entire shell has fine growth lines. White with brown or red streaks; aperture iridescent within. **Habitat:** intertidal and offshore. **Distribution:** Mas, SO.

Family SKENEIDAE
The small or minute species in this superfamily have thin, umbilicate, non-porcellaneous shells which are top shaped or discoidal and bear a thin, chitinous, multi-spiral operculum. Very little is known about their biology. Those recorded from eastern Arabia have come mostly from shell-sand siftings.

Daronia subdisjuncta (H. Adams, 1868): **61**
15mm. Thin, loosely coiled, later stage of body whorl disconnected from spire, penultimate whorl almost enveloping rest of spire. Aperture oval, umbilicus large and deep. Early whorls strongly keeled at shoulder and spirally ribbed, last whorl well rounded and encircled by flat-topped, spiral ribs which do not continue into umbilicus. All white. **Habitat:** muddy sand and stones offshore. **Distribution:** GO.

***Lodderena* n. sp**: **62**
0.8mm. Thin, discoidal, with sunken apex, spire of 2 whorls and a protoconch of 1 smooth whorl. About 9 spiral, beaded keels, the 2 at periphery larger than others. Aperture almost circular, umbilicus wide and deep. White. **Habitat:** intertidal. **Distribution:** Mas, SO.

Family VITRINELLIDAE
A poorly known and poorly defined family, the vitrinellids are mostly tiny and have low, flattened shells which are translucent when fresh but become opaque later. An umbilicus is usually present and there is a chitinous operculum. The shells are usually found beached.

Leucorhynchia crossei Tryon, 1888: **63**
4mm. Solid, glossy, low spired, aperture circular, sutures very fine, apex flat; umbilicus deep, largely obscured by thick, lip-like callus extending from base of columella. Surface smooth. White. **Habitat:** deep water. **Distribution:** GO.

Lodderia novemcarinata (Melvill, 1906): **64**
3mm. Solid, silky, discoidal, wide deep umbilicus; protoconch of 2 glossy whorls. Conspicuously spirally keeled, usually 9 keels on last whorl; fine and close axial striae. White. **Habitat:** deep water. **Distribution:** SEG, Mas, GO.

Morchiella moreleti (Fischer, 1877): **65**
1.3mm. Thick lipped, almost discoid, with a large umbilicus and slightly heterostrophe protoconch. Aperture directed downwards. White. Provisionally placed in this family. **Habitat:** deep water and offshore. **Distribution:** NWG, GO.

Stomatia cf *rubra* 60

Daronia subdisjuncta 61

Leucorhynchia crossei 63

Lodderena n. sp. 62

Lodderia novemcarinata 64

"*Cyclostrema*" *archeri* 67

"*Cyclostrema*" *charmophron* 69

"*Cyclostrema*" *euchilopteron* 71

"*Cyclostrema*" *eumares* 72

"*Cyclostrema*" *eupoietum* 73

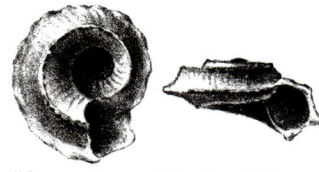

"*Cyclostrema*" *gyalum* 75

"*Cyclostrema*" *henjamense* 76

"*Cyclostrema*" *placens* 77

"*Cyclostrema*" *prominulum* 78

Woodringilla solida (Laseron, 1954): **66**
2mm. Dome-shaped, featureless except for delicate, zigzagging spiral sculpture overall. **Habitat:** intertidal. **Distribution:** Mas, GO.

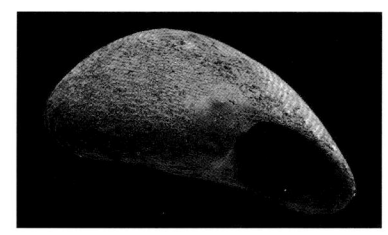

Woodringilla solida **66**

Morchiella moreleti **65**

"Cyclostrema" quadricarinatum **79**

"Cyclostrema" quinquecarinatum **80**

"Cyclostrema" solariellum **81**

"Cyclostrema" supremum **82**

Family "CYCLOSTREMATIDAE"
At least some species formerly classified under this family name should now be considered members of the family Turbinidae (subfamily Liotiinae) or of other families. Although the Cyclostrematidae may be non-existent, as a family, the name is retained here nominally for them because most of the many species of *"Cyclostrema"* from eastern Arabia described by Melvill and Standen have not been studied critically. Their shells have a "family likeness", being very small, mostly coiled on the flat, and often spirally ridged, but there the likeness ends. The species are listed but not described here. Descriptions and illustrations of several of them are given by Melvill, 1906a.

"Cyclostrema" archeri Tryon, 1888: **67**
1mm. Described, as *C. annellarium*, by Melvill & Standen, 1903b:292, from 156 fathoms off Muscat (GO). The accompanying figure illustrates Melvill and Standen's taxon.

"Cyclostrema" carinatum
H. Adams, 1873: **68** NOT ILLUSTRATED
Listed by Melvill, 1906a:21, from NG and GO.

"Cyclostrema" charmophron Melvill, 1906: **69**
0.75mm. Described from 156 fathoms off Muscat (GO).

"Cyclostrema" cingulatum (Dunker, 1859): **70** NOT ILLUSTRATED
Listed by Melvill, 1906a:21, from the Arabian Gulf and GO.

"Cyclostrema" euchilopteron Melvill & Standen, 1903: **71**
2mm. Described from 156 fathoms off Muscat (GO).

"Cyclostrema" eumares Melvill, 1904: **72**
1.75mm. Described from 156 fathoms off Muscat (GO).

"Cyclostrema" eupoietum Melvill, 1904: **73**
2mm. Described from 156 fathoms off Muscat (GO).

"Cyclostrema" exiguum (Philippi, 1849): **74** NOT ILLUSTRATED
Listed by Melvill, 1906a:22, from 156 fathoms off Muscat (GO).

"Cyclostrema" gyalum Melvill, 1904: **75**
5mm. Described from 156 fathoms off Muscat (GO).

"Cyclostrema" henjamense Melvill & Standen, 1903: **76**
6mm. Described from Henjam Island, (NG).

"Cyclostrema" placens (Melvill & Standen, 1901): **77**
1.50mm. Listed by Melvill, 1906a:25, from 156 fathoms off Muscat and deeper water nearby (GO); listed by Smythe, 1979a:64, from SEG.

"Cyclostrema" prominulum Melvill & Standen, 1903: **78**
2mm. Described from 156 fathoms off Muscat (GO).

"Cyclostrema" quadricarinatum Melvill & Standen, 1901: **79**
2.25mm. Described from deep water in the Gulf of Oman (GO). Listed by Biggs, 1973:352, and Smythe, 1979a:64, from U. A. E. (SEG) and by Glayzer et al, 1984:318, from Kuwait (NWG).

"Cyclostrema" quinquecarinatum Melvill, 1906: **80**
2.55mm. Described from 156 fathoms off Muscat (GO). Listed by Smythe, 1979a:64, from U. A. E. (SEG).

"Cyclostrema" solariellum Melvill, 1893: **81**
1.50mm. Listed by Melvill, 1906a:24, from NG and GO; listed by Glayzer et al, 1984:318, from Kuwait (NWG).

"Cyclostrema" supremum Melvill & Standen, 1903: **82**
4mm. Described from near Fao and Bundar Abbas (NG).

Family TURBINIDAE
Extensive, worldwide family of small to large shells which are mostly thick, solid, globose, few-whorled and strongly ornamented but of subdued colour and pattern. The body whorl is well rounded in some genera, keeled in others. The columella is usually smooth and lacking teeth or folds. The aperture is nacreous within. Unlike that of the Trochidae, the spiral operculum is calcareous and often thick and heavy, its inner surface bearing a thin, chitinous layer, its outer surface being smooth or variously ornamented and often brightly coloured; its appearance may be diagnostic for some species. Most turban shells browse small algae from rocks. There are few species in Oman and Gulf waters but some are abundant and conspicuous members of the intertidal fauna. The systematic position of some of the smaller species included here is uncertain.

Subfamily LIOTIINAE
Liotia echinacantha Melvill & Standen, 1903: **83**
7.5mm. Described from Muscat (GO) and occurs off Masirah (Mas). Almost certainly not a *Liotia*.

Liotia echinacantha **83**

Superfamily: **TROCHOIDEA**

Liotia romalea Melvill & Standen, 1903: **84**
5mm. Listed by Melvill, 1906a:28, from Sheyk Shoeyb Island (NG), Muscat and 156 fathoms off Muscat (NG). Almost certainly not a *Liotia*.

Cyclostrema ocrinium Melvill & Standen, 1901: **85**
3mm. Fragile, glossy, spire depressed-conical, sutures channelled, umbilicus deep. Strong spiral keels, 5 on last whorl, crossed by fine, close-set axial ridges; outer lip crenulate. A true *Cyclostrema* and a turbinid. **Habitat:** in mud and sand offshore. **Distribution:** NWG, SEG, GO.

Subfamily TURBININAE

Bothropoma cf **bellula** (H. Adams, 1873): **86**
2mm. Thin, dull, flat spired, all but last whorl coiled in one plane, apertural lip thickened; umbilicus wide and deep. Coarse, spiral ribs above and below periphery crossed by very fine, close-set, axial threads (absent from sunken protoconch). White tinged pink above and around umbilicus, or with pink spots. **Habitat:** intertidal. **Distribution:** NWG, GO.

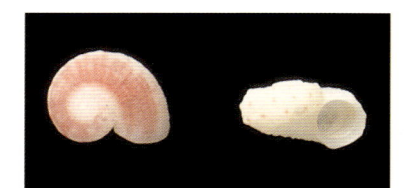

Bothropoma cf **munda** (H. Adams, 1873): **87**
4.5mm. Thick, dull, whorls appearing somewhat straight sided in profile, apex domed; umbilicus small but deep. Regular, close-set, spiral ridges (strongest just above periphery), their intervals crossed by very fine, oblique threads; thick, coarsely beaded ridge around umbilicus. White variously blotched, lined and spotted grey or reddish brown; aperture and umbilicus white. Operculum has central pit. **Habitat:** offshore. **Distribution:** NWG, GO.

Bothropoma cf **pilula** (Dunker, 1860): **88**
6mm. Thick, dull, turbinate, aperture about half total length, spire whorls rounded, apex flat topped, sutures deep, base rounded; umbilicus small but deep. Close-set spiral cords, strongest at periphery; thick, coarsely beaded ridge around umbilicus. White variously blotched, lined and spotted pink or brown; aperture and umbilicus white. **Habitat:** intertidal under stones. **Distribution:** NWG, GO, Mas, SO.

Lunella coronata (Gmelin, 1791): **89**
40mm. Thick, domed or almost flat topped, spire whorls often eroded, last whorl large and broadly expanded. Shallow channel extends from lip below columella; umbilicus small, shallow, smooth sided, sometimes filled in. Last whorl has heavy or light spiral rows of nodules centrally, subsutural nodules larger. Dull green, orange-yellow, brown or reddish grey; aperture nacreous. Operculum faintly granulose, greenish with white centre. Varies greatly in size and ornament. **Habitat:** intertidal on rocks. **Distribution:** all.

Liotia romalea 84

Bothropoma cf *bellula* 86

Cyclostrema ocrinium 85

Bothropoma cf *pilula* 88

Bothropoma cf *munda* 87

Lunella coronata 89

SEASHELLS OF EASTERN ARABIA

Superfamily: **TROCHOIDEA**

Turbo bruneus Röding, 1798: **90**
40mm. Thick, heavy, much taller than broad, apex pointed, body whorl large, sutures deep. Lip below columella extends downwards slightly; umbilicus narrow but deep. Last whorl keeled at shoulder; smooth spiral cords most prominent above and below periphery. Ivory mottled reddish brown; aperture nacreous. Operculum finely granulose, purple becoming paler towards margin. **Habitat:** intertidal on rocks. **Distribution:** GO, Mas, SO.

Turbo chemnitzianus Reeve, 1848: **91**
40mm. Thick, dull, almost as broad as tall, apex rounded, sutures deep; Thickening at base of columella; no umbilicus. Whorls strongly keeled at periphery; scaly spiral cords, strongest at and below periphery. Alternating brown and white, axial streaks on all whorls, scales usually paler; aperture nacreous. **Habitat:** intertidal on rocks. **Distribution:** Mas.

Turbo jonathani Dekker, Moolenbeek & Dance, 1992: **92**
38mm. Thick, heavy, moderately glossy, as broad as tall, apex rounded, sutures deeply channelled. Lip below columella thickened and recurved; no umbilicus. Smooth spiral cords strongest below sutures and above periphery; whole shell covered by oblique, flattened ridges. Reddish brown fading to pinkish brown with paler blotches and streaks; aperture nacreous. Operculum has large, flattened tubercles, greenish or greyish becoming orange or white towards inner margin. **Habitat:** intertidal and offshore under boulders. **Distribution:** GO, SO.

Turbo radiatus Gmelin, 1791: **93**
50mm. Thick, heavy, much taller than broad, apex pointed but always eroded, last whorl large, sutures deep; columella smooth and evenly curved; no umbilicus; edge of aperture below columella extends downwards slightly. Last whorl strongly angled at suture and looks flat sided; coarse spiral cords develop scaly nodules above and below periphery. White mottled reddish brown; aperture nacreous. Operculum shiny, greenish white. May grow to 70mm. **Habitat:** intertidal on rocks. **Distribution:** NWG, SEG, GO, Mas.

Subfamily HOMALOPOMATINAE

Homalopoma tapparonei (Caramagna, 1888): **94**
3.5mm. Thick, dull, globose-turbinate, spire a low dome, apex flat topped, sutures deep, base convex; umbilicus narrow but deep. Columella thickened and straight. Strong, spiral cords as wide as their intervals, crossed by crowded, oblique threads. White, cords intermittently purple-brown, occasionally fusing as blotches; aperture nacreous. Operculum with almost central nucleus, white. **Habitat:** intertidal under stones. **Distribution:** GO, Mas, SO.

Leptothyra rubens Melvill & Standen, 1903: **95**
4.5mm. Listed by Melvill, 1918:152, from Musandam and from 156 fathoms off Muscat (GO).

Turbo chemnitzianus **91**

Turbo bruneus **90**

Leptothyra rubens **95**

Turbo radiatus **93**

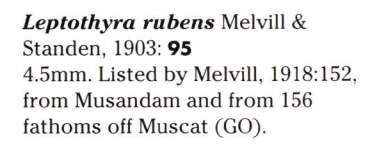

Homalopoma tapparonei **94**

Turbo jonathani **92**

Superfamily: **TROCHOIDEA**

Gabrielona roni **96**

Family PHASIANELLIDAE
Shells of this small, algae-browsing family are small to moderate in size, thin, smooth and shiny, elongate-ovate or globose, umbilicate or not. True pheasant shells, genus *Phasianella*, are recognisable by their spiral capillary lines. Amalgamated here with the one *Phasianella* species occurring in the region are species of *Tricolia* and *Gabrielona* which lack such lines on their shells and are usually placed in a separate family, the Tricoliidae. The calcareous operculum is convex externally and white.

Gabrielona roni Moolenbeek & Dekker 1993: **96**
1.8mm. Thin, semitranslucent to opaque, globose; aperture large, obliquely ovate; umbilicus a mere chink with wide, deep, associated channel; columella smooth, glassy. Fine spiral striae on last whorl. Early whorls white; upper part of last whorl has about 10 spiral lines of white and brown spots; spiral, brown lines below periphery; columella white. Polished, white operculum. Colour and pattern vary greatly. **Habitat:** In algae, shallow water. **Distribution:** SO.

Phasianella solida (Born, 1778): **97**
15mm. Thick, lightweight, elongate-ovate, about 6 whorls, apex blunt, sutures lightly impressed, aperture almost half total height; no umbilicus; columella gently curved. Yellowish brown or orange mottled white and brown; interrupted, spiral, red, capillary lines on all later whorls, often with brown or white crescents between. Polished, white operculum. Colour and pattern variable. **Habitat:** intertidal on algae and beached. **Distribution:** NWG, SEG, GO, Mas, SO.

Tricolia fordiana (Pilsbry, 1888): **98**
4mm. Thin, fragile, globose to elongate-ovate, about 5 whorls, last whorl much taller than spire, suture shallow, apex rounded; umbilicus narrow, columella smooth. Surface glossy and smooth. Colours include reddish brown, orange, pink, yellowish, amber and white with patterns of spots, stripes and mottlings in extraordinary variety. Operculum polished (except for granulated upper portion) and white. **Habitat:** intertidal among muddy stones, sand and grass. Syn: *Phasianella minima* Melvill, 1896. **Distribution:** NWG, SEG.

Tricolia ios Robertson, 1985: **99**
4mm. Thin to solid, translucent to opaque, elongate-ovate, about 5 whorls, last whorl much taller than spire, sutures shallow, apex scarcely raised; no umbilicus. Fine spiral striae. Colour pattern very variable; usually white and brown tinged red, orange or olive-green; conspicuous

Tricolia fordiana **98**

Tricolia ios **99**

Phasianella solida **97**

Tricolia variabilis **100**

Nerita adenensis **101**

flame-like markings and pale subperipheral band. Polished, white operculum. **Habitat:** intertidal on algae. **Distribution:** Mas, SO.

Tricolia variabilis (Pease, 1861): **100**
3.5mm. Thin to solid, elongate to globose, sutures well impressed; no umbilicus, columella smooth. Smooth or faintly spirally striate. Pink, red, brown, yellow, grey in complex patterns of stripes, bands, dots, zigzags. Polished, white operculum. An incredibly variable species, the form prevalent in the region is low spired and diagonally striped. **Habitat:** intertidal on algae. **Distribution:** Mas, SO.

Superfamily NERITOIDEA
A mainly tropical superfamily, its component species have coiled or limpet-like shells and occur in marine and non-marine biotopes. Some occur in brackish water in eastern Arabia. A distinctive operculum with a hooked process on its inner side is characteristic of the principal family.

Family NERITIDAE
All species have very solid, globular shells with a low or depressed spire, a large body whorl and a callused columellar pad which may be smooth, granulated or ridged; there is no umbilicus. The shell surface may be smooth or spirally ribbed and the aperture may be denticulate at its outer edge. The calcareous operculum is half-moon shaped (as is the aperture); smooth or granulose on the outer surface, it is smooth on the inner surface which also bears a hook-like projection; usually greyish externally. Abundant on rocky shores worldwide and well represented in the region, nerites are vegetarians; many species live under intertidal rocks but some are conspicuous on rocky surfaces above high-tide level.

Subfamily NERITINAE
Nerita adenensis Mienis, 1978: **101**
18mm. Thick, dull, globose, as tall as broad, low spire. Columella has 2 central teeth; upper callus pad wrinkled; outer lip smooth inside. Low spiral ridges. White, yellowish, reddish brown, grey or black, often spirally banded and flecked; columella, callus pad and aperture white, tinged yellow or orange. Operculum smooth, glossy. **Habitat:** intertidal on rocks. **Distribution:** SEG, GO, Mas, SO.

Nerita albicilla Linnaeus, 1758: **102**
23mm. Thick, dull, elongate-globose, spire depressed, aperture flaring. Columella has 4 weak, central teeth, smooth above; callus pad heavily granulated; weak denticles inside outer lip. Low spiral ridges and fine growth lines. Greenish, brown, reddish or black, usually banded or blotched; columella white, callus pad and aperture white to yellow-orange. Operculum granulated, dull. **Habitat:** intertidal on rocks. **Distribution:** all.

Nerita debilis Dufo, 1840: **103**
20mm. Thin or thick, dull, cap shaped, spire depressed, broader than tall, aperture flaring. Columella and callus pad smooth. Inner lip thickened and smooth inside; elongate tooth towards base of outer lip. Spiral ridges and fine growth lines. White, orange, beige or black variously patterned with blackish zigzags, spiral lines and dashes; columella, callus pad and aperture white or yellow-orange. Operculum finely granulated, glossy. Syn: *N. anodonta* Melvill, 1898. **Habitat:** intertidal on rocks. **Distribution:** GO, Mas, SO.

Nerita longii Récluz, 1841: **104**
30mm. Thick, dull, globose with slightly raised spire, protoconch always eroded. Columella has 4 strong teeth; callus pad is irregularly granulated; outer lip has close-set ridges inside and 2 large teeth at top. Low, spiral ribs and coarse growth lines. Greenish, yellowish brown or blackish, often mottled and dotted; apertural area white with dark rim, callus pad tinged orange. **Habitat:** intertidal on rocks. **Distribution:** NG, SO.

Nerita debilis **103**

Nerita albicilla **102**

Nerita longii **104**

Superfamily: **NERITOIDEA**

Nerita textilis on exposed rocks at Juzor Al Halaaniyaat

Plesiothyreus omanensis **110**

Smaragdia souverbiana **107**

Plesiothyreus evansi **109**

Nerita polita orbignyana Récluz, 1841: **105**
18mm. Thick, glossy, broader than tall, sutures distinct, flat topped. Columella has about 4 short, broad and narrow teeth; callus pad almost smooth; outer lip weakly ridged inside. Prominent growth lines. Shades of red, brown, orange yellow, green, grey and ivory with darker bands, zigzags and mottlings; apertural area white tinged orange. Columella smooth, glossy. **Habitat:** Intertidal on rocks. **Distribution:** NG, SEG, GO, SO.

Nerita textilis Gmelin, 1791: **106**
35mm. Thick, heavy, dull, much broader than tall, spire depressed. Columella has 2 or 3 weak teeth; callus pad densely granulated; outer lip strongly ridged inside. Coarsely wrinkled, spiral cords. White or orange-yellow dashes on black; apertural area white, callus pad tinged orange. Often heavily eroded. Operculum densely granulated. **Habitat:** intertidal on rocks. **Distribution:** GO, Mas, SO.

Subfamily SMARAGDIINAE
Smaragdia souverbiana (Montrouzier, 1863): **107**
4mm. Thin, glossy, taller than broad, spire blunt topped. Columella has row of small teeth centrally, callus pad smooth. Smooth all over. Bright green with spiral rows of white "arrow heads", fine, black or purple, axial lines overall; occasionally brown or orange spiral bands; callus pad orange-yellow. Operculum smooth, dull. **Habitat:** intertidal among stones and sand. **Distribution:** NWG, SEG, Mas, SO.

Family PHENACOLEPADIDAE
A small family of limpet-like species with thin, colourless shells bearing scaly radial ribs. The backwardly pointing apex is near the posterior end and the shell margin is slightly arched. A horseshoe-shaped muscle scar, similar in outline to that occurring in Patellidae, is present but may be difficult to see. Configuration of the radula separates this family from the true limpets. The species recorded from eastern Arabia usually occur beached.

Plesiothyreus elongatus (Thiele, 1909): **108** NOT ILLUSTRATED
7mm. Similar to *P. evansi* (see below) but smaller, less elevated, with straighter sides and apex farther from margin. Has crowded, radiating threads crossed by weakly scaled, concentric ridges. **Habitat:** beached. **Distribution:** Mas, SO.

Plesiothyreus evansi (Biggs, 1973): **109**
6mm. Thin, brittle, translucent, roundly oval in outline, flat-cap shaped in profile, margin slightly arched along longer axis; bulbous apex almost overhangs posterior margin. Scaly ridges radiate from apex to margins, visible as grooves internally; muscle scar deeply indented laterally. White. **Habitat:** offshore and beached. **Distribution:** NWG, SEG.

Plesiothyreus omanensis (Biggs, 1973): **110**
8mm. Thin, brittle, translucent, roundly oval in outline, flat-cap shaped in profile, margin very slightly arched along longer axis, posterior slope concave; globular protoconch near posterior margin. Fine concentric ridges with irregular scales which are larger and denser towards shell edge, visible

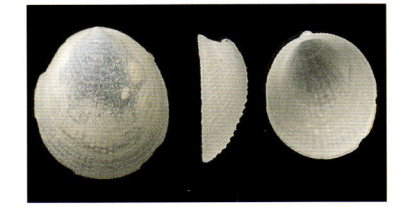

■ *Nerites, oysters and large chitons are familiar to anyone who has clambered around exposed rocks along the coasts of eastern Arabia. The shells of* Nerita textilis *are associated with rocky coasts around Oman where their jazzy colour pattern makes them stand out at a distance. For much of the day they stay firmly fixed to rocks exposed to the broiling heat of the sun, seemingly none the worse for their ordeal. In certain localities, Masirah Island in particular, they may grow much larger than elsewhere. Such outsized specimens are usually very eroded which suggests that they may be older than those of more normal growth.*

Nerita polita orbignyana **105**

Nerita textilis **106**

44 SEASHELLS OF EASTERN ARABIA

internally. Muscle scar has 2 distinct lobes each side linked by opaque rim parallel with shell edge. White. **Habitat:** offshore and beached. **Distribution:** NWG, SEG.

Plesiothyreus osculans (C. B. Adams, 1852): **111** NOT ILLUSTRATED
5mm. Similar to *P. omanensis* (see above) but smaller, more fragile, with straighter sides and closer spaced sculpture lacking scales. **Habitat:** offshore and beached. **Distribution:** GO, SO.

Plesiothyreus pararabicus (Christiaens, 1988): **112**
25mm. Thin, brittle, translucent, oval in outline, rather depressed and gently curved in profile, margin slightly arched along longer axis, posterior slope concave; apex placed about a third of total length from posterior end. Crowded, prickly ridges radiate from apex to margins, visible internally; muscle scar slightly indented laterally. White. **Habitat:** offshore and beached. **Distribution:** NWG, SEG, GO. Also recorded: *P. galathea* (Lamarck, 1819) and *P. scobinata* (Gould, 1859), both from NWG.

Superfamily LITTORINOIDEA
Although this superfamily includes many terrestrial gastropods only one marine family occurs in eastern Arabia, the Littorinidae.

Family LITTORINIDAE
The larger members of this family are known as periwinkles or winkles, their compact, ovate-conical shells being familiar objects on stony surfaces intertidally or high up on the shore where they feed mostly on algae. Plentiful locally, they are not as significant a feature of eastern Arabian coasts as they are of northern Atlantic coasts. Their shells are smooth or have spiral ribs or rows of nodules and a thin periostracum. Except for the tiny species of *Peasiella* none has an umbilicus but all have a smooth, concave columella, a thin outer lip and a thin, chitinous operculum with few spiral whorls. Material of the family Eatoniellidae is represented but it is still under review and comprises species which may all be new to science.

Subfamily LITTORININAE
Littoraria (Littoraria) glabrata (Philippi, 1846): **113**
20mm. Solid, dull, ovate-conical, sutures incised; pointed protoconch of 3 smooth whorls. Spire whorls gently rounded, periphery of last whorl angular. Later whorls spirally grooved. Cream to pinkish with oblique or zigzag brown stripes or reticulations; aperture orange, base of columella brown. Very variable colour pattern. **Habitat:** rocks at high-tide mark. **Distribution:** GO, Mas, SO.

Littoraria (Littorinopsis) intermedia (Philippi, 1846): **114**
17mm. Thin, dull, elongate-ovate, sutures deep; pointed protoconch of 3.5 smooth whorls. Spire whorls rounded, last whorl gently angled at periphery. All whorls evenly, spirally grooved; strong subsutural cord. Grey, orange or reddish brown often with spiral rows of dashes, paler mottlings and spots; aperture and columella reddish brown. **Habitat:** intertidal rocks. **Distribution:** NG, NWG, SEG, Mas.

Nodilittorina (Nodilittorina) arabica El Assal, 1990: **115**
9mm. Thick, dull, globose-ovate, sometimes weakly angular at periphery of last whorl, sutures deep; pointed protoconch of 3 smooth whorls. Early spire whorls spirally ridged, ridges on later whorls becoming coarsely nodular. Orange or shades of brown; columella whitish or brown. **Habitat:** intertidal rocks. **Distribution:** NWG, SEG.

Plesiothyreus pararabicus **112**

Littoraria (Littorinopsis) intermedia **114**

Littoraria (Littoraria) glabrata **113**

Nodilittorina (Nodilittorina) arabica **115**

Superfamily: **LITTORINOIDEA, RISSOOIDEA**

Nodilittorina (Nodilittorina) millegrana 116

Nodilittorina (Nodilittorina) natalensis 117

Truncatella marginata 120

Nodilittorina (Nodilittorina) millegrana (Philippi, 1848): **116**
10mm. Thick, dull, globose-ovate or ovate-conical, slightly tabulated below the deep sutures; pointed protoconch of about 2 smooth whorls. Early spire whorls spirally ridged, ridges on later whorls becoming coarsely nodular, ridges alternately strong and weak. Whitish or yellowish orange; columella and aperture orange brown. **Habitat:** intertidal rocks. **Distribution:** NWG, SEG, GO, SO.

Nodilittorina (Nodilittorina) natalensis (Philippi, 1847): **117**
10mm. Thick, dull, high spired, straight sided, sutures deep: pointed protoconch. Coarse blunt nodules (3 spiral rows on last whorl), coarse growth lines. Lower part of columella broadly flared. Greyish with whitish nodules; aperture chocolate with paler edge and paler band emerging at base. Entire shell often eroded and encrusted. **Habitat:** intertidal rocks. **Distribution:** Mas, SO.

Peasiella infracostata (Issel, 1869): **118**
4mm. Thick, dull, compressed top shaped, flat based, spire whorls broadly tabulated, angled at periphery, sutures incised; flat protoconch of 2.5 smooth whorls, usually eroded. Thick keel may be present at periphery of last whorl; outer lip has 5 or 6 notches corresponding to keel and sharp spiral ridges on base; deep umbilicus; fine axial growth lines overall. Amber with reddish brown, oblique stripes. **Habitat:** intertidal under stones. **Distribution:** GO, Mas, SO.

Peasiella isseli (Semper in Issel, 1869): **119**
5mm. Description as for *P. infracostata* (above) but slightly larger, much more compressed, whorls more angled at periphery, keel of last whorl sharper and basal ridges less prominent towards umbilicus. Colour pattern similar but oblique stripes wider apart. **Habitat:** intertidal under stones. **Distribution:** NWG, SEG, GO, Mas.

Superfamily RISSOOIDEA
The small to tiny gastropods placed in this superfamily belong to many families some of which occur only in fresh-water habitats. As diverse in external appearance as they are in behaviour, their relationships are only imperfectly understood. Many species have elongated shells but others are discoidal. The status of many small or tiny species belonging to the families Barleeidae, Cingulopsidae and Rissoidae is still under review. Preliminary studies indicate that most of them may be new to science.

Family TRUNCATELLIDAE
Most species in this small, herbivorous family live at high-water mark or in brackish water but some have colonised the land. The slender shells are strongly ribbed and nearly always lose their early whorls by maturity. Only one eastern Arabian species is recorded.

Truncatella marginata Küster, 1855: **120**
7mm. Thin, dull, narrowly elongate, sutures deep, early whorls usually missing so that spire is characteristically truncate; protoconch of 3 smooth, rounded whorls. Outer lip projects slightly, no umbilicus. Regularly spaced axial ribs on all whorls. White. **Habitat:** pools at high-water mark and beached. **Distribution:** SWG, Mas.

Family IRAVADIIDAE
Small gastropods with usually solid, conic shells, non-umbilicate shells which are smooth or spirally sculptured shells. The periostracum is sometimes well developed and there is an operculum. Shells are usually found in beach drift.

Iravadia quadrasi (Boettger, 1893): **121**
2.5mm. Solid, ovate with rounded whorls having a coarse cancellate sculpture, aperture circular and outer lip thickened; protoconch of about 2 large, smooth whorls. **Habitat:** intertidal. **Distribution:** SEG.

Lucidinella densilabrum (Melvill, 1912): **122**
3.5mm. Elongate-ovate with well rounded whorls, slightly channelled sutures; columellar lip sharply angled towards base; protoconch of 2.5 flattened whorls. Crowded spiral threads. Pale brown to greyish white. **Habitat:** intertidal. **Distribution:** SEG.

Pseudonoba alphesiboei (Melvill, 1912): **123**
5mm. Columnar in outline, sometimes slightly curved, outer lip thickened; protoconch flattened. Close spiral striae overall. White. Large for the genus. **Habitat:** intertidal and offshore. **Distribution:** NG, NWG.

Peasiella infracostata 118

Peasiella isseli 119

Peasiella isseli 119

Iravadia quadrasi 121

Lucidinella densilabrum 122

Pseudonoba alphesiboei 123

SEASHELLS OF EASTERN ARABIA

Superfamily: **RISSOOIDEA**

Pseudonoba aristaei (Melvill, 1912): **124**
3.5mm. Listed by Glayzer et al, 1984:319, from Kuwait (NWG).

Pseudonoba columen (Melvill, 1904): **125**
4mm. Described from 156 fathoms off Muscat (GO).

Pseudonoba elspethae (Melvill, 1910): **126**
3.5mm. Spire whorls rounded, aperture prolonged obliquely downwards; protoconch large and bulbous. White. **Habitat:** intertidal and offshore. **Distribution:** GO.

Pseudonoba ictriella (Melvill, 1910): **127**
4mm. Attenuate-fusiform with rounded whorls, aperture prolonged downwards; protoconch of 2 flattened whorls. Delicately spirally striate. Greyish. **Habitat:** intertidal. **Distribution:** NWG.

Family RISSOIDAE
Small or tiny gastropods with conical shells, smooth or variously sculptured, having an aperture usually shorter than the spire. The operculum is usually oval, may be thin or thick and may bear a peg on its inner surface. The animals feed on algae and diatoms and are common intertidally and beached. Eastern Arabian members of the family have been little studied and even some of the relatively large species of *Rissoina* have not been identified satisfactorily.

Alvania cf *mahimensis* Melvill, 1893: **128**
2.5mm. Solid, dull, ovate, aperture small, sutures deep; protoconch of 2 rounded whorls. Axial ribs cut by spiral grooves producing a cancellate sculpture, groove below suture deepest. White with a spiral red band below sutures and another around base. **Habitat:** intertidal. **Distribution:** GO, Mas, SO.

Benthonellania charope (Melvill & Standen, 1901): **129**
1.75mm. Described from deep water north-west of Muscat (GO).

Merelina sp: **130**
1mm. Translucent, glossy, sculptured with widely spaced, hook-like lamellae and with a large, finely spirally striate protoconch, this distinctive species is still unidentified. **Habitat:** intertidal. **Distribution:** SO.

Rissoina cerithiiformis Tryon, 1887: **131**
4.5mm. Thick, elongate-conic, spire whorls straight-sided, sutures deep; protoconch of 3 whorls, the first smooth, the next spirally striate, the third weakly keeled and bearing axial threads. Outer lip thickened. Regular axial ribs and spiral grooves produce a delicately cancellate sculpture. **Habitat:** intertidal. **Distribution:** GO.

Rissoina pachystoma Melvill, 1896: **132**
4mm. Thick, elongate-conic with deep sutures and narrow aperture; dome-shaped protoconch. Outer lip greatly thickened. Widely spaced, straight axial ribs, thick spiral rib at base. Greyish white. **Habitat:** intertidal and beached. **Distribution:** Mas.

Pseudonoba elspethae **126**

Pseudonoba ictriella **127**

Pseudonoba aristaei **124**

Pseudonoba columen **125**

Alvania cf *mahimensis* **128**

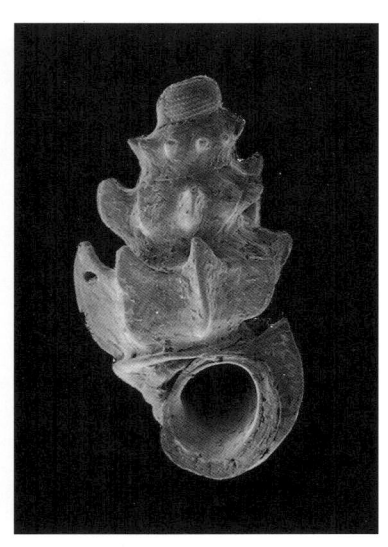

Merelina sp. **130**

Benthonellania charope **129**

Rissoina cerithiiformis **131**

Rissoina pachystoma **132**

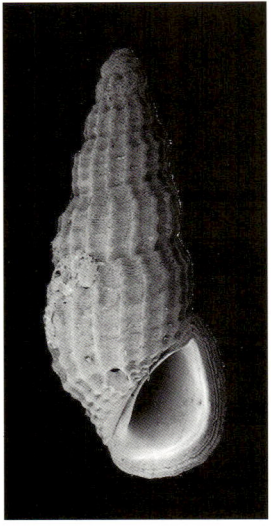

Superfamily: **RISSOOIDEA**

Rissoina phormis Melvill, 1904: **133**
3.5mm. Thick, elongate-conic with thickened outer lip slightly protruding forwards at base; protoconch of 2 rounded, glassy whorls. Cancellate sculpture overall. **Habitat:** intertidal. **Distribution:** GO.

Rissoina* cf *pulchella (Brazier, 1877): **134**
3.5mm. Thick, elongate-ovate, spire whorls slightly convex; dome-shaped protoconch of about 3 rounded whorls. Fine, straight axial ribs narrower than their intervals, fine spiral threads between ribs at base. White. **Habitat:** intertidal. **Distribution:** GO.

***Rissoina* sp**: **135**
12mm. Thick, dull, elongate-ovate, spire whorls slightly convex, sutures deep; pointed protoconch of about 3 rounded, glassy whorls. Thickened outer lip protrudes forwards. Slightly curved axial ribs crossed by weaker spiral ridges which become stronger on lower half of whorls and are strongest on outer lip. White tinged pink. **Habitat:** intertidal in sand. **Distribution:** Mas.

Stosicia annulata (Dunker, 1860): **136**
3.3mm. Thick, semitranslucent, elongate-ovate, outer lip thickened behind; flat-topped protoconch of about 3 smooth whorls. Bold, sharp, spiral keels (6 on last whorl), otherwise smooth. White. **Habitat:** intertidal pools and in shell sand. **Distribution:** NG, NWG, SEG, GO, Mas.

Voorwindia tiberiana (Issel, 1869): **137**
1.3mm. Thin, glossy, globose-ovate, whorls rounded, sutures deep, umbilicus narrow; blunt-topped protoconch. Very fine, spiral striae overall. White. **Habitat:** intertidal and offshore. **Distribution:** NWG, SEG, GO, Mas.

Zebina insignis (A. Adams & Reeve, 1850): **138**
10mm. Solid, glossy, ovate, spire whorls rounded, sutures well defined; protoconch not seen. Aperture less than half total length; edge of outer lip lirate, its lower part protruding forwards. Regular, fine, spiral threads overall. White. Apex often missing. **Habitat:** beached. **Distribution:** Mas.

Zebina tridentata (Michaud, 1830): **139**
9mm. Thick, glossy, ovate with a straight-sided spire which is flatter on the apertural than on the dorsal side, sutures barely visible; protoconch of 1 smooth whorl. Early spire whorls have weak axial riblets. Aperture half total length; thick parietal callus continuous with thickened outer lip. Lower part of outer lip has 3 blunt denticles. White (but dorsal side of last whorl often tinged orange-violet when occupied by a crustacean). **Habitat:** offshore presumably among echinoderms. **Distribution:** GO, Mas, SO.

Family BARLEEIDAE
Tiny snails with thin, smooth, conical-ovate, reddish-brown shells. There is a thick operculum bearing a peg and a thick, longitudinal ridge. They occur on algae intertidally. Represented in eastern Arabia by an unidentified species from Masirah.

***Barleeia* sp**: **140**
1.3mm. Thin, smooth, ovate, spire whorls slightly convex, sutures impressed, aperture almost circular, dome-shaped apex. Brown with a white zone around the tiny umbilicus. **Habitat:** intertidal. **Distribution:** Mas.

Family ORBITESTELLIDAE
Tiny gastropods known from widely scattered localities but recognised only recently in the Indian Ocean. Their shells (less than a millimetre across) are among the smallest in the world. In the intertidal zone they occur under rocks and among coralline algae.

Superfamily: **RISSOOIDEA**

Orbitestella bermudezi (Aguayo & Borro, 1946): **141**
0.5mm. A species known previously only from the western Atlantic, it has been recorded by Moolenbeek, 1994:5, from Masirah and Al Halaaniyaat (Kuria Muria islands) (SO).

Boschitestella donaldi Moolenbeek, 1994: **142**
0.86mm. Fragile, translucent, discoidal, flat spired, umbilicus very wide; initially dimpled protoconch of half a whorl. Sharp keel around periphery crossed by regular, axial ridges; finer and more numerous axial ridges around umbilicus; fine spiral threads overall. White.
Habitat: in tidal pools with rocks.
Distribution: GO.

Boschitestella eloiseae Moolenbeek, 1994: **143**
0.55mm. Differs from *B. donaldi* (see above) by its larger and differently formed protoconch and by having fewer spiral threads.
Habitat: in tidal pools with rocks.
Distribution: SO.

Family CAECIDAE
The curious, tiny shell of members of this family develops successively from a) the spirally coiled shell of a free-swimming, veliger larva and b) a slightly curved tube from which the larval shell drops off. After some growth a thin wall (or septum) forms near the growing end and the tube behind it drops off. Further growth produces the tubular shell of the final stage. A spiky process (or mucro) may project from the septum. There is a thin, multispiral operculum. Some species are known to feed on diatoms.

Caecum sp. (a): **144**
1mm. A thin, gradually enlarging, curved tube blocked off by a smooth septum at the rear end. Smooth, dull, white. **Habitat:** intertidal.
Distribution: Mas.

Caecum sp. (b): **145**
1.5mm. A thin, gradually enlarging, curved tube encircled by sharp-edged, concentric rings and blocked off by a septum bearing a mucro at the rear end. Dull, white. **Habitat:** intertidal. **Distribution:** GO.

Family PELYCIDIIDAE
Minute gastropods with solid, elongate shells which are usually found in beach drift. The systematic position of this family is uncertain.

Pelycidion cf *xanthias* (Watson, 1888): **146**
0.8mm. Listed by Smythe, 1979a: 65, from U. A. E. (SEG) and by Glayzer et al, 1984: 319, from Kuwait (NWG). Shell from Abu Dhabi illustrated.

Orbitestella bermudezi 141

Boschitestella donaldi 142

Boschitestella eloiseae 143

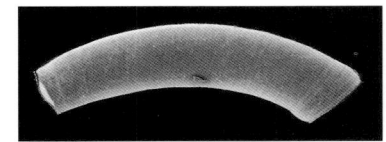

Caecum sp. (a): 144

Caecum sp. (b): 145

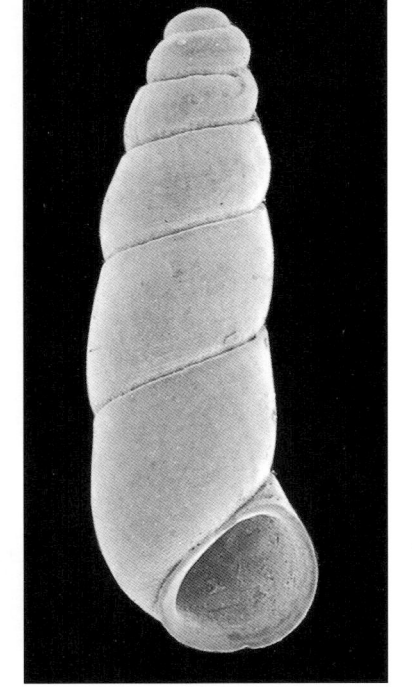

Pelycidion cf *xanthias* 146

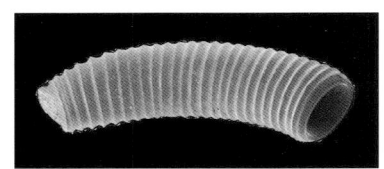

Superfamily CERITHIOIDEA

Several of the component families of this superfamily, such as the Cerithiidae, have long-spired shells with a siphonal notch. Some, such as the Planaxidae and Modulidae, have short-spired shells. The Turritellidae display a long spire and a thin-edged lip without a notch, while the Vermetidae have a similarly thin-edged lip on a shell which grows haphazardly and is usually cemented to a substrate. A circular, spirally coiled, chitinous operculum is usually present. Widely distributed in warm seas and well represented in eastern Arabia, most members of this very large and varied group browse algae and may occur in prodigious numbers.

Family PLANAXIDAE

Planaxis snails have a strong, ovate shell which is sometimes covered with a persistent, fibrous periostracum but is often eroded on the spire. The posterior canal is a distinct groove, the outer lip has fine ridges inside, and there is no umbilicus; the operculum is chitinous with few spiral turns. Planaxids browse algae between tide marks where they may congregate in large numbers. Recent research shows that species formerly placed in the family Fossaridae should be included in the Planaxidae. The shell of *Fossarus* has a prominent protoconch, strong spiral sculpture and a narrow or broad umbilicus; the chitinous operculum has an open spiral and a raised nucleus. Several so-called species of *Fossarus* have been recorded from eastern Arabia but the identifications of most of them are untrustworthy.

Planaxis niger Quoy & Gaimard, 1834: **147**
15mm. Thick, moderately glossy or dull, ovate-conical, spire longer than aperture, sutures impressed. Spiral grooves on all whorls but obsolete at centre of last whorl. Aperture flaring, widely spaced ridges inside. Uniformly brown except for paler early whorls. **Habitat:** intertidal on mud and rocks. **Distribution:** GO.

Planaxis sulcatus (Born, 1780): **148**
20mm. Solid, glossy or dull, ovate, whorls slightly convex, last whorl angled towards base, spire whorls usually eroded. Shallow spiral grooves on all whorls, deeper on base; aperture spirally ridged inside, its edge sharp. Greyish white with oblique, reddish brown stripes or uniformly brown; columella white; aperture brown with pale ridges. **Habitat:** intertidal on mud and rocks. **Distribution:** all.

Fossarus ambiguus (Linnaeus, 1758): **149**
3mm. Solid, dull, globose, spire short; protoconch of about 4 sculptured whorls (described below). Columella slightly curved, aperture semi-lunate, outer lip thin; umbilicus narrow and deep. Strong spiral cords (about 5 on last whorl) and distinct growth lines. Yellowish white. Syn: *F. aptus* Melvill, 1912. **Habitat:** Intertidal and offshore. **Distribution:** NWG, GO, Mas, SO.

Planaxis niger 147

Fossarus trochlearis 150

Planaxis sulcatus 148

Fossarus ambiguus 149

Superfamily: **CERITHIOIDEA**

▪ *Widely distributed in eastern Arabia,* Fossarus ambiguus *also occurs in many other parts of the globe. The reason for its wide distribution is revealed by the electron scanning micrograph which shows that the first whorl of the shell's large protoconch has a granulated surface, the next three whorls bearing axial and spiral ribs — hall-marks of a long larval development.*

Fossarus trochlearis (A. Adams, 1854): **150**
6mm. Differs from *F. ambiguus* (see above) by its larger size and by having bold keels (3 on last whorl) instead of spiral cords, the outer lip having correspondingly strong crenulations. **Habitat:** intertidal. **Distribution:** NWG, GO, Mas, SO.

Family MODULIDAE
Sturdy conical or depressed-conical shells in which the last whorl is broadly expanded and angulate, a tooth at the base of the columella points downwards, and the operculum is thin, corneous and multi-spiral. The few species live in sandy places intertidally and offshore.

Modulus tectum (Gmelin, 1791): **151**
25mm. Solid, squat shell with broadly expanded last whorl, about 6 depressed spire whorls and deep suture. Smooth columella ends in ridge-like tooth; tiny umbilicus. Coarse oblique nodules crossed by irregular spiral ridges. White speckled brown. **Habitat:** offshore in sand. **Distribution:** SEG, GO.

Family CERITHIIDAE
Known as creepers or ceriths, the larger members of this family are among the commoner gastropods around the coasts of Oman and the Gulf. Most species have a narrowly elongate, many-whorled, thick and dull shell ornamented with ribs or nodules and a short, recurved siphonal canal. There is no umbilicus. The thin, chitinous operculum is usually oval and has few turns. Some live among rocks and algae, often in silty or muddy places, others prefer the vicinity of coral reefs where they frequent patches of sand or rubble. All are herbivorous and some may congregate in large numbers. Their shells may vary considerably, even within a single habitat, and this aggravates the already difficult job of identifying some of them. Apart from a few small, deep-water species most of those recorded from the region occur intertidally.

Subfamily CERITHIINAE
Argyropeza divina Melvill & Standen, 1901: **152**
6mm. Thin, translucent, glazed, narrowly elongate, deep sutures; pointed protoconch of 3 glossy whorls. Columella straight; no umbilicus. First post-nuclear whorl has 3 spiral ridges; remaining spire whorls have 2 keeled, spiral rows of prickly nodules; 2 basal, spiral ridges. White tinged amber. **Habitat:** deep water in mud. **Distribution:** GO.

Argyropeza verecundum (Melvill & Standen, 1903): **153** NOT ILLUSTRATED
Similar to **A. divina** but smaller, more slender and with a more curved columella. From 156 fathoms off Muscat. Syn: *Cerithium pervicax* Melvill, 1904. **Habitat:** deep water. **Distribution:** GO.

Bittium anembatum (Melvill, 1904): **154** NOT ILLUSTRATED
5mm. Solid, dull, fusiform, sutures incised; protoconch of 4 smooth, spirally keeled whorls. Spire whorls encircled by rows of weakly beaded cords. Aperture rounded-ovate with a short siphonal canal. Off-white. **Habitat:** deep water. **Distribution:** GO.

Bittium atramentarium Melvill & Standen, 1901: **155** NOT ILLUSTRATED
Described originally from Karachi specimens, this was listed by Smythe, 1979a:65, from Dubai (SEG).

Bittium (Dahlakia) proteus Jousseaume, 1930: **156**
10mm. Thin, dull, elongate, more than twice as long as wide, blunt topped. Slightly oblique, low axial ribs crossed by fine, irregular, spiral striae. Columella smooth, outer lip thin, wavy edged. Reddish brown with whitish ribs; aperture reddish brown. An older name may be *Litiopa bucciniformis* Hornung & Mermod, 1928. **Habitat:** beached. **Distribution:** GO.

Bittium tenthrenois Melvill, 1896: **157** NOT ILLUSTRATED
Described originally from Bombay, this was listed by Smythe, 1979a:66, from Dubai (SEG).

Cerithium caeruleum Sowerby, 1855: **158**
35mm. Thick, almost biconic, sutures shallow. Fine, unevenly spaced striae cross low or elevated nodules; nodules usually prominent at periphery of whorls, smaller and more numerous subsuturally. Varix opposite outer lip; short, narrow siphonal canal; outer lip weakly corrugated or smooth. Yellowish with grey-blue or brownish, spiral blotches, dashes and dots; aperture bluish white. **Habitat:** intertidal under rocks. **Distribution:** NG, NWG, GO, Mas, SO.

Bittium (Dahlakia) proteus **156**

Argyropeza divina **152**

Modulus tectum **151**

Cerithium caeruleum **158**

SEASHELLS OF EASTERN ARABIA

Superfamily: **CERITHIOIDEA**

Cerithium columna **159**

Cerithium nodulosum adansonii **160**

Cerithium rueppelli **161**

Cerithium scabridum **162**

Cerithium columna Sowerby, 1834: **159**
35mm. Thick, slightly more than twice as long as wide, sutures indistinct. Coarse spiral cords and fine threads cross large, pointed nodules, nodules being smaller and more numerous on last whorl. Thick varix opposite outer lip; short, oblique siphonal canal not overhung by corrugated outer lip; aperture weakly lirate. White or cream, mottled, speckled and lined reddish brown; aperture white. **Habitat:** intertidal in sand. **Distribution:** NG, SEG, GO, Mas, SO.

Cerithium nodulosum adansonii Bruguière, 1792: **160**
60mm. Thick, dull, about 2.5 times as long as wide, sutures shallow. Coarse spiral threads cross large, rounded nodules at periphery of whorls. Thick varix opposite outer lip; posterior canal distinct; short, oblique siphonal canal partly overhung by corrugated outer lip. White or cream often mottled and spotted with brown; aperture white. **Habitat:** intertidal in sand. **Distribution:** GO, Mas, SO.

Cerithium rueppelli Philippi, 1848: **161**
30mm. Solid, dull, about 2.5 times as long as wide: pointed protoconch. Spiral cords (2 strong and 2 weak per spire whorl) cross strong axial ribs, nodular at intersections, nodules prominent mid-whorl. Deep posterior canal and short, recurved siphonal canal; outer lip sharply corrugated. White with brown spots on cords, spots larger on cord below each suture. **Habitat:** intertidal. **Distribution:** SEG.

Cerithium scabridum Philippi, 1848: **162**
20mm. Solid, about 3 times as long as wide; pointed protoconch of 1.5 whorls. Spiral cords (2 or 3 dominant ones per spire whorl) cross strong axial ribs, nodular at intersections; flattened spiral threads between cords. Short, recurved siphonal canal; outer lip corrugated. White with brown spiral bands or white with reddish brown spots on cords and nodules; greenish periostracum. Syn: *C. adenense* Sowerby. **Habitat:** subtidal in sand. **Distribution:** NWG, SEG, GO, Mas, SO.

Superfamily: **CERITHIOIDEA**

Cerithium tuberculatum
(Linnaeus, 1758): **163** NOT ILLUSTRATED
This name, occasionally used in the literature of eastern Arabia, may have been intended for *C. echinatum* (Lamarck, 1822) but cannot be applied with certainty to any species now.

Cerithium egenum Gould, 1849: **164**
10mm. Thick, dull, about 3 times as long as wide, almost straight sided, sutures shallow; apex usually eroded. Spiral row of nodules each mid-whorl, occasional low varices, entire surface has low spiral threads; posterior canal distinct; short, oblique siphonal canal. White or cream with axial brown flames and blotches; aperture white. **Habitat:** intertidal under rocks and beached. **Distribution:** Mas.

Clypeomorus bifasciatus bifasciatus (Sowerby, 1855): **165**
18mm. Thick, dull, twice as long as wide, almost straight-sided, sutures indistinct. Low axial ribs crossed by low spiral cords bearing low nodules, grooves between cords. Siphonal canal short and oblique; outer lip thickened. White or cream usually with broad, dark brown, spiral bands; outer lip dotted brown; aperture white or brown. **Habitat:** intertidal on and under rocks. **Distribution:** NG, GO, SO.

Clypeomorus bifasciatus persicus Houbrick, 1985: **166**
16mm. Less compact, more prominently noduled and usually smaller than *C. b. bifasciata*. Easily distinguished by its black and white chequered appearance and sculpture of large, white beads. Sometimes brown with white nodules; spiral grooves may have brown tint. **Habitat:** intertidal under rocks. **Distribution:** NWG, SEG.

Colina pinguis A. Adams, 1855: **167**
16mm. Thin, glossy, elongate-fusiform, last whorl narrower than penultimate; pointed protoconch usually missing from adults. Low, rounded axial ribs on lower half of whorls and regular spiral grooves overall. Aperture flares at base; columella smooth. Shades of brown mottled and spotted white; sutures dotted dark brown; brown-barred outer lip. Varies greatly in size and obesity. **Habitat:** intertidal among algae in pools or under rocks. **Distribution:** NWG, GO, Mas, SO.

Plesiotrochus cf penitricinctus (Cotton, 1932): **168**
10mm. Solid, glossy, pagodiform, twice as long as wide; apical whorls not seen. Prominent, rounded keel and a lesser keel below on each whorl; spiral ridges at base. Cream with axial, flame-like, brown streaks; aperture white. Originally described from Australian shells, the identification is tentative. **Habitat:** intertidal and beached. **Distribution:** Mas.

Clypeomorus bifasciatus bifasciatus **165**

Clypeomorus bifasciatus persicus **166**

Cerithium egenum **164**

Colina pinguis **167**

Plesiotrochus cf *penitricinctus* **168**

Plesiotrochus souverbianus Fischer, 1878: **169**
10mm. Description as for *P.* cf *penitricinctus* (see above) but lacks double spiral cords at periphery and has finer spirals around base. **Habitat:** intertidal among algae. **Distribution:** GO.

Plesiotrochus souverbianus **169**

Superfamily: **CERITHIOIDEA**

Rhinoclavis (Proclava) kochi
(Philippi, 1848): **170**
35mm. Solid, glossy, straight sided, about 3.5 times as long as wide; pointed protoconch; aperture with deep posterior canal, reflexed siphonal canal and blunt fold basally. Flattened spiral cords cross irregular axial ribs giving beaded appearance. Often a low varix opposite outer lip. Pale brown with fine brown spiral lines and cream spiral band below each suture; columella white. **Habitat:** intertidal in sand. **Distribution:** NWG, GO, Mas.

Rhinoclavis (Proclava) sordidula
(Gould, 1849): **171** NOT ILLUSTRATED
According to Houbrick, 1978:73, this is present in the Arabian Gulf.

Rhinoclavis (Rhinoclavis) articulata (Adams & Reeve, 1850): **172** NOT ILLUSTRATED
According to Houbrick, 1978:52, there is a record for the Arabian Gulf.

Rhinoclavis (Rhinoclavis) aspera
(Linnaeus, 1758): **173** NOT ILLUSTRATED
Listed by Melvill (1928:101) from Muscat (GO) but not found since.

Rhinoclavis (Rhinoclavis) fasciata
(Bruguière, 1792): **174**
65mm. Thick, glossy, straight sided, flatter ventrally, about 4.5 times as long as broad, early spire whorls sometimes with varices. Close-set, flat axial ribs (stronger towards sutures) crossed by 2 or more incised spiral lines per whorl. Aperture has weak posterior canal and strongly reflexed siphonal canal. Columella has strong fold centrally. White to shades of brown, blotched, lined and spotted with pale and darker brown; aperture white. **Habitat:** subtidal in sand and beached. **Distribution:** NWG, SEG, GO, Mas, SO.

Rhinoclavis (Rhinoclavis) sinensis
(Gmelin, 1791): **175**
40mm. Solid, glossy or dull, about 2.5 times as long as wide; blunt protoconch of 1.5 whorls; aperture with deep posterior canal and acutely reflexed siphonal canal. Spiral cords strong (up to 10 on last whorl), nodular below each suture, often spinose. Columella has a thick fold. Cream or white, blotched, spotted and lined with shades of brown; columella white, tinged brown basally. **Habitat:** intertidal under rocks. **Distribution:** NG, GO, Mas, SO.

Family DIALIDAE
Tiny, elongate snails often locally abundant on weed in silty places near coral formations. The shells are usually smooth and lack both anterior notch and siphonal canal. They differ from those of the Litiopidae (see below) by having a smooth protoconch. There is a chitinous operculum.

Diala semistriata (Philippi, 1849): **176**
2mm. Thin, glossy, elongate-conic, narrow and channelled sutures; protoconch of about 1.5 smooth whorls. Fine spiral threads at base, otherwise almost smooth. Columella straight, no umbilicus. Brown, occasionally dark brown or cream, with brown and white spiral band below sutures. **Habitat:** intertidal and offshore. **Distribution:** NWG, SEG, GO, Mas.

Rhinoclavis (Proclava) kochi **170**

Rhinoclavis (Rhinoclavis) fasciata **174**

Rhinoclavis (Rhinoclavis) sinensis **175**

Diala sulcifera martensi (Issel, 1869): **177**
3mm. Description as for *D. semistriata* (see above) but has strong spiral ridges on all whorls and is white. **Habitat:** beached. **Distribution:** NWG, SEG.

Family LITIOPIDAE
Tiny snails with thin, elongate shells differing from those of the Dialidae (see above) by having a strongly sculptured protoconch. Prominent varices occur on the shells of some species. There is a chitinous operculum.

Alaba virgata (Philippi, 1849): **178**
4.5mm. Thin, translucent, glossy, elongate-conical, sutures channelled; protoconch of about 3.5 rounded whorls. Irregular, pronounced varices on spire whorls. Columella straight. Whitish with thin, axial brown streaks. **Habitat:** intertidal and offshore. **Distribution:** NWG, SEG, GO, Mas.

Litiopa melanostoma (Rang, 1829): **179**
5mm. Known intertidally from Dhofar (SO), this species has a circumtropical distribution and its distinctive protoconch shows that there is a long, free-swimming larval stage.

Styliferina goniochila A. Adams, 1860: **180**
3mm. Fragile, translucent, glossy, broadly elongate, sutures impressed; pointed protoconch of about 5 whorls. All whorls rounded and smooth. Columella straight. Colourless with spiral row of white blotches around periphery of last whorl. **Habitat:** intertidal and offshore. **Distribution:** NWG, SEG, GO.

Styliferina goniochila 180

Alaba virgata 178

Litiopa melanostoma 179

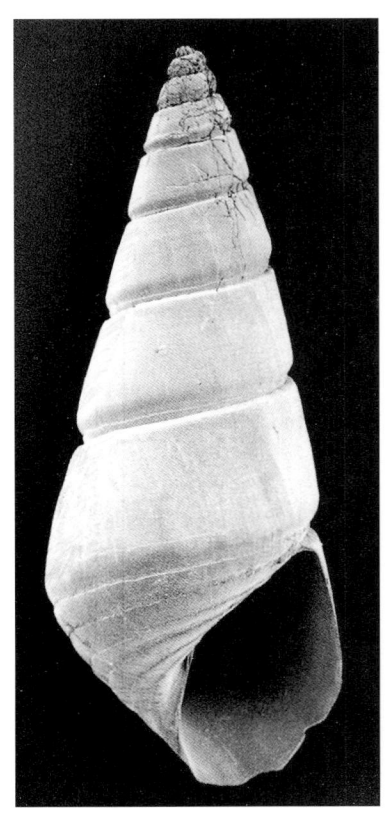

Diala semistriata 176

Diala sulcifera martensi 177

Superfamily: **CERITHIOIDEA**

Cerithidium cerithinum **181**

Cerithidium diplax **182**

Obtortio pupoides **183**

Cerithidea cingulata **184**

Family DIASTOMATIDAE
Tiny snails with thin or thick, elongate shells with convex whorls and a broad siphonal canal. They live among weeds in shallow water and probably feed upon minute algae. The aperture may have a wide, shallow siphonal canal. After revision this little-known family may accommodate several other species occurring in eastern Arabia.

Cerithidium cerithinum (Philippi, 1849): **181**
2mm. Thin, narrowly or broadly elongate, sutures impressed; protoconch of 2 rounded whorls. Axial ribs crossed by spiral threads, nodulose at intersections. Columella straight. White to brown. Syn: *Fenella reticulata* (A. Adams, 1860), *Bittium perparvulum* (Watson, 1886). **Habitat:** tide pools and offshore. **Distribution:** NG, NWG, GO, Mas.

Cerithidium diplax (Watson, 1886): **182**
4mm. Thin, translucent, narrowly elongate, sutures impressed; pointed protoconch of 1.5 smooth whorls. Widely spaced axial ribs crossed by widely spaced spiral threads, noduled at intersections. Columella straight. White or cream. Syn: *Obtortio elongella* (Melvill, 1910). **Habitat:** intertidal and offshore. **Distribution:** NWG, GO, Mas.

Obtortio pupoides (A. Adams, 1860): **183**
4mm. Thin, translucent, narrowly elongate, sutures impressed; blunt protoconch of about 3 smooth whorls. Spire whorls have coarse axial riblets crossed by broad, flat, spiral threads, nodulous at intersections becoming smooth basally. Columella curved. White or brown, rarely with brown subsutural band. **Habitat:** intertidal and offshore. **Distribution:** NG, NWG, SEG, GO, Mas.

Family SCALIOLIDAE
These tiny, colourless snails are remarkable for attaching sand grains to their shells, a practice which makes them difficult to detect and to identify. The specimens found so far in eastern Arabia may belong to several species but none has yet been identified positively so they are ignored here.

Family POTAMIDIDAE
Popularly known as horn shells, mudcreepers or mudwhelks, members of this family live in the mud of river estuaries and mangrove swamps, where they feed on microscopic plants. The columella is strongly twisted and the short siphonal canal gutter-like, the outer lip sometimes surrounding these features like a curved blade. The thin, chitinous operculum is brown and multispiral. Horn shells are poorly represented in eastern Arabia as the region offers few suitable habitats for them, although in some places they may occur in astronomical numbers.

Cerithidea cingulata (Gmelin, 1791): **184**
30mm. Solid, dull, 3 times as long as wide; pointed protoconch usually eroded; deep posterior canal; short, twisted siphonal canal. Nodular axial ribs, coronate at sutures; sometimes 1 or 2 spiral grooves per whorl, base spirally grooved. White and dark brown spiral lines; pale brown periostracum. **Habitat:** intertidal in sand and among mangroves. **Distribution:** all.

Potamides conicus (Blainville, 1826): **185**
15mm. Thick, moderately glossy, straight sided, aperture squarish; blunt protoconch of 2 glossy whorls; short, broad, recurved siphonal canal. Coarse, obliquely axial ribs, 2-4 nodulous spiral cords per spire whorl (subsutural cord thickest). White with dark brown and orange spiral bands, nodules paler. Syn: *Pirenella caillaudii* (Potiez & Michaud, 1838). Very variable. **Habitat:** intertidal on muddy sand. **Distribution:** all.

Telescopium telescopium (Linnaeus, 1758): **186**
60mm. Thick, moderately glossy, elongate-conical, about 16 whorls, sutures poorly defined; protoconch eroded. 3 broad and 1 narrow, smooth spiral ribs; flat base has few spiral ribs and groove around

Superfamily: **CERITHIOIDEA**

Potamides conicus **185**

Telescopium telescopium **186**

Terebralia palustris **187**

Burnt remnants of a meal including shells of *Terebralia palustris*

strongly twisted and channelled columellar pillar. Dark brown; columella pale brown. A few, mostly adventitious (beach) records from eastern Arabia; probably not native.
Habitat: mud among mangroves.
Distribution: NG, GO.

Terebralia palustris (Linnaeus, 1767): **187**
90mm. Thick, heavy, elongate, straight sided, sutures distinct; protoconch eroded. Oblique, flattened, axial ribs crossed by 3 spiral grooves per whorl; 1 low varix per whorl; rounded base has many spiral ribs; Columella strongly twisted. Dark brown or greenish, varices paler; aperture chocolate; greenish brown periostracum.
Habitat: mud among mangroves.
Distribution: SEG, GO, SO.

Family TURRITELLIDAE
Elongate, slender, many-whorled shells ornamented with spiral ridges and usually brown or yellow. There is no umbilicus and the thin outer lip is seldom complete even in mature examples. A thin, chitinous, multispiral operculum is present. There are many species worldwide, living as herbivores in sandy or muddy places, but few occur in eastern Arabia. Identification is difficult because shells vary greatly in ornament, colour pattern and spire angle, even at one locality.

Turritella cingulifera Sowerby, 1825: **188** NOT ILLUSTRATED
Listed by Melvill & Standen, 1901:378, from Linjah (NG) and GO, but occurrence of the species in eastern Arabia needs confirmation.

SEASHELLS OF EASTERN ARABIA

Superfamily: **CERITHIOIDEA**

Turritella cochlea 189

Turritella columnaris 190

Turritella fultoni 191

Turritella maculata 192

Turritella cochlea Reeve, 1849: **189**
45mm. Thin, semitranslucent, spire angle about 13°. Each whorl rounded (but seeming straight sided) with strong, sharp, median keel and equally strong keel just above succeeding suture; irregularly spaced spiral ridges cover rest of each whorl. Thin-edged aperture almost circular. Yellowish white with brown mottlings and stripes. Keels vary greatly in strength. Syn: *T. auricincta* von Martens, 1882 (sensu Glayzer et al, 1984:319). **Habitat:** intertidal in sand. **Distribution:** all.

Turritella columnaris Kiener, 1844: **190**
90mm. Solid, semitranslucent, spire angle about 10°. Each whorl overhangs succeeding whorl to make sutures deep; strong spiral ridges on lower part of whorls, lesser ridges above. Thin-edged aperture squarish, columella straight. Whitish heavily mottled with brown fading to orange-brown. Length varies greatly. **Habitat:** muddy sand among stones. **Distribution:** NG, GO, SO.

Turritella fultoni Melvill, 1898: **191**
50mm. Solid, opaque, spire angle about 16-20°. Rounded whorls separated by deep sutures, each whorl encircled by strong ridges with lesser ridges between; fine, straight or curved, axial striae between ridges. Thin-edged aperture broadly ovate, columella straight. White, or rarely with flame-like, orange-brown markings on the whorls. **Habitat:** offshore and beached. Syn: *T. illustris* Melvill, 1904. **Distribution:** NG, NWG, SEG, GO.

Turritella maculata Reeve, 1849: **192**
65mm. Thin (becoming thicker in large specimens), semitranslucent or opaque, spire angle about 20°. Each whorl has 2 strong, rounded keels, occasional spiral ridges and many, fine, spiral threads. Aperture thin-edged, squarish. Violet-white with brown mottlings and stripes (fading to yellow); fine, brown lines sometimes encircle whorls. Keels sometimes obsolete. Syn: *T. torulosa* Kiener, 1843-44 (sensu Bosch & Bosch, 1982:41, and others; the name is now applied to West African specimens.). A pure white form was named, unnecessarily, var. *chionia* by Melvill, 1928:102. **Habitat:** intertidal. **Distribution:** all.

Turritella spectrum Reeve, 1849: **193** NOT ILLUSTRATED
Listed by Smythe, 1979a:65, from U. A. E. (SEG) but the name is probably a synonym for one of the other species names above.

Turritella vittulata Adams & Reeve, 1850: **194** NOT ILLUSTRATED
Listed by Melvill & Standen, 1901:378, from PG and GO.

Family VERMETIDAE
Worm shells, which may not be recognisable immediately as gastropod molluscs, have irregularly coiled, tubular shells cemented to hard surfaces. The animals capture floating organic matter by mucus threads or by ciliary action. In most species an operculum is present, which is just as well because it is one of the few reliable features in identification. The names given to the two eastern Arabian species described here are tentative only. Shells of other unidentified species have been found, including some of a large worm shell found on beaches at Juzor al Halaaniyat (Kuria Muria Islands).

Dendropoma meroclista Hadfield & Kay, 1972: **195**
1.5mm (maximum tube diameter, length variable). Solid, irregularly tubular, occurring in crowded masses, usually encrusted with algae. Aperture circular but almost closed by a convex shell dome. Irregular axial ribs and fine concentric striae. Dark purple, aperture white sometimes tinged green. Operculum a plug fitting under convex dome of aperture. **Habitat:** Coral reefs. **Distribution:** Mas, SO.

Serpulorbis variabilis Hadfield & Kay, 1972: **196**
40mm (maximum diameter of coiled whorls), 8mm (tube diameter). Thick, irregularly tubular and coiled in a flattened spiral which is often hollow at centre, usually occurring in small, entangled groups, algae-encrusted or covered with attached sand grains. Strong axial ribs crossed by strong and weak spiral ribs. White or tan, sometimes mottled with darker streaks, aperture white and pale brown. No operculum. **Habitat:** on rocks and flat surfaces. **Distribution:** GO, Mas, SO.

Dendropoma meroclista 195

Serpulorbis variabilis 196   *Natural size*

Superfamily: **STROMBOIDEA**

Superfamily STROMBOIDEA
For details see family description below.

Family STROMBIDAE
Medium-sized to large, thick shells with an expanded outer lip which may be simple, elongated in the direction of the apex, or furnished with several short or long digitations. The lower (i.e. the front) end of the outer lip has an indentation (the stromboid notch) below which the animal protrudes one of its stalked eyes. Some species (*Tibia*) have an elongated spire and a long siphonal canal. All eastern Arabian species have a sharply pointed chitinous operculum, sometimes with a serrated edge, which the animal uses to help it make lunging movements. A family of warm-water sand dwellers which browse on algae, it includes *Lambis truncata sebae*, the largest gastropod found in eastern Arabia.

Strombus (Canarium) erythrinus Dillwyn, 1817: **197**
28mm. Thick, elongate, with about 8 angulate and axially ribbed, weakly varicose, spire whorls; last whorl with nodules sometimes extending nearly to base. Outer lip thickened, its edge smooth but sharp; aperture lirate, columella lirate except centrally; stromboid notch very shallow. Last whorl spirally ribbed. White or yellow banded and mottled with brown; aperture chocolate, outer edge of columella white. **Habitat:** sandy places among rocks. **Distribution:** GO, Mas.

Strombus (Canarium) fusiformis Sowerby, 1842: **198**
35mm. Thin, smooth, elongate, about 9 whorls, last whorl longer than spire. Each whorl has spiral groove just below suture, base of last whorl spirally grooved; 3 or 4 varices per whorl, 2 or 3 blunt nodules at shoulder of last whorl. Thickened outer lip extends above last whorl; stromboid notch shallow; aperture and columella lirate. White with dark or pale brown blotches and spots; aperture tinged brown and purple. **Habitat:** in sand. **Distribution:** all.

Strombus (Gibberulus) gibberulus gibberulus Linnaeus, 1758: **199**
55mm. Thick, smooth, last whorl has humped appearance; spire short, earlier whorls stepped, varicose;

Strombus (Canarium) urceus urceus **201**

Strombus (Canarium) erythrinus **197**

Strombus (Canarium) mutabilis mutabilis **200**

Strombus (Canarium) fusiformis **198**

Strombus (Gibberulus) gibberulus gibberulus

Superfamily: **STROMBOIDEA**

Strombus (Conomurex) decorus masirensis **202**

Strombus (Conomurex) persicus **203**

Strombus (Dolomena) plicatus sibbaldii **204**

apex pointed. Outer lip thickened, aperture weakly lirate, columella smooth, thickened at top; stromboid notch moderately deep. Last whorl smooth on apertural side, outer lip spirally grooved outside. Pale brown with thin, white, spiral bands, darker brown below suture of last whorl; aperture rose-pink. **Habitat:** in sand. **Distribution:** NG, GO, SEO, Mas, SO.

Strombus (Canarium) mutabilis mutabilis Swainson, 1821: **200**
27mm. Thick, smooth, last whorl square sided with few blunt nodules, spire whorls shouldered, sutures deep, apex pointed. Outer lip thickened, its edge smooth but sharp; aperture and columella finely lirate; stromboid notch moderately deep. Occasional thin band at suture; outer lip spirally grooved outside, otherwise smooth. Orange to brown variously mottled, banded and spotted; aperture pink or orange, columellar lirae white. **Habitat:** in sand among rocks. **Distribution:** GO, Mas, SO.

Strombus (Canarium) urceus urceus Linnaeus, 1758: **201**
55mm. Solid, glossy, narrow aperture much longer than spire, pointed apex; stromboid notch broad and shallow. Whorls stepped, angled and more or less noduled at shoulder, otherwise smooth. Thick columellar lip lirate above and below; aperture lirate. Cream blotched, banded and reticulated with shades of brown and orange; aperture and columella orange to brown. Known from eastern Arabia by 3 beached shells only. **Habitat:** offshore and beached. **Distribution:** GO, Mas.

Strombus (Conomurex) decorus masirensis Moolenbeek & Dekker, 1993: **202**
60mm. Heavy, spire short, pointed. Sutures well defined. Narrow aperture almost length of shell. Columella smooth, straight. Thin-edged lip has broad, shallow stromboid notch. Shallow posterior and siphonal canals. Prominent nodules at shoulder of last whorl. Fine spiral striae cover shell. Creamy white with spiral and axial rows of brownish flecks and zigzags. Columella orange; aperture orange-white. **Habitat:** intertidal in sand. **Distribution:** Mas, SO.

Strombus (Conomurex) persicus Swainson, 1821: **203**
50mm. Heavy, dull; short, pointed spire, last whorl triangular in outline. Thick, smooth columellar callus. Thin-edged lip with shallow stromboid notch. Shallow siphonal canal. Smooth keel at shoulder of last whorl. Fine axial striae cover shell. White with pale brown streaks and interrupted spiral bands. Columella white; aperture white or pale pink. Syn: *S. beluchiensis* Melvill, 1901, *S. mauritianus* Lamarck, 1822. **Habitat:** intertidal in sand. **Distribution:** NG, NWG, SEG, GO.

Strombus (Dolomena) plicatus sibbaldii Sowerby, 1842: **204**
40mm. Sturdy, lightweight with almost square-sided last whorl and narrowly elongate, slightly concave spire. Outer lip rounded, thickened at edge. Row of nodules at shoulder, coarse axial ribs on spire whorls. Aperture and columella lirate; upper half of last whorl smooth, lower half spirally ridged. White, blotched and banded with beige; columellar lirae tinged brown; aperture white. Syn: *S. yerburyi* Smith, 1891. **Habitat:** offshore in sand. **Distribution:** NG, SEG, GO, Mas, SO.

Superfamily: **STROMBOIDEA**

Strombus (Tricornis) oldi 205

■ Lambis truncata sebae, *the largest and heaviest gastropod in eastern Arabia, is familiar as an ornamental object, although its vivid colouring fades when exposed continually to strong sunlight. In some places it is common, even in shallow water, and Omani fishermen usually know where to find it. Long ago, they used to accumulate piles of this solidly constructed species to form the walls of crude seaside huts. Before reaching maturity the shell lacks the finger-like extensions typical of the* Lambis *group and may then be mistaken for a species of an entirely different group, such as Conus. Its pointed spire distinguishes it from the other subspecies,* L. truncata truncata, *which has a blunt-topped spire.*

Strombus (Tricornis) oldi
Emerson, 1965: **205**
100mm. Thick, heavy, glossy, last whorl swollen; flaring outer lip and stepped spire seldom more than half length of last whorl. Large tubercles above each suture and at shoulder of last whorl; spiral rows of prominent tubercles on outer lip. White, heavily blotched and banded chocolate brown; aperture cream with large chocolate blotch. Outer lip usually damaged. **Habitat:** in sand offshore and beached. **Distribution:** Mas.

Strombus (Tricornis) tricornis
Lightfoot, 1786: **206**
115mm. Thin, lightweight, last whorl with outer lip projecting above level of sharp apex; stromboid notch broad and shallow. Spire whorls angulated and noduled above suture; last whorl with large, blunt tubercles at shoulder, coarse, widely spaced spiral ribs and fine axial growth lines. Parietal wall broadly glazed. Pinkish brown, aperture pinkish. Martyn Day has found this Red Sea species at Musandam. **Habitat:** offshore in sand. **Distribution:** GO.

Lambis lambis (Linnaeus, 1758): **207** NOT ILLUSTRATED
Listed by Melvill & Standen, 1901:381 from Muscat but not found since, in spite of intensive collecting there.

Lambis truncata sebae (Kiener, 1843): **208**
350mm. Thick, heavy, dull outside but glossy inside, about 11 whorls, the last greatly expanded, outer lip bearing 7 to 9 open digitations at maturity. Pointed spire is characteristic of this subspecies of *L. truncata* (Linnaeus, 1758). Whorls heavily coronated; last whorl has large nodules at shoulder. Parietal wall heavily glazed. Pale brown under amber periostracum. Aperture and parietal wall pink, orange, purplish, yellow or white. **Habitat:** on sand near rocks and coral. **Distribution:** GO.

Strombus (Tricornis) tricornis 206

Superfamily: **STROMBOIDEA**

Lambis truncata sebae **208**

Tibia insulaechorab curta 211

Superfamily: **STROMBOIDEA**

Tibia delicatula **210**

Terebellum terebellum (Linnaeus, 1758): **209**
50mm. Thin, lightweight, torpedo shaped and wide open at base, about 5 whorls, suture very shallow. Flattened spiral thread above suture, whole surface smooth and glossy. Outer lip and columella straight. Orange, yellow, white, or purple, each colour disposed in spiral bands, zebra stripes, blotches, dots and streaks; aperture white or pinkish. **Habitat:** muddy sand. **Distribution:** all.

Tibia delicatula (Nevill, 1881): **210**
75mm. Solid, last whorl (excluding siphonal canal) 1.5 times as long as convex-sided spire whorls; short, pointed siphonal canal. Fine spiral grooves on early whorls becoming obsolete on later whorls. Outer lip has 3, 4, 5 or (very rarely) 6 spines; posterior canal continued along suture. Beige with paler spiral bands (4 on last whorl), orange-lined lip; aperture white. **Habitat:** deep water. **Distribution:** GO.

Tibia insulaechorab curta (Sowerby, 1842): **211**
150mm. Thick, heavy, glossy, last whorl (excluding siphonal canal) much shorter than concave-sided spire whorls. Early whorls axially ribbed, later ones smooth. Columella continues upwards as a sinuous, lumpy callus; outer lip thickened, bearing 4 to 6 blunt digitations; siphonal canal slightly curved. Coffee coloured or orange-brown, sometimes darker brown below suture; aperture and inside edge of columella white; thick, brown periostracum when fresh. **Habitat:** intertidal on sand. **Distribution:** NG, NWG, GO, Mas.

■ *In common with many other molluscs* Tibia delicatula *is thinly represented in the world's larger museums although it is abundant in nature. F. W. Townsend obtained many specimens in the Gulf of Oman, from 225 fathoms and from 200-400 fathoms. He noted that the water temperature where it occurred was always 62°F. A manuscript note of his at Manchester Museum gives the following details about the incidence of labial spines on specimens he dredged:*
lip with 3 processes - very rare, not 1 in 500
lip with 4 processes - the common type
lip with 5 processes - about 1 per cent
lip with 6 processes - 1 only out of many thousands.
In 1905 Melvill and Standen published new varietal names for the first three of these forms: tridenticulata, quatuordenticulata *and* quinquedenticulata, *a pointless exercise in zoological nomenclature.*

Terebellum terebellum **209**

Superfamily HIPPONICOIDEA
Contains four families of small gastropods with shells which are cap shaped or have a large last whorl.

Family HIPPONICIDAE
The hoof shells are more or less cap shaped and may be mistaken for true limpets. Some species attach themselves to rocks, forming a shelly plate as a base. Others attach themselves to molluscan shells or to the spines of sea urchins, a habit which affects the shape and sculpture of their own shells. They feed on suspended particles of food or, if attached to other molluscs, the faecal pellets of their host. Most start life as males but change into females later. The apex may be recurved and the shell surface is smooth or ornamented, the shell having a horseshoe-shaped muscle scar inside. There is no operculum.

Cheilea cicatricosa (Reeve, 1858): **212**
30mm. Thin or solid, dull, semitranslucent, aperture more or less circular, elevated spire recurved backwards. Coarse, concentric corrugations crossed by crowded, flattened, radial ridges. Internally smooth and glossy, a thin, gutter-like, process issuing from just below apex. White. **Habitat:** intertidal on other objects. **Distribution:** NG, GO.

Cheilea equestris (Linnaeus, 1758): **213**
30mm. Similar to description of previous species but has stronger radial ridges and lacks the coarse, concentric corrugations. **Habitat:** intertidal on other objects. **Distribution:** NG, GO, SO.

Hipponix conicus (Schumacher, 1817): **214**
10mm. Solid, dull externally, cap shaped in profile but varying in height, outline of margin oval to pentagonal, shape conforming to surface of its host; apex recurved backwards. Coarse, irregular, flattened radial ribs, usually indenting margin. White or grey blotched and streaked reddish brown, internally white or pinkish violet with yellowish muscle scar. **Habitat:** intertidal on other shells. **Distribution:** Mas.

Hipponix foliaceus Quoy & Gaimard, 1835: **215**
12mm. Thick, dull, flattened-cap shaped; protoconch of about 1.5 partially submerged whorls which usually project over posterior edge. Aperture wide open, its margin conforming to shape of surface to which it was attached. Irregular, concentric, overlapping lamellae crossed by fine radiating striae. Adductor-muscle scars horseshoe shaped and conspicuous. White. **Habitat:** intertidal on undersides of rocks. **Distribution:** GO.

Hipponix ticaonicus (Sowerby, 1847): **216**
16mm. Solid, dull, limpet-like, the aperture almost circular, elevated spire usually strongly recurved backwards, the apex tilted to one side. Coarse, crowded, radial ridges crossed by irregular concentric threads and prominent growth ridges. White. Covered by pale brown, fibrous periostracum when fresh. **Habitat:** offshore on other objects and beached. **Distribution:** Mas, SO.

Malluvium lissum (Smith, 1894): **217**
11mm. Solid, dull (but glossy inside), limpet-like, low- to high-domed, often misshapen, the apex recurved and overhanging posterior margin, protoconch of 2.5 closely wound whorls. Fine, irregular growth lines, otherwise smooth. Internally the posterior rim is shelf-like and there are 2 elongate muscle scars. White. **Habitat:** on other objects in deep water. **Distribution:** NG, GO.

Malluvium lissum **217**

Cheilea cicatricosa **212**

Cheilea equestris **213**

Hipponix ticaonicus **216**

■ *Living examples of* Malluvium lissum *form colonies on dead shells of other molluscs, such as* Tibia delicatula *and* Conus planiliratus, *a juvenile usually attaching itself to the dorsal surface of a larger one, presumably its parent. Usually they have the rounded outline normal for the species. They may frequent less accommodating surfaces than molluscan shells, however, as F. W. Townsend found in 1903 when he examined some on a sea-urchin dredged from 122 fathoms in the Gulf of Oman. Adapting themselves to the shape of the sea-urchin's spines, they had assumed a narrowly oblong form, sometimes forming a thick basal plate in the process. Invariably, however, the offspring attending them displayed the normal rounded outline.*

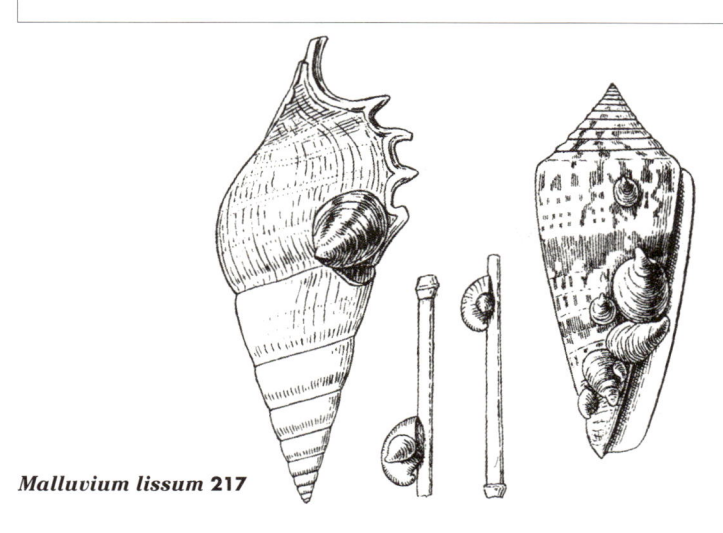

Malluvium lissum 217

Family VANIKORIDAE
A small family of species living in clean sand near coral. Shells of the few eastern Arabian representatives are mostly thin, fragile, dull and white. The operculum is thin and chitinous. A poorly understood family, it seems that differences in sculpture between shells otherwise similar may sometimes be nothing more than male and female features of a single species. The strong axial ribbing often present on the early spire whorls may represent a male phase. As with many other gastropod shells the protoconch may provide the most reliable clues to species relationships. Melvill and others record from eastern Arabia several so-called species, including some identified wrongly as members of the family Fossaridae (now Planaxidae). As the systematic position of most of these is uncertain they are not included here.

Vanikoro gueriniana 222

Hipponix foliaceus 215

Hipponix conicus 214

Vanikoro plicata 223

Berthais intertexta (Melvill & Standen, 1903): **218**
7mm. Thin, high spired with protoconch of about 3 smooth whorls. Spire whorls narrow, shouldered, with very deep sutures, bearing regular, low axial ribs (about 22 on last whorl) slightly elevated just below sutures, with spiral lirae between each pair of ribs giving weakly cancellate effect. Aperture elongate; peristome sharp edged. No umbilicus. White and dull overall. **Habitat:** deep water. **Distribution:** GO.

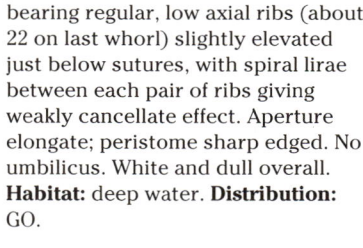

Macromphalus* cf *subreticulatus (Nevill, 1884): **219**
6mm. Solid, translucent, dull, broadly ovate, high spired, semi-lunate aperture about half total length; pointed protoconch of 3 smooth, glossy whorls. Spire whorls ledged at sutures, columella straight and smooth, umbilicus narrow and deep. Whorls coarsely, axially ridged, later ones encircled by strong cords and very fine, spiral threads. White. **Habitat:** beached. **Distribution:** NG, NWG.

Macromphalus thelacme (Melvill, 1904): **220**
3mm. Differs from *M.* cf *subreticulatus* (see above) by its smaller size, narrower outline, weaker sculpture and less pointed protoconch. **Habitat:** deep and shallow water. **Distribution:** NWG, SEG, GO.

Vanikoro cancellata (Lamarck, 1822): **221** NOT ILLUSTRATED
Listed by Melvill & Standen, 1901: 361, from Hindarabi Island and Linjah (NG) but occurrence of the species in eastern Arabia needs confirmation.

Vanikoro gueriniana (Récluz, 1844): **222**
10mm. Smaller than *V. plicata* (see below) with fine, sharp axial ribs and weak spiral threads. Protoconch amber. **Habitat:** intertidal among rocks. **Distribution:** GO.

Vanikoro plicata (Récluz, 1844): **223**
15mm. Thin, globose-ovate, short spire protruding prominently above last whorl, columella straight; protoconch of about 2.5 apparently smooth whorls. Umbilicus broad and deep. Strong, sharp, axial ribs becoming obsolete on last whorl; strong, close-set, spiral threads overall. White. **Habitat:** offshore and beached. **Distribution:** GO, Mas.

Berthais intertexta 218

Macromphalus thelacme 220

Macromphalus cf *subreticulatus* 219

Superfamily: **HIPPONICOIDEA, CREPIDULOIDEA**

Vanikoro tricarinata (Récluz, 1843): **224**
15mm. Thin, globose-ovate, spire scarcely raised above last whorl, columella straight; protoconch of about 2 whorls. Umbilicus narrow and deep. Beaded spiral cords encircle shell (stronger on early whorls), beads also aligned axially. White. **Habitat:** beached. **Distribution:** NWG.

Vanikoro tricarinata **224**

Vanikoro sp: 225
18mm. Solid, globose, spire scarcely protruding above last whorl; protoconch of about 2 whorls bearing fine spiral threads. Early whorls have equally strong spiral and axial cords which are roundly noduled at intersections; sculpture becomes progressively weaker on later whorls. Umbilicus narrow and deep. White. **Habitat:** offshore and beached. **Distribution:** GO, Mas.

Superfamily CREPIDULOIDEA
Like the Hipponicoidea the members of this family have limpet-like shells but they have, in addition, a spirally coiled apex and, internally, some of them have a ledge-like or cup-like process (septum). None has an operculum. Because they attach themselves to surfaces which may be irregular the shape of their shells varies greatly. Some are known to change sex with age. Using a large gill, they capture minute floating organisms.

Vanikoro sp. **225**

Calyptraea spinifera **228**

Family CREPIDULIDAE
Members of this family have shells bearing an internal ledge-like or cup-like process which is why most of them are known as cup-and-saucer limpets.

Calyptraea edgariana Melvill, 1898: **226**
20mm. Thin or thick, dull, misshapen and accommodated to objects upon which it settles, but essentially high domed and patelliform. Coarse, irregular radial ribs, usually encrusted with coralline growths. Internally glossy, a relatively small, laterally compressed cup issuing from below apex. Externally pale orange-brown, internally white suffused with orange-yellow or violet. **Habitat:** attached to objects offshore. **Distribution:** NG, SEG, GO, Mas, SO.

Calyptraea pellucida (Reeve, 1859): **227**
10mm. Thin, opaque, dull, circular in outline, low-cap-like, sharp-edged; sub-central protoconch of about 1.25 glossy, inrolled whorls. Fine concentric striae and occasional growth ridges. Internally glossy, a thin shelf spiralling down from below apex. White tinged yellow around apex; internally white tinged yellow, shelf brown edged. **Habitat:** intertidal on stones and other objects. **Distribution:** NWG, SEG, Mas.

Calyptraea spinifera (Gray, 1867): **228**
25mm. Thin, dull, irregularly limpet shaped, low to high domed, margin conforming to shape of attachment site; centrally placed protoconch of about 1.25 glossy whorls. Radiating rows of scaly ridges crossed by irregular growth ridges. Internally glossy, a plate-like process descending from below apex. White tinged orange around apex; internally orange-brown with white adductor-muscle scars. **Habitat:** intertidal under surfaces of rocks. **Distribution:** Mas, SO.

Crepidula fornicata (Linnaeus, 1758): **229**
30mm. Solid, dull externally, elongate-oval in outline, resembling single valve of a marine mussel with

Calyptraea edgariana **226**

Calyptraea pellucida **227**

Superfamily: **CREPIDULOIDEA**

a weak ridge from apex to margin and a broad, shelf-like structure occupying narrower half of underside. Apex slightly coiled, just overhanging margin. Cream spotted and flecked violet-brown, internally glossy, streaked and blotched orange to violet-brown, shelf whitish. An oyster pest around Atlantic coasts. Dead shells of this well-known Atlantic species have been found at Masirah and Muscat, probably introduced accidentally by human agency. **Habitat:** beached. **Distribution:** GO, Mas.

Crepidula walshi Reeve, 1859: **230**
20mm. Thin, dull (but glossy internally), opaque or semitranslucent, assuming shape of attachment site, outline irregular, usually longer than broad; pointed apex at posterior edge. Fine concentric striae and coarse growth lines. Broad, shelf-like plate issues from below apex. White. **Habitat:** intertidal and offshore usually in apertures of shells occupied by hermit crabs. **Distribution:** NWG, SEG, GO, Mas, SO.

Trochita dhofarensis Taylor & Smythe, 1985: **231**
40mm. Solid, dull, circular in outline, trochoidal but varying in height, aperture virtually area of base, sutures indistinct; protoconch of 1.25 glossy, inrolled whorls (usually eroded). Coarse, rough, rounded, obliquely axial ribs; base glossy and smooth, a thin ledge spreading out from apex occupying half basal area. Purplish brown with paler ribs; internally orange-brown to chocolate. Usually found worn and faded. **Habitat:** intertidal and offshore on rocks and in crevices. **Distribution:** SO.

Crepidula fornicata **229**

Crepidula walshi **230**

Trochita dhofarensis **231**

SEASHELLS OF EASTERN ARABIA

Superfamily: CREPIDULOIDEA, XENOPHOROIDEA

Family CAPULIDAE
Gastropods with limpet-like shells and commonly known as cap shells. The apex is usually conspicuously coiled and often extends beyond the posterior margin. Cap shells attach themselves to other shells, stones and rocks and the configuration of the attachment surface is usually repeated on their own surfaces, sometimes resulting in remarkable distortions. The animals are either parasites or ciliary feeders.

Capulus badius Dunker, 1881: **232**
10mm. Thin, matt (shiny when worn), narrowly cap-shaped with recurved or inrolled apex overhanging posterior margin. Margin conforming to shape of object to which it was attached. Fine radiating ribs and irregular growth ridges. Rosy pink, darker anterior to apex, paler below apex, rarely uniformly white. **Habitat:** offshore and beached. **Distribution:** Mas.

Family TRICHOTROPIDAE
Essentially a cold-water family, the hairy shells (so called because their spiral ridges or keels are covered by bristle-like periostracal "hairs" when fresh) are thinly represented in eastern Arabia. A chitinous operculum is present. The animals feed on organic particles among stones and silt.

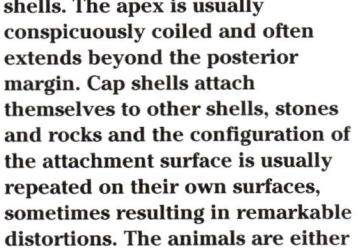

Lippistes tropaeum Melvill, 1912: **233**
6mm. Thin, translucent, ovate-fusiform, all whorls sloping below deep sutures; large protoconch of 2.5 smooth, glossy whorls. 3 ridge-like keels per whorl. Columella forms acute angle with base; narrow, deep umbilicus. Milky white. **Habitat:** deep water. **Distribution:** GO.

Separatista helicoides (Gmelin, 1791): **234**
13mm. Thin, silky, short spired, last whorl greatly expanded and ultimately disconnected from previous whorl, gutter-like siphonal canal; protoconch not seen. Spiral keels on all whorls (3 at periphery of last whorl), another keel around the broad, deep umbilicus. Fawn. **Habitat:** beached. **Distribution:** Mas.

Trichotropis crassicostata Melvill, 1912: **235** NOT ILLUSTRATED
Described from shells found at the same time and place as *T. pulcherrima* (see below) to which they seem closely allied. Available shells are too fragmentary to describe or illustrate adequately.

Trichotropis pulcherrima Melvill & Standen, 1903: **236**
4mm. Thin, dull, attenuate-fusiform, spire twice as long as the broadly ovate aperture, sutures impressed; protoconch of 2 smooth whorls. Spire whorls 2-keeled, last whorl 4-keeled. Periostracum developed into obliquely axial ridges. White to straw coloured. **Habitat:** offshore. **Distribution:** GO.

Trichotropis townsendi Melvill & Standen, 1901: **237**
5mm. Lightweight, fragile, dull, pyramidal with well rounded whorls and deep sutures; flat-topped protoconch of 1.5 smooth whorls. Aperture large; columella straight; flaring, corrugated outer lip; no umbilicus. Widely spaced axial and spiral cords produce broadly cancellate sculpture; thin axial threads overall. White. Described and illustrated from type specimens which appear juvenile. **Habitat:** deep water. **Distribution:** GO.

Superfamily XENOPHOROIDEA
Most of the few small to medium-sized members of the only family in this superfamily have a thin, trochoidal shell with a pronounced keel around the last whorl and usually lack an umbilicus. Spiny processes project from the keel of one species. A thin, corneous operculum is present. Virtually all species attach foreign material to the shell surface, particularly at the periphery of the last whorl, a habit apparently developed to help the mollusc escape detection. As with members of the Strombidae the animal leaps or loops over the sea bed. Most species live in warm or temperate waters where they feed upon forams or algae. Although some occur in shallow waters others may be dredged from considerable depths.

Family XENOPHORIDAE
For description see superfamily description above.

Xenophora (Stellaria) chinensis (Philippi, 1841): **238**
45mm. Thin, low spired, nearly twice as wide as high, slightly convex whorls; base slightly convex, extended by broad peripheral flange. Sparse attached debris at sutures and margin. Umbilicus broad, exposing all whorls. Strong spiral ridges crossed by axial ridges,

Trichotropis pulcherrima **236**

Capulus badius **232**

Lippistes tropaeum **233**

Trichotropis townsendi **237**

Separatista helicoides **234**

70 SEASHELLS OF EASTERN ARABIA

beaded at intersections. Yellowish brown to yellowish white. Syn: *Phorus calculiferus* (Reeve, 1842). **Habitat:** deep water. **Distribution:** NG.

Xenophora (Stellaria) solaris (Linnaeus, 1764): **239**
90mm (excluding spines). Thin, translucent, trochoidal, much broader than long; base almost flat, fringed by long and short, hollow, downward-curving spines (spines are impressed into spire whorls). Sparse attached debris around apex. Umbilicus broad and deep; inner lip slightly thickened. Coarse, undulating ribs on all whorls; strong, lamellate, radial ribs on base. Pale yellowish brown. **Habitat:** in sand offshore. **Distribution:** NWG, NG, SEG, GO.

Xenophora (Xenophora) corrugata (Reeve, 1842): **240**
50mm (excluding attachments). Thin, translucent, trochoidal with convex or slightly concave base and stepped whorls; upper parts of whorls partly hidden by adherent shell fragments and debris. Umbilicus closed when adult, open when juvenile. Coarse, oblique ridges above sutures; spiral and radial ribs around umbilical area, radial ribs elsewhere on base. Outer lip thin. Yellowish brown to white. Syn: *X. caperata* Philippi, 1855. **Habitat:** on sand. **Distribution:** NG, NWG, SEG, GO, Mas.

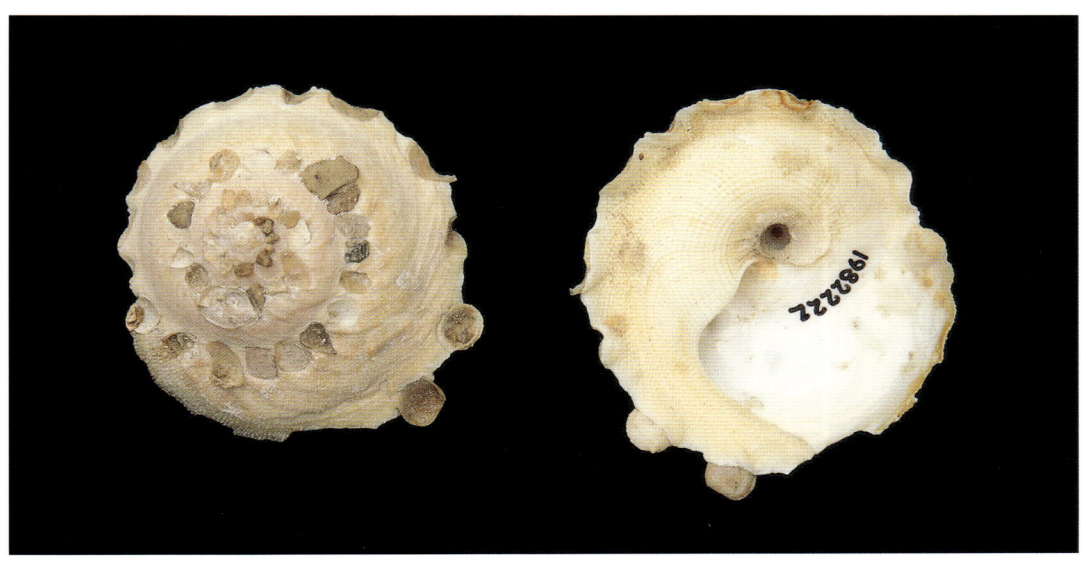

Xenophora (Stellaria) chinensis **238**

Xenophora (Xenophora) corrugata **240**

Xenophora (Stellaria) solaris **239**

SEASHELLS OF EASTERN ARABIA 71

Superfamily: **XENOPHOROIDEA, CYPRAEOIDEA**

Xenophora (Xenophora) pallidula (Reeve, 1842): **241**
65mm (excluding attachments). Thin, lightweight, trochoidal, broader than long, whorls slightly convex; base almost flat or slightly concave. Adherent shell fragments usually obscure much of upper surface, may radiate from periphery like wheel spokes. Umbilicus small or quite closed. Wavy, oblique riblets cover upper surface; wavy growth ridges on base. White or yellowish white. Illustrated from shells dredged off SE India. **Habitat:** on sand or mud in deep water. **Distribution:** GO.

Superfamily CYPRAEOIDEA
As well as the strikingly patterned and polished members of the well-known family Cypraeidae (described below) this large superfamily includes the cowry-like triviids, the less familiar but often colourful ovulids of the family Ovulidae and the Lamellariidae with internal, virtually colourless shells. These families show considerable differences in appearance and life styles and their placement together in this superfamily is provisional.

Xenophora (Xenophora) pallidula **241**

Cypraea annulus **242**

Cypraea caputserpentis **244**

Family CYPRAEIDAE
With few exceptions each species in this large family has a smooth, globular or cylindrical shell which is highly polished all over and has a long, narrow aperture bearing small, blunt teeth on each side. Cowries, as they are generally known, are not highly coloured but many are strikingly patterned. Often, however, the animal's mantle, which secretes the shell, is colourful and bears simple or elaborately branched papillae. The dorsum of some species displays a line (the sulcus) from top to bottom where the mantle lobes meet. There is no operculum. Abundant in the tropics, cowries are omnivorous and mostly conceal themselves by day under rocks and coral or in crevices and caves. Some of the many species recorded from eastern Arabia vary greatly in size and pattern. Nearly all have a smooth, highly polished dorsum. The posterior end is usually broader than the anterior and is always uppermost in the accompanying illustrations. For descriptive purposes the flattened apertural side is here called the "base", the outline of the shell when resting on its base describing its shape, each edge bearing a more or less thickened margin. Cowry shells may fade noticeably with time and exposure to light. Illustrations best convey the differences between cowry species, so the descriptions of them are brief.

Cypraea annulus Linnaeus, 1758: **242**
25mm. Thick, heavy, ovate and humped centrally; margins thickened. Teeth short and coarse. Grey, bluish or pale yellowish-brown, the humped portion encircled by a bright orange line. Mantle dark grey with greyish-yellow papillae. **Habitat:** under rocks and coral. **Distribution:** SEG, GO, Mas, SO.

Cypraea camelopardalis Perry, 1811: **243** NOT ILLUSTRATED
A freshly dead specimen of this Red Sea species has been found near Salalah (SO).

Cypraea caputserpentis Linnaeus, 1758: **244**
30mm. Thick, heavy, ovate with well developed margins confluent with dorsum which is conspicuously humped. Teeth large and spreading onto almost flat base. Margins chocolate-brown, ends blotched whitish, dorsum brown reticulated with whitish spots, sulcus paler brown; base whitish brown. **Habitat:** under rocks. **Distribution:** Mas, SO.

Cypraea carneola Linnaeus, 1758: **245**
45mm. Thick, heavy, ovate with well developed, lumpy margins confluent with dorsum; anterior end slightly produced. Base convex and uneven; teeth small and deep set. Dorsum purplish-brown or beige with 4 paler bands, purple rim above margins; margins brown speckled with yellow; base whitish to beige; teeth

Superfamily: **CYPRAEOIDEA**

Cypraea carneola **245**

Cypraea caurica **246**

purple. Varies greatly in size. **Habitat:** among rocks and coral. **Distribution:** NG, SEG, GO, SO.

Cypraea caurica Linnaeus, 1758: **246**
45mm. Thick, elongate to ovate, with well rounded dorsum, produced ends and lumpy margins. Aperture much wider anteriorly; large teeth spread over base. Dorsum bluish-white with 3 darker transverse bands overlain by brown spots and blotches; margins purplish brown to beige with large, dark brown spots; base and interstices of teeth purplish brown or beige, teeth paler. Varies greatly in size, shape and colour pattern. **Habitat:** among rocks and coral. **Distribution:** NWG, SEG, GO, Mas, SO.

Cypraea chinensis Gmelin, 1791: **247**
30mm. Thick, elongate-ovate, with gently rounded dorsum and lumpy margins. Labial teeth small, not spreading over base; much larger columellar teeth reach margin. Dorsum cream with orange-yellow blotches; margins cream spotted violet; orange-yellow between cream teeth. Reddish animal has dark red mantle. **Habitat:** under rocks. **Distribution:** Mas, SO.

Cypraea cicercula Linnaeus, 1758: **248**
18mm. Thin, globular, extremities produced and joined dorsally by a grooved sulcus, a small but deep dorsal depression below posterior extremity; entire dorsum usually granular. Aperture moderately curved, the many teeth spreading over convex base. Pale yellow with small brown spots, posterior depression dark brown; teeth and extremities pale brown. Known by a worn shell from Khor Fakkan

Cypraea caurica with mantle exposed on underside of a rock at Masirah

collected by H. Kauch. **Habitat:** offshore and beached. **Distribution:** GO.

Cypraea clandestina Linnaeus, 1767: **249**
18mm. Thin, elongate-ovate, labial margin slightly lipped, apex marked by a depression. Teeth prominent, not extending over convex base. Dorsum cream with broad, transverse, greyish-orange bands overlain by very fine zig-zag, brown lines; base and teeth cream. Black animal has grey and black mantle. **Habitat:** under rocks and in rock crevices. **Distribution:** Mas, SO.

Cypraea coloba Melvill, 1888: **250**
25mm. Thick, rounded-ovate, with well produced, lumpy margins, labial margin lipped, apex a slight depression, extremities weakly produced. Teeth do not extend over convex base, labial teeth well

Cypraea chinensis **247**

Cypraea clandestina **249**

Cypraea cicercula **248**

Cypraea coloba **250**

developed. Dorsum cream irregularly spotted brown, base and margins orange with large purple spots, bright orange between teeth. Possibly only a form of *C. chinensis*. **Habitat:** shallow water under rocks. **Distribution:** Mas.

SEASHELLS OF EASTERN ARABIA 73

Superfamily: **CYPRAEOIDEA**

Cypraea cribraria Linnaeus, 1758: **251**
30mm. Moderately thick, ovate to elongate-ovate, labial margin lipped and slightly wrinkled, large depression at apex. Columellar teeth small, deep-set; labial teeth larger, reaching margin. Dorsum reddish brown with large and small, sometimes coalescing, white spots; margins, base and teeth white. Animal's mantle deep red. **Habitat:** under rocks. **Distribution:** GO, Mas.

Cypraea erosa Linnaeus, 1758: **252** NOT ILLUSTRATED
Listed by Melvill & Standen, 1901:382, from the Arabian Gulf but its presence in eastern Arabia needs confirmation.

Cypraea felina fabula Kiener, 1846: **253**
20mm. Thick, elongate-ovate, low domed, extremities not produced, margins only slightly lipped. Fine teeth slightly invade rounded base. Dorsum cream with 4 underlying greenish-grey bands, overlain by coalescing, dark brown spots strongly developed at margins; two large spots at each extremity; base and teeth cream or orange-yellow, lacking spots. Syn: *C. listeri* Gray, 1825. **Habitat:** under rocks. **Distribution:** GO, Mas, SO.

Cypraea grayana **258**

Cypraea cribraria **251**

Cypraea gracilis **257**

Cypraea felina fabula **253**

Cypraea fimbriata Gmelin, 1791: **254** NOT ILLUSTRATED
Listed from the Arabian Gulf by Melvill & Standen, 1901:382, but its presence in eastern Arabia needs confirmation. Easily confused with *C. gracilis*.

Cypraea gangranosa Dillwyn, 1817: **255** NOT ILLUSTRATED
Recorded from Oman without more precise locality information. Its presence in eastern Arabia needs confirmation.

Cypraea globulus Linnaeus, 1758: **256** NOT ILLUSTRATED
Recorded from Oman but without more precise locality information. Its presence in eastern Arabia needs confirmation.

Cypraea gracilis Gaskoin, 1849: **257**
20mm. Thin, elongate-ovate, lightly margined, aperture much broader anteriorly. Teeth widely spaced, deep set, not reaching base. Dorsum blue-grey spotted and blotched reddish-brown; brown mid-dorsal blotch or transverse band; 2 purple-brown blotches at each extremity, brown spots on margins; base and teeth yellowish. Reddish-brown animal has orange-red mantle with red papillae. Syn: *C. notata* Gill, 1858. **Habitat:** intertidal under rocks. **Distribution:** NWG, SEG, GO, Mas, SO.

Superfamily: **CYPRAEOIDEA**

Cypraea grayana Schilder, 1930: **258**
70mm. Thick, humped, with prominently lipped margins and protruding extremities; apex often prominent. Poorly developed teeth deep-set, not reaching flat base. Dorsum bluish white with large, faint, brown blotches overlain by irregular axial lines interrupted by brown reticulations; margins with large, brown spots; extremities brown; base pinkish, teeth brown. Grey-black animal has greenish-grey mantle. Syn: Smythe, 1979a:67, lists **C. arabica** Linnaeus, 1758, and **C. histrio** from U. A. E. (SEG) but these are probably misidentifications for **C. grayana**. **Habitat:** intertidal under rocks, rock ledges and in crevices. **Distribution:** SEG, GO, Mas, SO.

Cypraea helvola Linnaeus, 1758: **259**
25mm. Thick, roundly ovate, usually with expanded, lipped margins, pitted posteriorly; aperture very narrow. Central columellar teeth poorly developed, labial teeth extending around margin anteriorly. Dorsum lilac-grey covered with white spots which are overlain by orange-brown spots and blotches (fused and darker towards margins); base and teeth orange-brown; extremities lilac. Animal and mantle purplish brown. **Habitat:** among stones and coral. **Distribution:** Mas, SO.

Cypraea isabella Linnaeus, 1758: **260**
40mm. Almost straight sided, with poorly developed margins and non-protruding extremities. Teeth very fine, short, deep set, absent from almost flat base. Dorsum yellow-grey to reddish brown, with faint transverse bands overlain by thin, axial, black streaks; 2 orange-brown blotches at each extremity; base and teeth white. Brown-black animal has jet-black mantle. **Habitat:** among rocks and coral. **Distribution:** GO, Mas, SO.

Cypraea kieneri Hidalgo, 1906: **261**
20mm. Thin, almost straight sided, with undeveloped margins; anterior extremity produced. Labial teeth strong, columellar teeth weaker, absent from flat base. Dorsum creamy white with 3 large, irregular, grey-blue zones; margins and base spotted brown; pair of dark brown blotches at each extremity; teeth white. **Habitat:** under rocks and in rock crevices. **Distribution:** SO.

Cypraea helvola **259**

Cypraea isabella **260**

Cypraea kieneri **261**

Superfamily: **CYPRAEOIDEA**

Cypraea lamarckii Gray, 1825: **262**
40mm. Thick, pear shaped, well rounded, margins poorly developed but distinctly pitted; anterior extremity slightly upturned. Teeth well developed, absent from convex base. Dorsum greenish to reddish brown covered with small bluish-white dots; margins and extremities with irregular brown spots; base and teeth white. **Habitat:** among rocks, mud and silt. **Distribution:** SO.

Cypraea lentiginosa Gray, 1825: **263**
30mm. Solid, ovate, slightly humped, with lipped margins and protruding front extremity. Labial teeth nearly reach margin, columellar teeth absent from convex base. Dorsum greyish cream with faint, transverse, purple-brown bands overlain by pale or dark brown spots; margins have dark brown spots; base and teeth white. **Habitat:** among muddy rocks and coral. **Distribution:** all.

Cypraea limacina Lamarck, 1810: **264**
20mm. Solid but lightweight, elongate-ovate, extremities scarcely produced, outer lip margin evenly pitted along its length, dorsum smooth or noduled all over. Teeth long but not crossing base. Dark brown fading to slate grey with white base and white spots; teeth orange-brown often bordered with darker brown lines. Animal black to carmine. **Habitat:** shallow water under stones near reef. **Distribution:** Mas.

Cypraea lynx Linnaeus, 1758: **265**
45mm. Thick, elongate-ovate, with lumpy, sometimes lipped margins and produced extremities. Teeth well developed, deep-set within narrow aperture, absent from base which is flat or slightly concave. Dorsum cream with a bluish cast with purple, grey, orange and dark brown dots and blotches; margins white with dark brown blotches; ill-defined, pale brown sulcus; base and teeth white, reddish-orange between teeth. Grey-brown animal has a grey mantle with white papillae. **Habitat:** among coral and rocks. **Distribution:** GO, Mas, SO.

Cypraea macandrewi Sowerby, 1870: **266**
20mm. Lightweight, elongate-ovate with lipped labial margin, produced extremities and rounded base. Columellar teeth much finer and shorter than labial teeth, aperture narrow. Dorsum reddish brown sparsely covered with white spots some of them within brown rings; base mostly unspotted; columellar teeth white, labial teeth stained brown. Beached shells of this mainly Red Sea species have been found at Muscat and Masirah. **Habitat:** under stones. **Distribution:** GO, Mas.

Cypraea lamarckii **262**

Cypraea lentiginosa **263**

Cypraea macandrewi **266**

Cypraea limacina **264**

Cypraea lynx **265**

Superfamily: **CYPRAEOIDEA**

Cypraea mauritiana **268**

Cypraea marginalis pseudocellata
Schilder & Schilder, 1938: **267**
25mm. Thick, ovate with lipped, pitted margins. Dorsum olive-brown covered with white spots, some ringed purple-brown; greyish sulcus; violet base has purple spots and dashes; teeth paler than their intervals. Grey animal has greyish-brown mantle with branched papillae. **Habitat:** among stones. **Distribution:** NG, GO, Mas, SO.

Cypraea mauritiana Linnaeus, 1758: **268**
90mm. Heavy, ovate, humped, with smooth, sometimes gently wavy margins, scarcely produced extremities, almost flat base. Teeth large, deep set. Dorsum yellowish brown heavily reticulated with chocolate brown; margins, base and teeth chocolate brown, paler between teeth. Dark brown animal has black mantle. **Habitat:** among rocks and coral. **Distribution:** GO, Mas, SO.

Cypraea marginalis pseudocellata **267**

SEASHELLS OF EASTERN ARABIA

Superfamily: **CYPRAEOIDEA**

Cypraea moneta Linnaeus, 1758: **269**
30mm. Thick, heavy, ovate but usually modified by well-developed margins to give broadly pentagonal outline, low humped; extremities produced, base flat. Teeth small, deep set. Dorsum and margins yellow, often with 2 or 3 greyish, transverse bands; central part of base and teeth white. Varies greatly in shape and colour. Widely distributed by human agency. **Habitat:** under stones. **Distribution:** GO, Mas, SO.

Cypraea nebrites Melvill, 1888: **270**
30mm. Thick, rounded-ovate with strongly lipped, corrugated margins and rounded base. Teeth coarse, labial ones nearly reaching margin. Dorsum greenish brown with closely spaced ivory spots and distinct sulcus; large, dark brown blotch either side with faint connecting band; brown spots and dashes on margins, sometimes invading apricot coloured base but absent from teeth. Foot and mantle grey. **Habitat:** under rocks. **Distribution:** GO, Mas, SO.

Cypraea ocellata Linnaeus, 1758: **271**
28mm. Thick, rounded-ovate, slightly humped with well-developed, angulated, wrinkled margins and produced extremities. Teeth fine, slightly spreading over flattish base. Orange-brown to brown covered with small white spots some having central, dark brown spots; greyish sulcus well defined; margins creamy heavily spotted brown; base creamy, labial teeth usually lined brown. Grey animal and greyish-brown mantle. **Habitat:** under rocks and in crevices. **Distribution:** GO, Mas, SO.

Cypraea onyx succincta Linnaeus, 1758: **272**
48mm. Moderately thick but lightweight, pear shaped with produced anterior extremity. Teeth short and widely spaced, edging a wide aperture. Brown, darker on margins; sulcus and teeth orange-brown. Has 2 narrow, transverse bands dividing dorsum into 3 almost equal parts. Animal and mantle black. Syn: *C. persica* Schilder & Schilder, 1939. **Habitat:** in fragments of larger shells. **Distribution:** GO, Mas, SO.

Cypraea onyx succincta **272**

Cypraea moneta **269**

Cypraea ocellata **271**

Cypraea nebrites **270**

Superfamily: **CYPRAEOIDEA**

Cypraea pulchella pericalles Melvill & Standen, 1904: **273**
32mm. Thin, lightweight, pear-shaped, flat-based, with well-produced extremities and slight margins. Teeth fine, long, ridge-like, extending over entire base. Greyish-white dorsum has small pale brown dots and faint transverse bands overlain by large, irregular, darker brown blotches; dark brown spots on margins and extremities; teeth dark brown on white base. **Habitat:** on mud in deep water. **Distribution:** NG, GO.

Cypraea pulchra Gray, 1824: **274**
45mm. Solid, elongate-ovate, with scarcely raised margins and slightly produced extremities. Teeth fine, deep set in very narrow aperture. Dorsum pale pinkish brown with 3 faint, transverse bands; 2 large, chocolate-brown blotches at each extremity; base ivory or pink, teeth chocolate brown. Animal and mantle black. Known locally in Oman as "Four-eyes". **Habitat:** among branched coral offshore. **Distribution:** NG, SEG, GO.

Cypraea staphylaea Linnaeus, 1758: **275**
22mm. Solid, rounded-ovate, very swollen, lipped margins poorly developed, extremities slightly produced; dorsum covered with small pimples; sulcus deeply incised. Teeth form flat-topped ridges crossing entire base. Dorsum violet-grey, pustules white; base and margins white; extremities chocolate-brown; teeth pale orange; dorsum of fresh shells may be almost black. Greyish-brown animal has brownish-black mantle. **Habitat:** among coral blocks. **Distribution:** Mas, SO.

Cypraea staphylaea 275

Cypraea talpa Linnaeus, 1758: **276**
65mm. Solid, elongate-ovate, margins lightly developed, extremities almost smooth and produced; base almost flat; apex well indented. Teeth fine, short, deep set in very narrow aperture. Pale yellowish brown with 4 underlying, broad, transverse, darker brown bands; extremities, margins, and teeth chocolate brown, whitish between teeth. Black animal has smooth, greyish-black mantle. **Habitat:** under rocks or coral. **Distribution:** GO, Mas, SO.

Cypraea pulchella pericalles 273

Cypraea pulchra 274

Cypraea talpa 276

SEASHELLS OF EASTERN ARABIA

Superfamily: **CYPRAEOIDEA**

Cypraea tigris 279

Cypraea teulerei 278

Cypraea teres 277

Cypraea teulerei Cazenavette, 1845: **278**
45mm. Thick, heavy, rounded-ovate with thickened, lumpy margins and heavily callused posterior extremity. No columellar teeth; labial teeth absent or vestigial. Colour and pattern variable and marginal callus may cover most of dorsum; cream overlain by chocolate-brown blotches, stripes and zigzags (rows of chevrons or crescents on juveniles); sulcus usually prominent; large, brown spots on margins; base whitish. **Habitat**: under rocks or coral blocks. **Distribution**: Mas.

■ *Until the early 1970s* Cypraea teulerei *was one of the rarest of all cowries, the few specimens present in collections previously being of uncertain provenance - the Mocha area of the Red Sea was usually cited on specimen labels - and they were probably all collected some time during the nineteenth century. Through the collecting activities of the Bosch family Masirah Island became known as the principal locality for this distinctive cowry which is no longer considered a rarity. How specimens first reached Europe is likely to remain an unsolved mystery. Possibly all those known in older collections came from a single source. Whoever found them may have a claim to be considered a pioneer of shell collecting in eastern Arabia in general and Masirah Island in particular.*

Cypraea teres Gmelin, 1791: **277**
35mm. Solid, elongate-ovate, margins well developed and lumpy, extremities smooth and produced; base almost flat; apex indented. Teeth fine, short, in narrow aperture. Posterior columellar callus well developed, triangular. Cream with violet-brown, marginal blotches, dorsum streaked and blotched brown with broad transverse band centrally, base cream or white. Animal deep carmine. **Habitat**: shallow water among sponges or dead coral. **Distribution**: Mas.

Cypraea tigris Linnaeus, 1758: **279**
85mm. Thick, heavy, well rounded, high domed, margins smooth or slightly lumpy, anterior extremity protruding. Labial teeth short and rounded, columellar teeth longer and finer, all set in a concave base.

80 SEASHELLS OF EASTERN ARABIA

Superfamily: **CYPRAEOIDEA**

Cypraea turdus winckworthi **280**

Cypraea ziczac **282**

Cypraea vitellus **281**

Creamy white variously spotted and blotched with browns and blues; sulcus sometimes visible as dark line; base and teeth white. Varies greatly in size, thickness and colour pattern. **Habitat:** under rocks, among coral or in sand. **Distribution:** GO, Mas, SO.

Cypraea turdus winckworthi Schilder & Schilder, 1938: **280**
40mm. Heavy, rounded-ovate with heavily callused, usually lipped margins and produced, smooth extremities; base almost flat. Teeth short, coarse and deep set in broad aperture. Dorsum white, cream or greenish usually stippled with irregular brown spots which are larger on margins; sulcus seldom visible; base and teeth white. Description applies only to the form, prevalent in eastern Arabia, of *C. turdus* Lamarck, 1810. **Habitat:** under rocks and stones. **Distribution:** all.

Cypraea vitellus Linnaeus, 1758: **281**
50mm. Thick, heavy, ovate to pear shaped, very swollen, with lumpy margins and dorsum lumpy, anterior extremity produced and base rounded. Teeth coarse, deep set. Dorsum pale brown (sometimes with underlying, brown, transverse bands) with white spots; margins have fine, transverse, white lines; base and teeth white. Grey animal has grey mantle with yellow papillae. **Habitat:** in mud and sand. **Distribution:** GO, Mas, SO.

Cypraea ziczac Linnaeus, 1758: **282**
20mm. Thin, lightweight, pear shaped with slightly lipped margins and produced anterior extremity. Columellar teeth fine and deep-set, labial teeth coarser and longer. Dorsum fawn with 3 broad, transverse, white bands, each crossed by fawn chevrons or crescents; extremities bordered by elongate, brown spots; brown spots on margins continue onto orange base; teeth orange. Colour pattern varies greatly. **Habitat:** under coral. **Distribution:** GO, Mas.

Family OVULIDAE
Species in this diverse family are predominantly small and have thin, translucent shells, narrowly elongate to globular or pear shaped. The spire is usually submerged in the last whorl, the anterior and posterior extremities of some species forming beaks or long canals. Most shells are smooth or delicately striate with smooth or toothed outer lip; the posterior end of the columella may have a blunt, tooth-like process (the funiculum); rarely an umbilicus is present. Shells may be brightly coloured but have simple patterns. The animal occupants, too, have different colour patterns. Most species associate with corals, sea-fans, sponges and kelp offshore and are difficult to collect and study without diving for them. Melvill & Standen, 1901:384-5, and Melvill, 1928:103, list several ovulids additional to those described below, but the identity of most of them is questionable. Nonetheless, more species await discovery in eastern Arabia.

Calpurnus (Procalpurnus) lacteus (Lamarck, 1810): **283**
15mm. Solid but lightweight, broadly ovate, without produced extremities. Outer lip thickened, with ripple-like teeth along its length; a well-developed ridge extends along length of opposite wall. Very fine, transverse striae, otherwise smooth. Greyish white, outer lip milky white. **Habitat:** intertidal. **Distribution:** GO, Mas.

Crenavolva cf *renovata* (Iredale, 1930): **284**
10mm. Thick, glossy, narrowly elongate; sides of crescent-shaped extremities divergent; Aperture straight. Outer lip gently sinuous, toothed most of its length; funiculum has up to 5 short knobs. Dorsum very finely striate transversely. Reddish, dorsum paler where widest and at ends; outer lip and extremities pinkish rose. **Habitat:** offshore on soft coral. Syn: *Ovulum dentatum* A. Adams & Reeve, 1848. **Distribution:** GO.

Crenavolva striatula traillii (A. Adams, 1856): **285**
9mm. Thin, glossy, narrowly elongate; extremities crescent shaped; posterior columellar extremity pointed. Aperture straight. Outer lip toothed posteriorly; funiculum has up to 4 short knobs. Dorsum finely striate transversely. Reddish-brown to white, dorsum paler where widest; outer lip and extremities rose or orange. **Habitat:** offshore on soft coral. **Distribution:** NWG, SEG, GO.

Calpurnus (Procalpurnus) lacteus **283**

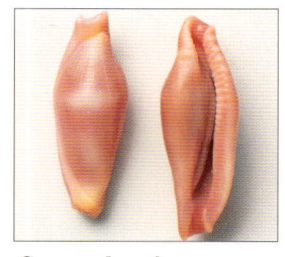

Crenavolva cf *renovata* **284**

Crenavolva striatula traillii **285**

Superfamily: **CYPRAEOIDEA**

Delonovolva labroguttata
(Schilder, 1969): **286**
25mm. Thin, narrowly spindle shaped; extremities recurved, pointed; aperture flaring anteriorly, narrow posteriorly. Thickened outer lip gently curved most of its length then sharply angled anteriorly. Fine spiral striae stronger each end. Orange-mauve, violet, pink or yellow with pale central band; conspicuous, evenly spaced spots matching main colour on white lip. **Habitat:** offshore on sea fans. **Distribution:** Mas.

Margovula tinctilis Cate, 1973: **287**
NOT ILLUSTRATED
15mm. Similar to ***Pseudosimnia* cf *marginata*** (Sowerby, 1828) (see below) but smaller, more pear shaped, with stronger teeth. Fine, widely spaced striae encircle shell. Whitish clouded on dorsum with pale brown, sometimes banded. **Habitat:** deep water. **Distribution:** GO.

Ovula ovum (Linnaeus, 1758): **288**
70mm. Sturdy but lightweight, ovate with roundly beaked, blunt extremities; outer lip thickened and wrinkled but not toothed. Dorsum smooth, sometimes malleated or wrinkled, highly polished; funiculum inconspicuous. Exterior porcelain white; interior reddish brown. Black animal has black mantle with white papillae. **Habitat:** coral reefs. **Distribution:** GO, Mas.

Phenacovolva (Phenacovolva) rosea rosea (A. Adams, 1855): **289**
25mm. Sturdy, narrowly spindle shaped with well-produced, slightly recurved extremities; aperture very narrow posteriorly; thickened outer lip curves inwards sharply anteriorly. Funiculum poorly developed. Delicate transverse striae each end. Rosy to orange-yellow with paler lip. **Habitat:** offshore on sea fans and corals. **Distribution:** GO.

Phenacovolva (Pellasimnia) weaveri pseudogracilis Cate & Azuma in Cate, 1973: **290**
29mm. Thin, spindle shaped; extremities evenly attenuated, slightly reflexed; aperture equal width throughout. Funiculum a slight knob. Fine spiral striae stronger each end. White with interrupted purplish stripe parallel to dull red or yellowish margin of outer lip. Yellowish-white mantle has black lines and pointed white papillae. **Habitat:** offshore on coral. **Distribution:** NG, SEG, GO.

Phenacovolva (Turbovula) dancei Cate, 1973: **291**
25mm. Thin, spindle shaped; extremities slightly recurved and sharply pointed (when undamaged); aperture wide and flaring anteriorly. Funiculum weak, elongate. Fine spiral striae stronger each end, smooth centrally on old specimens. Pale violet to violet-orange with paler central band, white outer lip and extremities. **Habitat:** offshore and beached. **Distribution:** GO, Mas.

Primovula* cf *platysia Cate, 1973: **292**
6mm. Solid, with produced extremities, posterior one pointed. Outer lip flattened and weakly toothed; funiculum a thick knob. Regular spiral striae (visible also on callus near aperture). Lavender or pink with whitish callus, outer lip and funiculum. **Habitat:** offshore on coral. **Distribution:** SEG.

Primovula singularis Cate, 1973: **293**
8mm. Solid, with produced extremities, posterior one twisted and roundly pointed. Outer lip flattened and toothed; funiculum a thick, triangular lump. Regular spiral striae (obscured by callus near aperture). Lavender with 3 diffuse, greyish-white, central and terminal areas on dorsum, pinkish margin and white funiculum. **Habitat:** offshore on coral. **Distribution:** SEG.

Primovula tropica (Schilder, 1931): **294**
6mm. Solid, with slightly produced extremities. Columellar callus ridge-like; outer lip toothed throughout; funiculum thick, triangular. Regular, spiral striae (obscured by callus near aperture). Pinkish orange with paler central and terminal areas on dorsum. Mantle whitish with white and dark spots, siphon blue-black. **Habitat:** offshore on soft coral. **Distribution:** NWG, SEG, GO, Mas.

Primovula (Adamantia) concinna (Adams & Reeve, 1848): **295** NOT ILLUSTRATED
Listed by Glayzer et al, 1984:321, from Kuwait (NWG).

Phenacovolva (Phenacovolva) rosea rosea **289**

Primovula cf *platysia* **292**

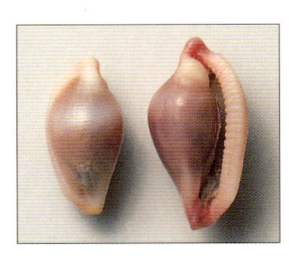

Primovula singularis **293**

Delonovolva labroguttata **286**

Phenacovolva (Pellasimnia) weaveri pseudogracilis **290**

Phenacovolva (Turbovula) dancei **291**

Superfamily: **CYPRAEOIDEA**

Primovula tropica **294**

Prionovolva pudica **296**

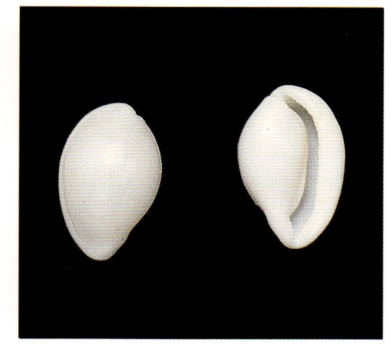

Pseudosimnia cf *marginata* **298**

Ovula ovum **288**

Trivirostra oryza **300**

Pseudocypraea adamsonii **297**

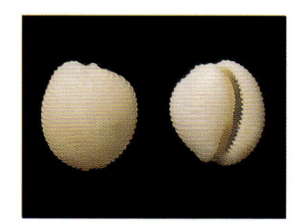

Niveria (Cleotrivia) globosa **299**

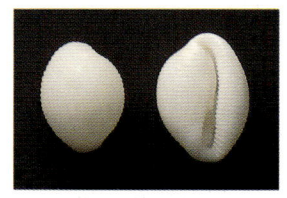

Trivirostra pellucidula **301**

Prionovolva pudica (A. Adams, 1855): **296**
15mm. Thin, translucent, ovate with slightly produced, rounded extremities; aperture much broader anteriorly. Thick columellar callus; outer lip thickened and toothed for most of its length; funiculum a raised ridge continuous with outer lip. Dorsum pale, pinkish brown, paler at widest part; outer lip and funiculum white. **Habitat:** unknown. **Distribution:** GO.

Pseudocypraea adamsonii (Sowerby, 1832): **297**
29mm. Solid, pear shaped with scarcely produced extremities, strongly humped, aperture very narrow, base slightly convex, labial margin lipped. Longitudinally, finely ridged and transversely ridged with uneven lines which end as strong teeth on broad outer lip. Off-white with large, irregular, pale brown blotches; outer lip white with pale brown blotches. **Habitat:** beached. **Distribution:** Mas.

Pseudosimnia cf *marginata* (Sowerby, 1828): **298**
15mm. Solid, glossy, pear shaped in outline, base almost flat; blunt extremities. Outer lip flattened, weakly and irregularly toothed; funiculum a rounded ridge. Fossula long and prominent. Only available shell is worn but has traces of minutely punctate, spiral grooves. Milky white. **Habitat:** beached. **Distribution:** Mas.

Family TRIVIIDAE

Shells of true triviids (subfamily Triviinae) resemble those of cowries but have transverse ridges covering the surface and entering the aperture, thus giving apertural edges a toothed appearance. With few exceptions the dorsum has a furrow from apex to base. When not white, the shells have subdued colours with no distinctive patterns. Shells of the subfamily Eratoinae are pear shaped with a rounded spire, have a toothed outer lip and small denticles on the inner lip. Most species have smooth shells and are yellow, greenish, or white. Members of this family feed on encrusting sea squirts (ascidians). Apparently poorly represented in Eastern Arabian waters but may have been overlooked.

Subfamily TRIVIINAE

Niveria (Cleotrivia) globosa (Sowerby, 1832): **299**
6mm. Solid, glossy, globose, aperture occupying total length; no spire. Entire shell encircled by close-set, spiral ridges which enter aperture and form many small teeth inside outer lip; dorsal furrow more or less prominent. White. **Habitat:** offshore and beached. **Distribution:** GO.

Trivirostra oryza (Lamarck, 1811): **300**
10mm. Thin, sturdy, elongate-globose with rounded base, the narrow aperture slightly curved posteriorly, extremities slightly produced. Fine, evenly spaced, transverse ribs cover entire shell surface, including shallow median groove extending length of dorsum, enter aperture and continue around columella; fine, axial striae between ribs. White. **Habitat:** beached. **Distribution:** GO, Mas.

Trivirostra pellucidula (Reeve, 1846): **301**
6mm. Thin, sturdy, globose with rounded base, the narrow aperture slightly curved posteriorly. Fine, evenly spaced, transverse ribs cover entire shell surface, enter aperture and continue around columella; unusually for a triviid there is no longitudinal dorsal furrow. White. **Habitat:** beached. **Distribution:** Mas, SO.

Trivirostra scabriuscula (Gray, 1827): **302** NOT ILLUSTRATED
Listed by Melvill & Standen, 1901:384, from the Gulf of Oman.

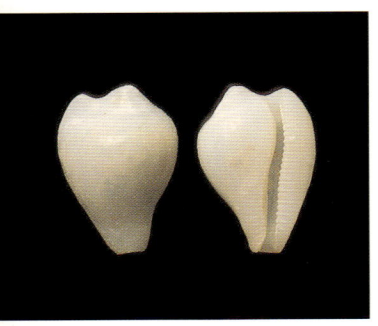

Superfamily: **CYPRAEOIDEA**, *Suborder:* **HETEROPODA**

Erato gallinacea 303

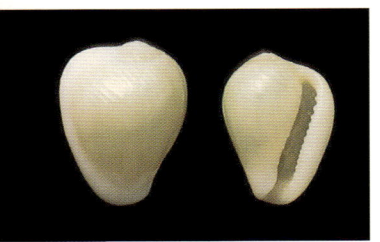

Sulcerato olivaria 305

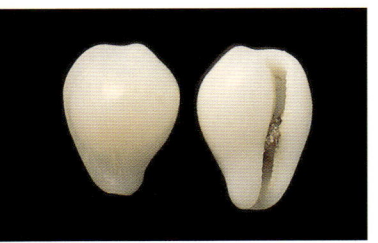

Sulcerato recondita 306

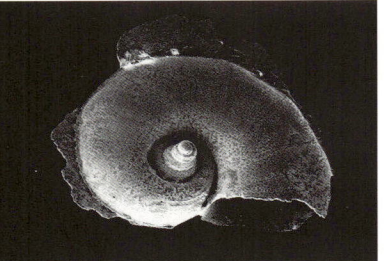

Atlanta turriculata 310

Lamellaria berghi 308

Subfamily ERATOINAE
Erato gallinacea (Hinds, 1844): **303**
5mm. Solid, glossy, globose-ovate, attenuated towards base, aperture slightly shorter than length of shell, spire nipple-like. Aperture slightly curved, many elongated denticles on outer lip, weak on lower half of columellar lip becoming obsolete above. Surface smooth. Milky white. **Habitat:** deep water. **Distribution:** GO.

Eratoena sulcifera (Sowerby, 1832): **304**
4.5mm. Solid, glossy, globose-ovate, aperture almost as long as shell, spire roundly pointed. Aperture narrow, straight; outer lip has about 20 denticles; columella with weak denticles along its length. Entire surface covered with axially aligned, elongated warts; groove along length of dorsum. Greyish white, paler on apertural side; purple stain on siphonal canal. **Habitat:** offshore and beached. **Distribution:** NWG.

Eratoena sulcifera 304

Sulcerato olivaria (Melvill, 1899): **305**
5mm. Solid, glossy, globose-ovate, aperture as long as shell, spire a smooth dome. Aperture narrow, straight; outer lip has about 15 denticles. Central dorsum smooth; unworn specimens under high magnification show diagonal threads elsewhere on shell ending in weak denticles along columella. Olive green; outer lip and apex white. **Habitat:** offshore and beached. **Distribution:** GO, Mas, SO.

Sulcerato recondita (Melvill & Standen, 1903): **306**
5mm. Solid, glossy, globose-ovate, aperture slightly shorter than length of shell, spire nipple-like. Aperture narrow, slightly curved. Many-denticled outer lip, denticles on columellar lip weak or absent. Surface smooth. Milky white. **Habitat:** deep water. **Distribution:** GO.

Family LAMELLARIIDAE
The thin, fragile shells of species in this family are usually found washed up on beaches. In life the animal's mantle envelops the shell, often totally concealing it. The spire is short and low, the last whorl greatly expanded, the aperture large. There is no operculum. The animals are known to feed upon sea squirts and sponges but even their empty shells are uncommon in Eastern Arabia. Their presence may not be suspected as they are not immediately recognisable as molluscs when alive.

Coriocella nigra 307

Coriocella nigra (Blainville, 1824): **307**
45mm. Thin, fragile, opaque, with short, low spire of about 3 whorls and greatly expanded, elongate last whorl; sutures shallow. Columella smooth, evenly curved, the outer lip sinuous. Smooth except for well-defined, irregular growth ridges. White with a thin, yellowish periostracum. Animal has yellowish or orange mantle with brown blotches or reticulations. **Habitat:** beached. **Distribution:** GO, Mas, SO.

Lamellaria berghi (Duclos, 1863): **308**
Recorded and illustrated by Melvill, 1918:142, from the Arabian Gulf without more precise locality.

Suborder HETEROPODA
Family ATLANTIDAE
Heteropods are small, pelagic molluscs with mostly discoidal, crested shells which help them float around, crest upright, in the surface waters of the sea, preying on tiny invertebrates. Unlike pteropods, with which they may be confused, heteropods have an operculum and their soft parts are too large for the shell. Seldom washed up on beaches, their shells abound in sediments on the sea floor.

Atlanta peronii Lesueur, 1817: **309**
5mm. Fragile, glossy, translucent, discoidal and coiled in one plane, the periphery is sharp and, in perfect specimens, has a large crest; protoconch multi-spiral. Fine growth striae. Crest usually imperfect. Milky white. **Habitat:** alive floating or dead in deep-sea sediments. **Distribution:** GO.

Atlanta turriculata Orbigny, 1836: **310**
5mm. Description as for ***A. peronii*** (see above) but spire elevated. **Habitat:** alive floating or dead in deep-sea sediments and beached. **Distribution:** Mas.

Atlanta peronii 309

Superfamily: **NATICOIDEA**

Mammilla melanostoma **311**

Mammilla sebae **312**

Mammilla syrphetodes **314**

Mammilla simiae **313**

Superfamily NATICOIDEA
Contains one family, the Naticidae (see below for details).

Family NATICIDAE
The sand-dwelling moon snails have thick or thin, mostly smooth and glossy shells with an unthickened outer lip. Some shells are globose or ovate, with a semilunar aperture; others are compressed and have a wide aperture; the sutures are seldom conspicuous. The umbilicus may be open or obscured by a thick calcareous plug (which may continue as a ridge into the umbilicus and is known as a "funicle"). In some species the operculum is smooth, chitinous and yellow to brown; in others it is calcareous, spirally ribbed and white. A thin, fibrous periostracum is usually present. To consume the flesh of molluscs, including other naticids, the naticid animal bores smooth-edged, circular holes into the protecting shells. Eggs of most species are cemented into flattened, coiled ribbons of sand or mud, these ribbons being common sights on sand in shallow water.

Subfamily POLINICINAE
Mammilla melanostoma (Gmelin, 1791): **311**
40mm. Solid but not heavy, ovate, spire short, blunt topped, straight sided; aperture much longer than wide. Umbilicus wide and deep, partly obscured by parietal callus. Fine growth lines. White or grey with 3 brownish-grey, sometimes interrupted bands on last whorl; parietal callus and umbilicus dark brown; paler brown inside aperture. Operculum chitinous, wine-red. **Habitat:** in sand. **Distribution:** GO, Mas.

Mammilla sebae (Récluz, 1844): **312**
40mm. Thin, semitranslucent, silky, spire nipple-like, very short; aperture less than twice as long as wide. Umbilicus very deep but mostly obscured by parietal and columellar callus. Fine growth striae crossed by much finer spiral striae. White with 3 spiral rows of widely spaced, brown blotches; umbilicus and covering callus dark brown; external pattern may show inside white aperture. Operculum chitinous (incompletely plugging aperture). Orange-yellow periostracum. **Habitat:** offshore and beached. **Distribution:** GO.

Mammilla simiae (Deshayes, 1838): **313**
22mm. Thin, semitranslucent, silky, ovate, spire very short, blunt topped; aperture twice as long as wide. Umbilicus deep but almost or quite obscured by parietal and columellar callus. Fine growth striae. White with 2 broad, spiral, orange-brown bands; occasional orange-brown blotches between bands; parietal and columellar callus chocolate-brown, sutures brown, paler brown inside aperture. Operculum chitinous. **Habitat:** offshore and beached. **Distribution:** NG, Mas.

Mammilla syrphetodes Kilburn, 1976: **314**
40mm. Thin, fragile, translucent, silky, broadly ovate, width of aperture about two-thirds of length; spire elevated and more prominent than in other Mammilla species. Umbilicus deep but narrow; thin callus on parietal wall and columella. Fine growth striae and finer spiral striae. White with 2 spiral rows of nebulous, widely spaced, orange-brown spots on last whorl; callus pale brown, umbilicus pale, colourless within. Operculum chitinous. One shell found on Batinah coast. **Habitat:** offshore and beached. **Distribution:** GO.

Superfamily: **NATICOIDEA**

Natica cernica **318**

Natica gualteriana **319**

Natica pulicaris **320**

Neverita (Glossaulax) didyma by its egg ribbon

Neverita (Glossaulax) didyma (Röding, 1798): **315**
40mm. Moderately heavy, spire low and blunt; aperture large, semi-lunar; last whorl slopes sharply below suture. Umbilical area almost or completely obscured by massive, tongue-like funicle. Fine, irregular growth lines; funicle has central groove. Bluish-white, grey or violet-brown (bleaches to violet-pink); apical whorls darker; paler and sometimes brown-streaked around umbilical area; funicle orange or dark brown; aperture shades of brown. Operculum chitinous. **Habitat:** in sand. **Distribution:** NWG, SEG, GO, Mas, SO.

Neverita (Neverita) peselephanti (Link, 1807): **316**
35mm. Thick, heavy, flat sided, spire low and pointed, aperture semi-lunar, last whorl not sloping below suture. Umbilicus very deep; broad funicle ends with reflected, flattened funicle. Very fine growth lines; funicle finely ridged. Orange, cream or pale yellow; umbilical area, apical whorls and inside aperture white. Operculum chitinous. **Habitat:** in sand. **Distribution:** NWG, SEG, GO, Mas, SO.

Polinices mammilla (Linnaeus, 1758): **317**
35mm. Heavy, ovate with low, bulbous spire. Semilunar aperture more than half height of last whorl; umbilical area filled with massive funicle. Fine, irregular growth lines all over, except funicle which is smooth. White, funicle sometimes tinged yellow. Operculum chitinous. Syn: *P. tumidus* (Swainson, 1840) of some modern authors. **Habitat:** in sand. **Distribution:** all.

Subfamily NATICINAE
Natica cernica Jousseaume, 1874: **318**
10mm. Thin, translucent, globose, spire short; aperture slightly flaring. Comma-shaped umbilicus narrow but deep, partly obscured by funicle which ends in broad callus. Oblique fine grooves below sutures, otherwise smooth. Spire whorls brown to greyish; white band below sutures; last whorl with 2 broad, spiral, pale brown zones containing darker spots and flecks; umbilical area white. Operculum calcareous, with sub-central nucleus and 2 strong grooves parallel to margin. Syn: *N. tranquilla* Melvill & Standen, 1901, is probably a pale form. **Habitat:** in sand offshore. **Distribution:** GO.

Natica gualteriana Récluz, 1843: **319**
15mm. Solid, globose with low spire and impressed sutures; upper part of whorls obliquely puckered, lower half with fine growth striae. Umbilicus deep, partly obscured by prominent funicle. White with 2 or 3 spiral zones of spots or short bars; sometimes greyish brown with unbroken spiral bands; umbilical area white. Calcareous operculum has marginal groove. **Habitat:** intermedial sand. **Distribution:** GO, Mas.

Natica pulicaris Philippi, 1852: **320**
20mm. Solid, globose with low, flat-topped spire; last whorl tabulated at suture and puckered, smooth below. Umbilicus deep, largely obscured by parietal callus. Cream or white with 2 or 3 pale or dark brown bands, also speckled with pale brown, sometimes patterned with large blotches; pale or dark brown around umbilicus; external pattern shows within aperture which is often stained violet. Operculum calcareous. **Habitat:** intertidal sand. **Distribution:** NG, NWG, Mas, SO.

Neverita (Glossaulax) didyma **315**

Neverita (Neverita) peselephanti **316**

Polinices mammilla **317**

SEASHELLS OF EASTERN ARABIA

Superfamily: **NATICOIDEA**

Natica pseustes Watson, 1881: **321**
12mm. Solid, globose with low spire; fine, oblique growth striae. Umbilicus deep, almost obscured by thick, columellar callus. White or cream patterned with crowded, wavy, dark or pale brown lines which form blotches below sutures, at periphery and around umbilicus; columellar callus dark or pale brown; white inside aperture. Operculum calcareous, with many grooves parallel to margin. Syn: *N. telaaraneae* Melvill, 1901. **Habitat:** in sand. **Distribution:** GO, Mas.

Natica vitellus (Linnaeus, 1758): **322**
18mm. Thick, silky, globose with low spire; very fine, oblique growth striae. Umbilicus deep, largely obscured by columellar callus; base of columella thickened and recurved. White with 1 or 2 narrow or broad, spiral, brown bands; apex reddish brown; white around base. Operculum calcareous, with 4 grooves parallel to margin. Shells may attain twice the size elsewhere in the Indo-Pacific. Syn: *Natica ponsonbyi* Melvill, 1899. **Habitat:** in sand, mostly offshore. **Distribution:** NG, SEG, GO.

Natica (Naticarius) alapapilionis (Röding, 1798): **323**
25mm. Solid, silky, globose-ovate with prominent spire; very fine, oblique growth striae, puckered at suture. Umbilicus deep and very wide; funicle slender and prominent; parietal callus reaches only top of aperture. Pale brown or greyish brown with 4 narrow, white bands each containing regularly spaced, dark brown, rectangular spots; umbilical area and columella white; reddish brown inside aperture. Operculum calcareous with many grooves parallel to margin. **Habitat:** in sand. **Distribution:** NG, GO, Mas.

Natica (Naticarius) manceli Jousseaume, 1874: **324**
17mm. Solid, globose-ovate with short spire; prominent grooves below sutures, smooth elsewhere. Umbilicus deep and wide; funicle broad and low. Creamy white to fawn with 2 spiral rows of sparse, brown streaks or spots; end of funicle either brown or white. Operculum calcareous with many grooves parallel to margin. Yellowish periostracum. Syn: Arabian Gulf form, described here, was named *N. strongyla* by Melvill, 1897. **Habitat:** offshore in sand. **Distribution:** NG.

Notocochlis n.sp: **325**
15mm. Specimens previously identified as ***Tanea lineata*** (Röding, 1798), distinctively patterned with many sinuous and zigzag reddish brown lines, belong to a species which will be described as new to science elsewhere. They occur in the Gulf of Oman and at Masirah.

Subfamily SININAE

Eunaticina papilla (Gmelin, 1791): **326**
20mm. Solid but lightweight, dull, broadly ovate with short spire and deep sutures. Columella has oblique axis. Fine, irregularly spaced, spiral grooves cover entire surface. Umbilicus broadly open, partly concealed by columellar callus. Operculum chitinous. **Habitat:** offshore and beached. **Distribution:** NWG, SEG, GO, Mas.

Sigatica pomatiella (Melvill, 1893): **327**
18mm. Thin, translucent, globose with short spire and deep sutures; fine, spiral grooves near sutures, almost obsolete mid-whorl, fine growth striae. Umbilicus large and deep, slightly obscured by columellar callus. White. **Habitat:** offshore and beached. **Distribution:** NWG, SEG, GO, Mas.

Natica (Naticarius) alapapilionis **323**

Natica pseustes **321**

Notocochlis n.sp. **325**

Natica (Naticarius) manceli **324**

Natica vitellus **322**

Eunaticina papilla **326**

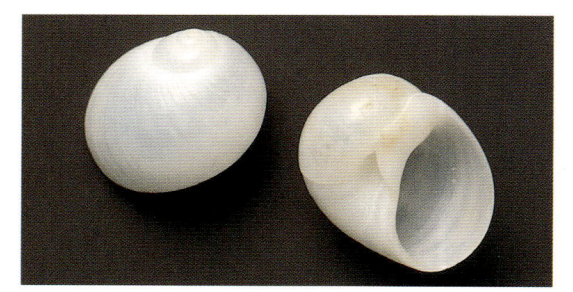

Sigatica pomatiella **327**

Superfamily: **NATICOIDEA, TONNOIDEA**

Sinum delessertii (Récluz, 1843): **328**
20mm. Thin, dull, ear shaped with low, nipple-like spire; evenly spaced, coarse, spiral threads crossed by growth ridges, early whorls smooth. Umbilicus a mere chink or covered by thin parietal callus. Orange-brown with white zone on last whorl below suture, at former sites of growing edge and around aperture. Operculum chitinous. **Habitat:** offshore and beached. **Distribution:** GO.

Sinum haliotoideum (Linnaeus, 1758): **329**
30mm. Fragile, translucent, dull, ear shaped, the flattened spire having almost lens-like profile. Crowded, fine, wavy, spiral threads crossed by coarse growth ridges which are strongest at sutures. Umbilicus closed by glossy parietal and columellar callus. White. **Habitat:** beached. **Distribution:** NWG, SEG, GO, Mas, SO.

Sinum laevigatum (Lamarck, 1822): **330**
35mm. Semitranslucent, dull, ear shaped with nipple-like spire. Flat-topped spiral ribs, 2 or 3 finer threads in their intervals, crossed by coarse, irregular growth ridges. Umbilicus usually closed by thin parietal callus. Last whorl white, tinged blue or violet, becoming pale brown on spire whorls; sub-sutural zone white. Operculum chitinous. **Habitat:** in sand and beached. **Distribution:** GO.

Sinum quasimodoides Kilburn, 1976: **331**
15mm. Thin, translucent, dull, ear shaped with nipple-like spire; aperture widest at anterior end. Evenly spaced, wavy, flat, spiral threads crossed by coarse growth ridges. Umbilicus a mere chink. White. Operculum chitinous. **Habitat:** offshore and beached. **Distribution:** NG, GO.

Superfamily TONNOIDEA
Well represented in eastern Arabia, this superfamily, comprising several carnivorous families differing widely from each other superficially, is distinguished for the large size of some of its species. Few species are small, none minute. Some have thin, brittle shells; others have thick, strong shells. A chitinous operculum is present in some families. A long, free-swimming larval stage has helped a few to establish a very wide, sometimes worldwide distribution.

Family TONNIDAE
Tun shells are medium to large with a thin, lightweight, globose shell, low spire and large aperture, the lip edge being wavy or toothed. The siphonal canal is short and broad, the columella smooth and twisted. Adults lack an operculum. Mostly from tropical areas where they feed on holothurians (or sea-cucumbers), they have a pelagic phase which may last for three months or longer. Most records for this family from eastern Arabia are of empty shells washed up on beaches; this suggests that the species prefer offshore habitats; inevitably their delicate shells would make them vulnerable to turbulent intertidal conditions.

Malea pomum (Linnaeus, 1758): **332**
55mm. Thick, globose, with short, low spire and deep sutures. Aperture narrow compared with other tun shells; outer lip thickened and bearing elongated teeth on its inner edge. Incurved columella has strong spiral folds. Umbilicus small

Sinum quasimodoides **331**

Sinum delessertii **328**

Sinum haliotoideum **329**

Sinum laevigatum **330**

Superfamily: **NATICOIDEA, TONNOIDEA**

or obscured. About 12 strong, smooth, spiral cords on last whorl. Cream with pale brown spots and blotches; parietal callus and outer lip white, inside aperture orange. Only a single worn shell has been found in eastern Arabia. **Habitat:** offshore, but usually found beached and worn. **Distribution:** SO.

Tonna cumingii (Reeve, 1849): **333**
80mm. Thin but sturdy, with short spire and deep sutures. Aperture wide, edge of outer lip fluted. Columella almost straight, umbilicus narrow and deep. Last whorl has 21 to 22 flattened, regularly spaced ribs separated by narrow grooves. Cream or pale brown with axial brown stripes, chevrons or crescents edged with white; pale reddish brown internally, edge of outer lip darker brown. **Habitat:** offshore, usually washed up dead on beaches. **Distribution:** GO, Mas, GO.

Tonna dolium (Linnaeus, 1758): **334**
100mm. Fragile, with short spire and deeply channelled sutures. Aperture wide, edge of outer lip thin and fluted. Lower part of columella prominently twisted. Umbilicus obscured in adults. Last whorl has up to 18 strong, sharp or rounded, spiral ridges and occasional lesser ones between. Ivory, cream or pale brown with evenly spaced, reddish brown spots on ridges. Protoconch and inside aperture brown. **Habitat:** offshore, usually beached. **Distribution:** GO, Mas.

Tonna dolium 334

Tonna cumingii 333

Malea pomum 332

SEASHELLS OF EASTERN ARABIA

Superfamily: **TONNOIDEA**

***Tonna olearium* 335**

***Tonna luteostoma* 336**

90 SEASHELLS OF EASTERN ARABIA

Superfamily: **TONNOIDEA**

Tonna olearium (Linnaeus, 1758): **335**
150mm. Fragile, with raised spire, pointed apex and deeply channelled sutures. Very large aperture slightly angulate above; edge of outer lip fluted. Columella strongly curved. Umbilicus obscured by parietal glaze. About 17 to 20 broad, flattened ribs on last whorl, occasional narrower threads between. Pale brown or yellowish with occasional, darker, spiral bands; protoconch reddish brown; inside aperture creamy white or brownish. A worn shell found at As Sib, Oman. Syn: some authors have used the name *T. zonata* (Green, 1830). **Habitat:** offshore, usually washed up dead on beaches. **Distribution:** GO.

Tonna luteostoma (Küster, 1857): **336**
150mm. Thick, sturdy, with low, almost flattened spire and deeply channelled sutures. Edge of outer lip fluted. Columella slightly curved. Umbilicus narrow and deep. About 16 strong, rounded, evenly spaced, spiral ridges on last whorl; large parietal callus. Ivory or cream with irregular brown blotches and axial streaks; inside aperture brown or orange. **Habitat:** offshore, usually beached. **Distribution:** GO, Mas.

Family FICIDAE
Fig shells have thin, fragile, swollen, pear-shaped or fig-shaped shells with an elongated aperture and well-developed siphonal canal; there is no umbilicus. The animal has a large foot, lacking an operculum; the tentacles and siphon are long and 2 large flaps envelop the shell on each side towards the head. Sand burrowers, they are seldom found alive in shallow water but their shells are often washed up on beaches around eastern Arabia.

Ficus gracilis (Sowerby, 1825): **337**
120mm. Thin, translucent, dull, low spired, twice as long as wide. Protoconch smooth; spire whorls with axial threads crossed by strong, flat-topped, spiral ridges; distinct parietal callus by upper corner of aperture. Whitish or pale brown variegated with darker brown, axial, wavy streaks; glossy and glazed white in aperture. **Habitat:** offshore and beached. **Distribution:** NG, GO, Mas.

Ficus gracilis **337**

Superfamily: **TONNOIDEA**

Ficus subintermedia (Orbigny, 1852): **338**
60mm. Thin but solid, dull, almost flat topped so that aperture is almost total height of shell. Sutures well impressed. Protoconch smooth; spire whorls with axial threads crossed by spiral threads (every fifth spiral thread below shoulder of last whorl conspicuously stronger than others); parietal callus by upper corner of aperture. Yellowish brown with reddish brown and whitish blotches form indistinct spiral bands; aperture pale brown glazed white at edge. **Habitat:** offshore and beached. **Distribution:** NWG, SEG, GO, Mas, SO.

Ficus variegata Röding, 1798: **339**
75mm. Sturdy, lightweight, dull, almost globular with very low spire and strongly recurved siphonal canal, sutures lined by callus deposit. Embryonic whorls smooth, all others with closely spaced spiral threads crossed by similarly spaced, weaker axial threads; parietal callus by upper corner of aperture. Pinkish or pale brown with about 6, paler, spiral zones containing reddish brown blotches; sutures white; aperture glossy, dark violet, glazed white towards edge. **Habitat:** offshore and beached. **Distribution:** SEG, Mas.

Family CASSIDAE
Mostly warm-water gastropods, helmet shells are represented in the region by a few species of moderate size, usually bearing a thin, chitinous operculum. All species have a capacious last whorl; most have varices, a toothed outer lip and columellar folds; some have pronounced knobs. Shells of males and females may differ in size and some species may hybridize with others. The sand-dwelling animals feed upon sea urchins and other invertebrates.

Ficus subintermedia **338**

Ficus variegata **339**

Cypraecassis rufa **340**

92   SEASHELLS OF EASTERN ARABIA

Superfamily: **TONNOIDEA**

Casmaria ponderosa unicolor **342**

Subfamily CASSINAE
Cypraecassis rufa (Linnaeus, 1758): **340**
115mm. Thick, heavy, broadly ovate in dorsal view, spire obscured by columellar callus and outer lip in apertural view. Early whorls exserted, otherwise spire almost flat; siphonal canal strongly recurved. Outer lip has widely spaced, elongated teeth; inner lip with finer teeth and long, thin folds. Dorsum has 3 to 4 spiral rows of blunt knobs; a series of axial bars encircles siphonal canal. Reddish brown interspersed with orange and apricot aperturally, the teeth white; white dorsally. **Habitat:** near coral reefs and beached. **Distribution:** Mas, SO.

Subfamily PHALIINAE
Phalium glaucum (Linnaeus, 1758): **341**
80mm. Thin, translucent but solid, globose, the spire short, well exserted; later spire whorls sharply angled and beaded at shoulders; sutures slightly channelled. Spire whorls reticulated; surface of last whorl lightly malleated. Umbilicus moderately broad, very deep, obscured by broad, thin, columellar shield. Teeth on outer lip longest near strongly recurved siphonal canal. Greyish apricot with brownish bars on outer lip; aperture dark brown. **Habitat:** offshore and beached. **Distribution:** GO, Mas.

Phalium glaucum **341**

Casmaria ponderosa unicolor (Pallary, 1926): **342**
70mm. Thick, glossy, ovate, spire short to moderately tall, whorls strongly shouldered or smoothly rounded. Fine growth ridges on all whorls, last whorl sometimes bearing a varix; outer lip thickened and regularly toothed along ventral side. No umbilicus. Bluish cream or greenish grey sometimes with up to 6 spiral rows of pale or dark brown blotches on last whorl. Operculum small, yellowish. Masirah specimens among largest known. **Habitat:** offshore and beached. **Distribution:** GO, Mas, SO.

Superfamily: **TONNOIDEA**

Semicassis bisulcata (Schubert & Wagner, 1829): **343** NOT ILLUSTRATED. 55mm. Thin or thick, moderately glossy, globose-ovate, last whorl rounded at shoulder. Varices none or as many as 6. Last whorl smooth or with crowded, spiral cords, sometimes malleated. Siphonal canal recurved; outer lip toothed. Thinly glazed columellar shield. Umbilicus narrow, deep. Cream, white or blue-grey, with or without 5 or 6 spiral rows of large or small, yellow to reddish-brown spots; last varix white with brown or mauve bars. Very variable in size and appearance. **Habitat:** deep water, sometimes beached. Available shells too worn for acceptable illustration here. **Distribution:** all.

Semicassis faurotis (Jousseaume, 1888): **344**
50mm. Thin, globose, without varices. Last whorl has well defined, regularly spaced, spiral grooves; early whorls have fine, axial riblets within spiral grooves. Thin columellar shield has 3 to 5 rounded folds on its left side. Outer lip toothed. Umbilicus deep, narrow. Early whorls dark purple (unique among Cassidae); last whorl cream or pinkish to pale brown with 5 spiral rows of rectangular, reddish brown blotches; dorsal side of outer lip has 5 broad, brown bars. Operculum covered with thin rods arranged fanwise. **Habitat:** offshore and beached. **Distribution:** all.

Family RANELLIDAE
The tritons are small to large, their shells are predominantly thick, heavy and coarsely ornamented with knobs and cords. Occasionally they have varices (which may vary slightly from the number given in the species descriptions below) and some have a long siphonal canal. The periostracum is usually conspicuous and may be thin or thick and bristly. The chitinous, brown operculum is often smaller than the aperture. A complex family, it is well represented in the region, most species occurring in rock pools or in rock crevices at low tide level or below; some bury themselves in sand. A long, free-swimming larval stage has ensured a very wide range for some species. Identification, especially when specimens are immature, may be difficult.

Gyrineum (Biplex) perca (Perry, 1811): **345**
36mm. Thick, dull, a dorso-ventral flattening giving leaf-like appearance, the deep sutures set at a conspicuous angle to the axis; trochoidal protoconch of 3.5 swollen, glossy whorls. Small, circular aperture. Coarse axial ribs crossed by coarse and fine spiral ridges extending onto webbed frills each side which end in rounded points. Yellowish white. **Habitat:** deep water. **Distribution:** GO.

Semicassis faurotis 344

Gyrineum (Biplex) perca 345

Semicassis faurotis 344

Superfamily: **TONNOIDEA**

Gyrineum (Gyrineum) bituberculare (Lamarck, 1816): **346**
45mm. Thick, elongate-ovate, with short, recurved siphonal canal and 2 thin varices per whorl, sinuously aligned on opposite sides. Outer lip weakly toothed. Columella has folds along its length. Between varices are 2 (sometimes 3) prominent nodules; surface has irregular spiral ridges and threads crossed by fine axial threads. Yellowish with reddish brown nodules; aperture white. **Habitat:** offshore. **Distribution:** NG.

Gyrineum (Gyrineum) natator (Röding, 1798): **347**
30mm. Thick, ovate, spire almost half total height, 2 thin varices per whorl, axially aligned on opposite sides. Short, oblique siphonal canal. Weak ridges on outer lip; 1 or 2 weak folds on columella. About 6 weak axial folds between varices, nodular where crossed by smooth spiral ribs. White heavily banded dark or pale brown; aperture white. **Habitat:** offshore and on intertidal rocks. **Distribution:** NG.

Gyrineum (Gyrineum) pusillum (Broderip, 1833): **348**
17mm. Sturdy, high spired with whorls deceptively flattened antero-dorsally because of prominent varices aligned each side; trochoidal protoconch of 3.5 swollen, glossy whorls. Columella has strong plicae above and below centre. Equally strong axial and spiral cords make pustulose sculpture. Outer lip has 7 to 8 teeth. Whitish with brown spiral bands, varices brown; aperture and columella purple. **Habitat:** near coral areas and beached. **Distribution:** GO, Mas, SO.

Gyrineum (Gyrineum) bituberculare **346**

Gyrineum (Gyrineum) natator **347**

Natural size

Gyrineum (Gyrineum) pusillum **348**

SEASHELLS OF EASTERN ARABIA

Superfamily: **TONNOIDEA**

Cymatium (Cymatium) ranzanii **349**

Subfamily CYMATIINAE
Cymatium (Cymatium) ranzanii (Bianconi, 1850): **349**
80mm. Thick, heavy, dull, the spire seeming shorter in mature specimens by development of large, flared aperture adjacent to wing-like varix opposite. Large, angulate protuberance on dorsal side of last whorl gives hump-backed appearance; about 7 broad, spiral ribs below protuberance. Spire whorls have strong, axial ribs crossed by finer spiral ribs. Ridge-like teeth inside the outer lip. Ivory with yellowish brown spiral bands; teeth have dark brown bars near edge of brown-and-white-barred outer lip; columella has large, chocolate-brown blotch; white inside aperture. May reach 220mm. **Habitat:** on intertidal rocks in muddy sand. **Distribution** GO, Mas, SO.

Cymatium (Lotoria) grandimaculatum (Reeve, 1844): **350**
60mm. Thick, heavy, elongate-ovate, low spired, the whorls shouldered, sutures slightly channelled, apex blunt. Up to 5 varices; later whorls with coarse spiral ribs and axially aligned nodules. Regular teeth inside; greatly thickened outer lip correspond to lirae within aperture; columella has a few poorly defined plicae; siphonal canal short and broad. Orange-brown with brown-and-white varices; 2 large brown blotches on parietal wall; teeth brown, aperture white. Grows much larger elsewhere within its range. **Habitat:** offshore and beached. **Distribution:** Mas.

Superfamily: **TONNOIDEA**

Cymatium (Lotoria) perryi **352**

Cymatium (Lotoria) grandimaculatum **350**

Cymatium (Lotoria) lotorium **351**

Cymatium (Lotoria) lotorium
(Linnaeus, 1758): **351**
80mm. Thick, heavy, irregularly ovate, short spired, sutures shallow. 2 strong varices per spire whorl. Siphonal canal short, sinuous, broad. 3 prominent nodes between varices; additional spiral cords thickened on varices. Aperture wavy edged; columella plicate for entire length. Orange with varices banded brown and whitish orange. Aperture white; inner lip chocolate edged. Columella brownish orange with 2 large chocolate blotches on upper half. A mature, living specimen found near Muscat. **Habitat:** intertidal rocks. **Distribution:** GO.

Cymatium (Lotoria) perryi
Emerson & Old, 1963: **352**
120mm. Thick, heavy, irregularly ovate, short spired, sutures deep. 3 or 4 strong varices per spire whorl. Siphonal canal sinuous, broad. Later spire whorls have 2 strong, nodulous, spiral ribs; last whorl has spiral row of 3 very large nodules. Reddish brown with darker brown spiral bands; varices brown and white; columella, parietal wall and teeth orange, aperture white. Lacks the columellar blotches of *C. (L.) lotorium* (see above). **Habitat:** rocks at low tide and below. **Distribution:** Mas.

Superfamily: **TONNOIDEA**

Cymatium (Monoplex) aquatile (Reeve, 1844): **353**
80mm. Thick, heavy, elongate-ovate, tall spired, deep sutures, 3 varices and short, recurved siphonal canal. Later spire whorls have 2 or 3 nodulous spiral cords, nodules very pronounced at shoulders. Inside outer lip are 2 concentric rows of paired teeth; columella densely lirate throughout its length. Various shades of orange-brown; varices dark brown and white; teeth and columellar lirae whitish, their interstices orange as is the aperture. **Habitat:** offshore and beached. **Distribution:** GO, Mas, SO.

Cymatium (Monoplex) nicobaricum (Röding, 1798): **354**
60mm. Very thick and heavy, elongate-ovate, high spired, with deep sutures and small aperture, 5 or 6 thick varices and short, broad siphonal canal. Spiral ribs nodulous and spirally grooved, 3 nodules at shoulder of last whorl very pronounced. Inside outer lip are 2 concentric rows of teeth; columella sparsely lirate throughout its length. Ivory white mottled grey and brown; varices white barred and spotted pale brown; teeth white, aperture orange; columellar lirae white, their interstices orange. **Habitat:** offshore, sometimes beached. **Distribution:** GO, Mas.

Cymatium (Monoplex) parthenopeum (von Salis, 1793): **355**
90mm. Thick, heavy, ovate to elongate-ovate, short or high spired with rounded whorls, deep sutures and short, broad, recurved siphonal canal. Coarse spiral ribs, the uppermost strongly or weakly noduled; 2 low varices. Paired teeth inside outer lip correspond to paired lirae in aperture; columella lirate throughout its length. Shades of brown; varices barred white and dark brown; aperture, teeth and columellar lirae white or orange, chocolate brown around teeth and between lirae. Periostracum fibrous, amber coloured. May grow to about 130mm. **Habitat:** intertidal rock pools and offshore. **Distribution:** all.

Cymatium (Monoplex) aquatile **353**

Cymatium (Monoplex) nicobaricum **354**

Cymatium (Monoplex) parthenopeum **355**

Superfamily: **TONNOIDEA**

Cymatium (Monoplex) pileare
(Linnaeus, 1758): **356**
100mm. Thick, heavy, elongate-ovate, high spired, with rounded whorls, deep sutures; short, broad, recurved siphonal canal; aperture narrow. Coarse spiral cords, the uppermost strongly or weakly nodulous, deeply grooved between cords, nodules sinuously aligned axially; 3 pronounced varices. Crowded, paired teeth correspond to paired lirae in aperture; columella lirate throughout its length. Shades of brown variegated with white spiral bands; teeth and lirae white, the aperture red. Brown periostracum. **Habitat:** intertidal rock pools and offshore.
Distribution: NG, SEG, GO, Mas, SO.

Cymatium (Ranularia) boschi
Abbott & Lewis, 1970: **357**
85mm. Thick, heavy, ovate, short spired, with strongly shouldered whorls and deep sutures; tapering siphonal canal; aperture much taller than spire. Coarse, nodulous spiral cords on later whorls, nodules axially aligned, spiral ribs between cords; 2 varices. A few deep-seated teeth inside wavy-edged outer lip; columella with few deep-seated folds. Shades of brown; aperture white, often tinted pink. **Habitat:** muddy rocks and crevices.
Distribution: GO, Mas, SO.

Cymatium (Ranularia) cynocephalum (Lamarck, 1816): **358**
60mm. Thick, moderately heavy, globose-ovate with short spire and deep sutures; short, recurved siphonal canal. Strong spiral cords have low nodules, pronounced at shoulder, spiral riblets between cords; 1 varix. Strong, elongate teeth inside outer lip; weak lirae on columella. Pale brown with darker brown and white bars on varix and edge of outer lip; large, chocolate-brown smudge on columella; teeth flushed pale brown or orange. Syn: *C. moritinctum* (Reeve, 1844).
Habitat: shallow water and beached.
Distribution: Mas, SO.

Cymatium (Monoplex) pileare 356

Cymatium (Ranularia) boschi 357

Cymatium (Ranularia) cynocephalum 358

Superfamily: **TONNOIDEA**

Cymatium (Ranularia) oboesum
(Perry, 1811): **359**
50mm. Thick, lightweight, globose with slightly raised spire, almost flat-topped last whorl, impressed sutures and long, attenuated siphonal canal. Widely spaced, low, nodulous spiral cords, coronate at shoulder of last whorl; no varices. Strong, ridge-like teeth on inner side of outer lip, uppermost tooth bifid; columella lirate throughout length. Ivory heavily blotched pale brown; teeth, columella and aperture white. **Habitat:** offshore and beached. **Distribution:** GO, Mas.

Cymatium (Ranularia) trilineatum
(Reeve, 1844): **360**
65mm. Very thick, heavy, ovate, short or high spired, with impressed sutures; long, recurved siphonal canal. Strong, nodulous spiral cords, the nodes forming sharp blades at shoulder, 1 dorsal node hump-like and larger than others; 2 varices. Strong, elongated teeth on inner side of outer lip; columella lirate throughout except for smooth central area. White heavily blotched dark brown; columella, teeth and aperture white. **Habitat:** offshore and beached. **Distribution:** NG, SEG, GO, Mas.

Cymatium (Ranularia) tripum
(Lamarck, 1822): **361**
75mm. Sturdy, lightweight, elongate-ovate, high spired with deeply channelled sutures and long, slightly curved siphonal canal. Low, spiral ribs with 3 or 4 finer spiral riblets in their intervals, nodes on uppermost ribs of each whorl axially aligned; 3 varices. Paired teeth on inner side of outer lip; columella irregularly lirate throughout length. Yellowish brown with brown-and-white-barred varices; columellar lirae and edge of inner lip white, brown between lirae, aperture ivory sometimes tinged purple. **Habitat:** offshore and beached. **Distribution:** SEG, GO, Mas.

Cymatium (Septa) rubeculum rubeculum (Linnaeus, 1758): **362**
40mm. Thick, elongate-ovate, spire about same height as aperture, siphonal canal short and broad. Coarse, uneven, beaded, spiral cords on all whorls; 5 or 6 varices. Strong teeth on inner side of outer lip correspond to lirae within aperture; columella has prominent lirae throughout length. Red or oranged-red varices have occasional bold white patches, often a single white spiral cord on last whorl; teeth and lirae white. **Habitat:** intertidal sand and rubble. **Distribution:** NWG, Mas.

Cymatium (Turritriton) labiosum
(Wood, 1828): **363**
28mm. Thick, squat, sometimes almost lozenge shaped in outline, the short spire much narrower than width of last whorl, sutures deep, siphonal canal short and recurved; no varices. Embryonic whorls smooth and glossy; rest of shell has coarse, wavy, spiral cords crossing strong axial ribs, producing rectangular pits; rows of rounded beads superimposed on cords give shagreened surface. Wavy-edged outer lip; columella smooth, almost truncate at lower end. Whitish with pits orange-brown or grey-brown; embryonic whorls dark brown; aperture white, columella white or tinted violet. **Habitat:** offshore and beached. **Distribution:** GO, Mas, SO.

Cymatium (Turritriton) vespaceum (Lamarck, 1822): **364**
35mm. Sturdy, lightweight, elongate-ovate, the narrow spire and attenuate siphonal canal giving it a spindle shape, apex pointed; sutures deep; 2 varices. Spiral cords, strongest and noduled at shoulder of whorls, crossed by axial riblets. Inner side of outer lip has broad, sometimes paired teeth; columella strongly lirate throughout its length. Chocolate-brown with paler brown or whitish spiral band on last whorl; outer lip banded white and brown; embryonic whorls brown; external pattern shows through aperture. **Habitat:** intertidal and offshore between stones. **Distribution:** NG, GO, Mas.

Linatella (Gelagna) n.sp: **365**
40mm. Thin, lightweight, globose, all whorls rounded, siphonal canal short and recurved, deep sutures; no varices. Regular, rounded, spiral cords with crowded, low, axial riblets between them. Regularly spaced lirae within aperture; columella lightly lirate above and below. Amber; apertural lirae

Cymatium (Ranularia) tripum **361**

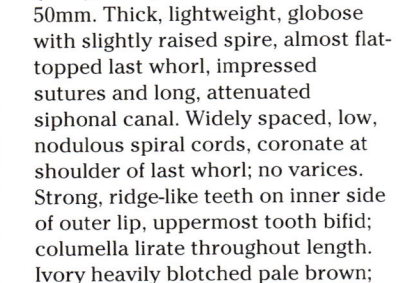

Cymatium (Septa) rubeculum rubeculum **362**

Cymatium (Turritriton) vespaceum **364**

Cymatium (Turritriton) labiosum **363**

Cymatium (Ranularia) oboesum **359**

Cymatium (Ranularia) trilineatum **360**

Superfamily: **TONNOIDEA**

chocolate-brown at edge of outer lip, rest of aperture white. Previously confused with *L. (G.) succincta* (Linnaeus, 1771), which has brown spiral cords, this species will be described elsewhere. **Habitat:** under intertidal rocks. **Distribution:** GO, Mas.

Linatella (Linatella) caudata (Gmelin, 1791): **366**
50mm. Thin, lightweight, globose-ovate, moderately high spired, spire whorls rounded or straight sided and keeled at shoulders, sutures deep, siphonal canal strongly twisted, outer lip thin; no varices. Regular, flat-topped, spiral cords, nodulous at shoulders, crossed by fine growth striae; wavy edge of outer lip corresponds to spiral cords; columella smooth or bearing impressions of underlying cords. Yellowish white with pale brown or grey-brown blotches and axial streaks; aperture and columella white. **Habitat:** under intertidal rocks. **Distribution:** GO, Mas.

Family PERSONIDAE
Members of this small family are easily recognised by the irregularly shaped whorls of their shells. The last whorl, in particular, is peculiarly distorted and enlarged. Varices are present on all whorls, aperture constricted by well developed teeth and folds.

Distorsio reticularis (Linnaeus, 1758): **367**
50mm. Thin, lightweight, irregularly spindle shaped, last whorl much larger than spire. Later whorls distorted, the last sometimes very humped; sutures uneven; aperture very constricted; outer lip strongly toothed. Columella indented centrally, its lower part with projecting folds; broad parietal callus with pustules on parietal wall; varices variable in strength and number. Spiral and axial ribs form cancellate pattern. Amber to yellowish white; aperture white. Often covered with thick, yellowish periostracum. Syn: *D. reticulata* (Röding, 1798) of some authors. **Habitat:** among rocks offshore. **Distribution:** GO.

Family BURSIDAE
Frog shells are superficially similar to those of the preceding family with which they may be easily confused. Small to large species with thick, heavy shells, their surface sculpture is even more rugged than that of the Cymatiidae and their varices are usually as well developed; the posterior canal may form an open-sided tube; one species has prominent spines. The chitinous operculum is usually thick and may not fill the aperture. They live in rock pools, in crevices or on and under rocks at low tide and below. As in the Ranellidae the larvae may have a very long free-swimming stage which has enabled some species to traverse entire oceans; this explains the large number of subspecies now accepted by most specialists.

Bufonaria (Bufonaria) crumena (Lamarck, 1816): **368**
55mm. Solid but lightweight, broadly ovate, spire less than 1/4 of total length; 2 axially aligned varices per whorl. Aperture large, outer lip ridged throughout; columella with crowded, deep-seated folds; siphonal canal short, broad, wide open; posterior canal short and broad. Up to 3 spiral rows of short, pointed nodules on last whorl, 1 row on each earlier whorl; rest of shell covered with closely spaced, beaded, spiral cords. White suffused brownish pink, brown spots between nodules; aperture white. **Habitat:** offshore and beached. **Distribution:** GO, Mas.

Bufonaria (Bufonaria) echinata (Link, 1807): **369**
55mm. Thick, lightweight, ovate and compressed dorso-ventrally; 2 axially aligned varices per whorl, later varices with spines, 3 or 4 on final varix. Aperture large, outer lip unevenly ridged; upper part of columella almost smooth, lower part with strong folds. Up to 3 spiral rows of unevenly spaced, pointed nodules on last whorl, 1 row on each earlier whorl; rest of shell covered with closely spaced, spiral threads and cords. Creamy to pale brown, darker brown between nodules. **Habitat:** on rocks intertidal and offshore. **Distribution:** NWG, NG, GO, Mas.

Distorsio reticularis **367**

Linatella (Gelagna) n.sp. **365**

Bufonaria (Bufonaria) crumena **368**

Linatella (Linatella) caudata **366**

Bufonaria (Bufonaria) echinata **369**

Bufonaria (Bufonaria) foliata (Broderip, 1844): **370** NOT ILLUSTRATED
Listed by Smythe, 1979a:67, from UAE (SEG) but record requires confirmation.

SEASHELLS OF EASTERN ARABIA 101

Superfamily: **TONNOIDEA**

Bufonaria (Bufonaria) gnorima **371**

Bufonaria (Bufonaria) rana **372**

Bursa (Bursa) davidboschi **374**

Bursa (Bufonariella) granularis **373**

Bufonaria (Bufonaria) gnorima (Melvill, 1918): **371**
60mm. Thick, heavy, broadly ovate with impressed sutures and short, broad siphonal canal; 2 low, axially aligned varices per whorl; protoconch of 4.5 smooth whorls. Aperture with unevenly ridged outer lip and strongly ridged lower half of columella. Posterior canal short and wide open. Spiral rows of coarse, granulose cords cover entire shell; spiral row of large nodules at shoulder of each whorl. Straw-coloured with yellowish to brown mottlings, darker between nodules; aperture white with orange-yellow margin. **Habitat:** offshore. **Distribution:** GO.

Bufonaria (Bufonaria) rana (Linnaeus, 1758): **372**
65mm. Thick, broadly ovate, with well-impressed sutures and broad, wide open, reflected siphonal canal; 2 varices per whorl, later varices with 1 or 2 spines. Long, narrow aperture with strongly ridged outer lip and flaring posterior canal; columella strongly ridged throughout length. Either 1 or 2 spiral rows of short, sharp nodules on each whorl; rest of shell covered with coarse, often beaded, spiral cords and threads. Cream, heavily mottled reddish brown; aperture white. Confirmation required of the occurrence of this Pacific species in Oman. **Habitat:** beached. **Distribution:** "Oman", without more precise locality.

Bursa (Bufonariella) granularis (Röding, 1798): **373**
50mm. Thick, heavy or lightweight, elongate-ovate, spire less than half total height; 2 axially aligned varices per whorl. Outer lip corrugated, with about 12 paired teeth just inside elongate aperture; columella lirate throughout length; siphonal canal short, narrow; posterior canal short, broadly open and callused at parietal junction. Spiral rows of bluntly pointed nodules on all whorls, nodules larger at shoulders; varices beaded. Red-brown to chocolate, darker between nodules, varices alternately dark and pale; aperture white, teeth orange. **Habitat:** intertidal rocks and crevices. **Distribution:** NG, GO, Mas, SO.

Bursa (Bursa) davidboschi Beu, 1986: **374**
45mm. Thick, solid, squat, spire less than half total height; 2 axially aligned varices per whorl. Aperture small, outer lip irregularly toothed, columella with irregular, deep-seated folds; posterior canal moderately long, partly open (canals visible on spire whorls). Coarse, nodulous, spiral cords and lesser ribs. Cream to yellowish brown with brown spotting on cords; aperture violet inside, its edges white. **Habitat:** under rocks. **Distribution:** GO, Mas, SO.

Tutufa (Tutufa) bardeyi (Jousseaume, 1894): **375**
250mm. Thick, heavy, high spired (the pointed apex usually eroded), last whorl greatly enlarged; 2 inconspicuous, thin varices per whorl. Each whorl has a spiral row of large, blunt nodules and irregular, coarse spiral cords. Outer lip of aperture wavy; columella smooth and sinuous; siphonal and posterior canals short, broad and wide open. Whitish to reddish brown, darker brown between nodules; aperture white or orange. **Habitat:** on sand or among rocks. **Distribution:** GO, Mas, SO.

■ The scientific discovery of the first shells of *Tutufa bardeyi* happened in a curious manner, at Aden. "It was not in the sea nor on the beaches that I found the first shell of this gigantic mollusc", says F. P. Jousseaume in a note published in 1921, "but at the gateway to the drawing room of the Italian consul Cecchi, who had placed one on each door post, where they resembled two sphinxes, guarding the entrance; at the first sight of these colossal shells, I could not conceal my surprise and my lust for them." The consul immediately offered both to him and said they had been brought to Aden by pearl fishers. Jousseaume said he could accept only one, an Italian museum being the right place for the other. The consul then let the delighted Jousseaume choose one.

Tutufa (Tutufa) bardeyi 375

Superfamily: **TONNOIDEA, CERITHIOPSOIDEA**

Tutufa (Tutufa) bufo (Röding, 1798): **376**
85mm. Thick, heavy, ovate, spire shorter than last whorl; 2 varices per whorl. Spiral rows of large, rounded nodules on all whorls, very prominent at shoulder of last whorl; coarse, pustulose spiral cords cover entire surface. Outer lip wavy edged, a series of blunt teeth just within aperture; columella smooth except for a few weak folds by siphonal canal; posterior canal short, deep and open; thick, broad parietal callus. Orange-red to brown mottled and blotched brown; parietal callus and aperture pale flesh-pink to white with bright red ring behind teeth. **Habitat:** among rocks. **Distribution:** Mas.

Tutufa (Tutufella) nigrita
Mühlhäusser & Blöcher, 1979: **377**
60mm. Thick, heavy, elongate-ovate, spire about half total height; 2 thick, non-aligned varices per whorl. Spiral rows of lumpy nodules on all whorls, prominent at shoulders; coarse, pustulose spiral cords cover entire surface. Outer lip wavy, a series of paired, elongate teeth just within aperture; columella lirate throughout length; posterior canal short, wide open; siphonal canal short, reflected, narrowly open. Yellow-orange variegated reddish brown (these colours alternating on varices); aperture yellow-orange, reddish brown between teeth; columellar lirae white, chocolate brown between lirae. **Habitat:** among coral and rocks in shallow water. **Distribution:** Mas.

Tutufa (Tutufella) oyamai Habe, 1973: **378**
80mm. Thick, heavy, elongate-ovate, spire about half total height, with deep sutures; 2 non-aligned varices per whorl. Spiral rows of prominent, compressed, paired nodules on all whorls; entire surface covered by spiral, beaded cords giving shagreened appearance. Aperture flared; outer lip irregularly corrugated; 11 to 14 low, paired, elongate teeth just within aperture; columella finely lirate throughout its length. Siphonal canal, short, narrow, recurved; posterior canal short and open. Fawn to red-brown, darker brown on nodules; aperture white, its edge sometimes pink. **Habitat:** on sand offshore and beached. **Distribution:** GO, Mas.

Superfamily CERITHIOPSOIDEA
For description see under next family.

Family CERITHIOPSIDAE
Many generic names have been proposed for the small, narrowly elongate or ovate shells in this family. Shell sculpture is well developed. The animals feed on sponges. Superficially the shells are similar in several respects to those of the Triphoridae (see below) but they all have dextral coiling, unlike those of many triphorids. Most of the Eastern Arabian species are still unidentified.

***Cerithiopsis* sp. (a): 379**
2.5mm. Elongate-conic, slightly convex-sided spire, sutures impressed; protoconch of 4 smooth whorls. Strong axial and spiral cords produce cancellate sculpture. **Habitat:** intertidal. **Distribution:** Mas.

Tutufa (Tutufella) nigrita **377**

Tutufa (Tutufa) bufo **376**

Tutufa (Tutufella) oyamai **378**

Superfamily: **CERITHIOPSOIDEA, TRIPHOROIDEA**

Cerithiopsis **sp. (b): 380**
2mm. Elongate-conical, spire (including protoconch) concave sided initially becoming convex sided later, aperture much narrower than maximum spire width; columnar protoconch of 4 whorls, first 2 bearing spiral rows of granules, next 2 axially ribbed. Spire whorls with strong, beaded, spiral cords. **Habitat:** intertidal. **Distribution:** GO.

Seila bandorensis (Melvill, 1893): **381**
8mm. Solid, dull, broadly needle-like, each whorl strengthened by 2 strong spiral ribs on either side of a weaker rib; protoconch not seen. Crowded, fine, axial threads between ribs; columella curved and smooth. Brown (but white when worn). **Habitat:** hanging by mucus threads under intertidal rocks. **Distribution:** NWG, SEG.

Seila hinduorum (Melvill, 1898): **382**
10mm. Solid, dull, needle-like, each whorl strengthened by 3 strong spiral ribs and a weaker pre-sutural rib; protoconch not seen. Very weak axial threads between ribs; columella sinuous and smooth. Pale brown with darker brown blotches. Listed by Smythe, 1979a:66, from UAE (SEG) but probably misidentified. **Habitat:** offshore in shell sand. **Distribution:** Mas.

Superfamily TRIPHOROIDEA
For description see under next family.

Family TRIPHORIDAE
Like the Cerithiopsidae (see above) members of this family feed on sponges and have small, narrowly elongate or ovate shells with bold sculpture. Additionally many of them have well developed siphonal and/or posterior canals. The principal feature separating the two families, however, is the way the shells coil, all the cerithiopsids being dextral while most of the triphorids are sinistral (at least one Eastern Arabian triphorid is dextral). About ten species are known from Eastern Arabia but few have been identified satisfactorily and the group requires a thorough revision.

Seilarex cf *verconis* Cotton, 1951: **383** NOT ILLUSTRATED
10mm. Thin, translucent, dextral, narrowly elongate-conical, spire whorls rounded, sutures deep; protoconch has fine spiral threads on first whorl followed by about 6 whorls with spiral thread and sharp axial riblets. Siphonal canal a wide notch. Spire whorls have sharp spiral cords with fine axial threads between them. White or pale buff. Illustration is of an immature shell

Cerithiopsis **sp. (a) 379**

1.5mm long. **Habitat:** intertidal. **Distribution:** Mas.

Triphora acicula Issel, 1869: **384**
NOT ILLUSTRATED
Listed by Glayzer et al, 1984:320, from Kuwait (NWG).

Triphora acuta (Kiener, 1841): **385**
NOT ILLUSTRATED
Listed by Smythe, 1979a:66, from U. A. E. (SEG) and by Glayzer et al, 1984:320, from Kuwait (NWG).

Triphora aegle (Jousseaume, 1884): **386** NOT ILLUSTRATED
Listed by Smythe, 1979a:66, from U. A. E. (SEG) and by Glayzer et al, 1984:320, from Kuwait (NWG).

Triphora cf *capensis* Bartsch, 1915: **387** NOT ILLUSTRATED
Listed by Glayzer et al, 1984:320, from Kuwait (NWG).

Triphora cingulata (A. Adams, 1854): **388** NOT ILLUSTRATED
Listed by Melvill and Standen, 1901:376, from NG and GO (but erroneously, see Melvill, 1928:102); by Smythe, 1979a:66, from U. A. E. (SEG) and by Glayzer et al, 1984:320, from Kuwait (NWG).

Triphora concatenata (Melvill, 1904): **389**
Described from 156 fathoms off Muscat (GO).

Triphora cf *distinctus* Deshayes, 1863: **390** NOT ILLUSTRATED
Listed by Glayzer et al, 1984:320, from Kuwait (NWG).

Triphora gemmata Blainville, 1828: **391**
5mm. Sturdy, glossy, sinistral, elongate-ovate, aperture about half width of shell at its widest, sutures impressed, siphonal canal sharply recurved backwards and almost forming a tube; protoconch: initial half whorl has hemispherical granules, next 4 whorls axially ridged. Spire whorls encircled by 2 rows of beads with fine spiral striae between rows. Amber with lowermost rows of beads dull white, reddish brown in their intervals. **Habitat:** intertidal and offshore on sponges. **Distribution:** GO, Mas.

Triphora idonea (Melvill & Standen, 1901): **392**
10mm. Described from Linjah (NG).

Triphora incolumnis Melvill, 1918: **393** NOT ILLUSTRATED
10mm. Described from Fao cable (NG); listed by Smythe, 1979a:66, from U. A. E. (SEG) and by Glayzer et al, 1984:320, from Kuwait (NWG).

Triphora interpres Melvill, 1918: **394**
11mm. Described from Musandam (GO); listed by Smythe, 1979a:66, from U. A. E. (SEG).

Superfamily: **TRIPHOROIDEA, EPITONIOIDEA**

Triphora maxillaris Hinds, 1843: **395** NOT ILLUSTRATED
Listed by Glayzer et al, 1984:320, from Kuwait (NWG).

Triphora tristoma (De Blainville, 1824): **396** NOT ILLUSTRATED
Listed by Glayzer et al, 1984:320, from Kuwait (NWG).

Viriola corrugata (Hinds, 1843): **397**
15mm. Solid, moderately glossy, broadly needle-like, spire straight sided, sutures impressed, siphonal canal sharply recurved backwards and open below, lower half of outer lip protruding forwards; first whorl of protoconch has hemispherical granules, later whorls have 2 spiral threads. 2 strong spiral keels with lesser keel between them; irregular axial riblets between keels. Amber with paler keels. **Habitat:** intertidal and offshore on sponges. **Distribution:** NWG, SEG, GO.

Superfamily EPITONIOIDEA
Contains two families, parasitic on coelenterates: the Epitoniidae, with mostly colourless, high-spired shells, and the pelagic Janthinidae, with globose or trochoidal shells.

Family EPITONIIDAE
Members of this large family of warm-water, parasitic species, popularly known as wentletraps, occur on or under sea-anemones and other coelenterates. Male when young but becoming female ultimately, they lay strings of eggs. Some are known to emit a purple dye for defensive purposes. Most have colourless, translucent shells with a round aperture and high spire, the whorls often buttressed by thin or thick, ribs (in a few the whorls connected only by the ribs). A dark brown, chitinous operculum is usually present. Many so-called species have been described from eastern Arabia but few occur intertidally or washed up on beaches. Some have been collected once only and several are known solely from immature or imperfect shells. Furthermore, many generic and subgeneric names have been proposed but are difficult to apply confidently. Kilburn (1985) has stressed the importance of the protoconch in the identification of wentletraps but this vulnerable feature is often missing. Rib counts also give valuable taxonomic characters (Robertson, 1983) but may be misleading if based on single or incomplete shells. So there is uncertainty about the status of most species and their classification. The arrangement adopted here partially follows that in the report on wentletraps from southern Africa and Mozambique by Kilburn (1985). The survey of eastern Arabian species by Melvill and Standen (1903a) has some useful illustrations. Identification of the smaller species described by them from eastern Arabia has been difficult to establish because the original specimens are often poorly preserved or seem to have become mixed up. That is why some of their original figures are reproduced here and why some of their taxa are excluded.

Acrilla cophinodes (Melvill, 1904): **398**
10mm. High spired with large, bluntly pointed, glossy protoconch of about 3.5 whorls. Well rounded spire whorls have crowded, oblique, low ribs interleaving at sutures. Smooth between ribs. No umbilicus. White. The few known examples are probably immature, the protoconch seeming disproportionately large on each of them. **Habitat:** deep water. **Distribution:** GO.

Amaea acuminata (Sowerby, 1844): **399**
30mm. Narrowly high spired, glossy, with smooth protoconch of about 2.5 whorls. Gently rounded spire whorls bear regular, low, slightly sinuous ribs (about 30 on 10th whorl and 50 or more on 12th) joined at shallow sutures; rounded rib encircles base (and continues at sutures). Aperture elongate. No umbilicus. Amber with paler spiral band mid-whorl; columella, basal and sutural ribs white. Syn: *A. minor* (Sowerby, 1873) may be merely a form with finer axial ribs. **Habitat:** intertidal and beached. **Distribution:** GO, Mas.

Amaea fimbriolata (Melvill, 1897): **400**
55mm. Narrowly high spired (protoconch not seen but presumably smooth). Spire whorls (14 whorls on largest specimen seen) enlarging regularly, superficially straight sided. Crowded, axial ribs (about 28 on last whorl), often wrinkled and layered on fresh specimens and hooked just below deep sutures; between ribs on later spire whorls about 6 strong, spiral threads have lesser threads between them. Strong cord encircles base. No umbilicus. White. **Habitat:** offshore in black mud and beached. **Distribution:** GO, Mas.

Amaea martinii (Wood, 1828): **401**
20mm. Narrowly high spired and glossy (no shells with protoconch available). Spire whorls have irregularly spaced, low, rounded ribs with occasional thicker ribs or varices (about 24 on last whorl); about 20 spiral threads between ribs. Aperture elongate. Small umbilicus. White. **Habitat:** offshore. **Distribution:** GO.

Viriola corrugata **397**

Acrilla cophinodes **398**

Amaea fimbriolata **400**

Amaea acuminata **399**

■ *"An exquisite species", said Melvill when describing* Amaea fimbriolata *as new to science in 1897. The shell he described, found at Karachi by F. W. Townsend, was only 19 millimetres long, but the following year he illustrated a much larger example, measuring 52 millimetres in length, one of two dredged off Muscat in 10 fathoms. In 1901 Melvill and Standen described this wentletrap as "the most elegant and refined, perhaps, of all the Mollusca collected by Mr. Townsend". Its only peer then would have been* Conus milneedwardsi, *a species of very different character. When large and perfectly preserved this wentletrap is an object of rare beauty. Unquestionably it may grow much longer than Townsend's largest example. One found recently by Richard Skinner on the beach at As Sib is 79 millimetres long - and to that may be added about 6 millimetres, representing the missing apex!*

Superfamily: **EPITONIOIDEA**

Amaea martinii **401**

Amaea xenicima **402**

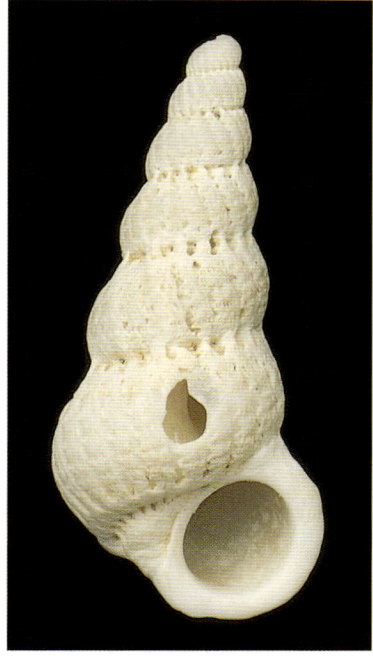

Cirsotrema corolla **403**

Cirsotrema varicosum **404**

Cycloscala hyalina **405**

Epitonium aculeatum **406**

Epitonium alatum **407**

Epitonium alizonae **408**

Epitonium amathusia **409**

Epitonium bonum **410**

Epitonium calideum **411**

Amaea xenicima (Melvill & Standen, 1903): **402**
14mm. Thin, high spired with pointed, smooth, glossy protoconch of about 3 whorls. Gently sloping, smooth, glossy spire whorls bear irregular widely spaced, curved or sinuous, very low ribs (about 13-16 on last whorl) which may join at sutures. Rounded rib encircles base and continues onto spire sutures. Faint spiral scratches between ribs. Aperture elongate; peristome thin. No umbilicus. Livid brown with paler axial streaks and greyish ribs. **Habitat:** deep water and offshore. **Distribution:** GO.

Cirsotrema corolla (Melvill & Standen, 1903): **403**
5mm. Thick, short spired with glossy protoconch (mostly missing from available specimen). Spire whorls have thick, slightly oblique ribs (about 16 on last whorl), raised up just below sutures, each rib edge rope like; base encircled by a strong cord. Outer lip thick and broad. No umbilicus. Chalky white. **Habitat:** deep water. **Distribution:** GO.

Cirsotrema varicosum (Lamarck, 1822): **404**
35mm. Solid, high spired with pointed protoconch and thickened outer lip. Well rounded spire whorls with occasional thick varices between which are flat, close-set, axial ribs, pointed at their tops and partly obscuring deep sutures. Surface scaly. Strong ridge encircles base. White. A worn shell collected by M. Day. **Habitat:** beached. **Distribution:** Mas.

Cycloscala hyalina (Sowerby, 1844): **405**
6.5mm. Broadly high spired, glossy, with protoconch of more than 3 whorls. Spire whorls all unconnected except at junctions of well developed, axial ribs (about 10 on first 2 whorls and 7 on last); angled at shoulders. Large umbilicus. White. **Habitat:** offshore. **Distribution:** NG, GO.

Epitonium aculeatum (Sowerby, 1844): **406**
23mm. High spired with evenly expanding, separated, whorls and oblique, strongly recurved ribs (9 on last whorl). Umbilicus obscured by converging ribs. White or tinged beige between white ribs. **Habitat:** sea anemones and beached. **Distribution:** NWG, SEG, GO, Mas.

Epitonium alatum (Sowerby, 1844): **407**
8mm. Short spired with rapidly expanding, slightly separated whorls and thick ribs which are upwardly hooked above and connected to each other at sutures. Umbilicus obscured by converging ribs. White. **Habitat:** sea anemones and beached. **Distribution:** NG, NWG.

Epitonium alizonae (Melvill, 1912): **408**
10mm. High spired with smooth, glassy protoconch of about 4.5 whorls. Very thin, wavy edged, unreflected, closely spaced ribs (about 36 on last whorl) with about 13 regular, spiral threads between ribs. No umbilicus. White except for chestnut protoconch and early spire whorls. **Habitat:** deep water. **Distribution:** GO.

Epitonium amathusia (Melvill & Standen, 1903): **409**
6.5mm. High spired, with pointed, glassy protoconch of about 3 whorls. Thin, sharp-edged, oblique, slightly curved ribs (about 18-20 on last whorl) interleaving at sutures; smooth between ribs. No umbilicus. White. **Habitat:** offshore. **Distribution:** NG.

Epitonium bonum (Melvill, 1906): **410**
14mm. Solid, high spired (protoconch missing from holotype, the only available specimen). Spire whorls bear thick, wedge-like ribs (about 16 on last whorl) interleaving at sutures; 1 or 2 varices per whorl. Fine, crowded, spiral threads between ribs. Thick rib encircles base. Peristome thick and expanded. No umbilicus. Chalky white. **Habitat:** deep water. **Distribution:** GO.

Epitonium calideum (Melvill & Standen, 1903): **411**
7mm. High spired with a glassy protoconch of about 2.5 whorls. Oblique, slightly reflected ribs (about 20 on last whorl) interleaving at sutures, slightly hooked just below sutures; each rib about one third the width of smooth intervals between them. No umbilicus. White. **Habitat:** deep water. **Distribution:** GO.

SEASHELLS OF EASTERN ARABIA

Superfamily: **EPITONIOIDEA**

Epitonium canephora **412** 
Epitonium continens **415** 
Epitonium gloriola **418** 
Epitonium idalia **420** 
Epitonium cerdantum **413** 
Epitonium deificum **416** 
Epitonium goniophora **419** 
Epitonium rissoinaeforme **430**
Epitonium laidlawi **423**

Epitonium glabratum **417**

Epitonium canephora (Melvill, 1906): **412**
5mm. High spired with a glossy protoconch of about 3 whorls. Oblique, reflected ribs (about 16 on last whorl) interleaving at sutures where they are also tabulated. About 16 low spiral threads between ribs. No umbilicus. White. **Habitat**: deep water. **Distribution**: NG, SEG, GO.

Epitonium cerdantum (Melvill & Standen, 1903): **413**
4mm. Narrowly high spired with a glossy, acutely pointed protoconch of about 4 whorls. Oblique ribs (about 14 on the last whorl) interleaving at sutures, raised into points just below. Smooth between ribs. No umbilicus. White. **Habitat**: deep water. **Distribution**: GO.

Epitonium concinnum (Sowerby, 1844): **414** NOT ILLUSTRATED
Listed by Smythe, 1979a:66, from U.A.E. (SEG).

Epitonium irregulare **421**

Epitonium continens (Melvill & Standen, 1903): **415**
5mm. High spired with glossy, acutely pointed protoconch of about 3 whorls; spire whorls expand rapidly after first one. Oblique, sharp-edged, unreflected ribs (about 14 on last whorl) joined at sutures. Smooth and glossy between ribs. No umbilicus. White. **Habitat**: deep water. **Distribution**: GO.

Epitonium deificum (Melvill & Standen, 1903): **416**
10mm. High spired with pointed, glossy protoconch of about 4 whorls. Spire whorls have thin, slightly oblique, upright ribs (about 17 on last whorl) diminishing and interleaving at sutures. Crowded, low, spiral threads, half as wide as their intervals, between ribs. No umbilicus. White. **Habitat**: deep water and beached. **Distribution**: NWG, SEG, GO.

Epitonium glabratum (Hinds, 1844): **417**
18mm. High spired, slender, glossy, with small, pointed protoconch of about 4 whorls. Spire whorls have widely spaced, thin, slightly recurved and poorly developed ribs (about 10 on last whorl); each rib enlarges at its top and joins rib of previous whorl at suture. Smooth between ribs. No umbilicus. Pale brown with white ribs. **Habitat**: deep water. **Distribution**: NG, GO.

Epitonium gloriola (Melvill & Standen, 1901): **418**
8mm. Broadly high spired with a presumably pointed protoconch (missing from available shells). Spire whorls bear regular, thin, axial ribs (about 22 on last whorl) which are saw-toothed in profile, about 9-11 "teeth" on each rib, some of them hooked. Low spiral threads between ribs correspond with "teeth". No umbilicus. Yellowish white with white peristome. **Habitat**: deep water. **Distribution**: GO.

Epitonium goniophora (Melvill & Standen, 1903): **419**
5mm. High spired, slender, glossy, smooth, with protoconch of about 3.5 whorls. Spire whorls have strong, oblique, reflected ribs (about 10 on last whorl) joined at sutures and angled just above centre of whorl. No umbilicus. All white. **Habitat**: deep water. **Distribution**: GO.

Epitonium idalia (Melvill, 1912): **420**
4mm. High spired with protoconch of about 3.5 glossy, smooth whorls. Spire whorls have thick, slightly recurved, oblique or wavy, ribs interleaving at sutures. Coarse, spiral threads between ribs. Peristome thick, especially at base. No umbilicus. White. **Habitat**: deep water. **Distribution**: GO.

Epitonium irregulare (Sowerby, 1844): **421**
30mm. Thin, high spired with rounded spire whorls and pointed glossy protoconch of about 3 whorls. Thin, close-set, spiral threads crossed by frilled, axial ribs (about 34 on last whorl) which meet at sutures. No umbilicus. Beige with a paler spiral band mid-whorl. **Habitat**: intertidal and offshore among muddy stones. **Distribution**: GO, Mas.

Superfamily: **EPITONIOIDEA**

Epitonium jomardi (Audouin, 1826): **422**
12mm. High spired with smooth protoconch of about 4 whorls. Spire whorls well rounded, sutures very deep. Fine, irregular, sometimes almost obsolete, sometimes thicker, oblique ribs (averaging about 28 on last whorl) meet irregularly at sutures; intervals smooth. Small umbilicus. Pale brown with darker brown spiral band on spire whorls (which tends to fade or disappear); ribs white. **Habitat:** offshore and beached. **Distribution:** GO.

Epitonium laidlawi (Melvill & Standen, 1903): **423**
13mm. Thin, high spired with protoconch of about 3 glossy whorls. Spire whorls bear thin, well developed, curved ribs (about 25 on last whorl) elevated at shoulder and interleaving with each other at sutures. Crowded spiral threads between and continued onto ribs. No umbilicus. Aperture elongate. White. **Habitat:** offshore. **Distribution:** GO.

Epitonium laxatum (Sowerby, 1844): **424** NOT ILLUSTRATED
Listed by Melvill & Standen, 1903:345, from NG and GO. A shell from Sheyk Shoeyb Island, so labelled in the Manchester Museum, is not this species so the identification is questionable.

Epitonium lyra (Sowerby, 1844): **425**
18mm. High spired, broad based; with deep sutures; no shells with protoconch available. Spire whorls bear fragile, often worn, oblique ribs (from 25 to 45 on last whorl) strongly reflected at sutures where they interleave; fine, irregular, spiral threads between ribs. Reflected inner lip partly obscures deep umbilicus. White with 2 strong spiral, pale brown bands; ribs and base white. **Habitat:** offshore. **Distribution:** GO.

Epitonium maculosum (Adams & Reeve, 1850): **426** NOT ILLUSTRATED
Listed by Melvill & Standen, 1901:354, from Kais Island, Arabian Gulf (NG).

Epitonium malcolmensis (Melvill, 1898): **427**
10mm. Broadly high spired with smooth, glossy protoconch of about 3 whorls. Later spire whorls unconnected and almost hidden under broad, strongly reflected, sinuously oblique ribs (about 10-16 on last whorl) which curl inwardly at edges and form claw-like extensions at sutures where they also join each other; fine, crowded, spiral threads between ribs. Umbilicus filled in by rib ends. White. **Habitat:** offshore. **Distribution:** GO.

Epitonium raricostum **428**

Epitonium replicatum **429**

Epitonium raricostum (Lamarck, 1822): **428**
25mm. Ovate-globose, last whorl expanding rapidly (no protoconch available for study). Spire whorls (of fresh specimens) bear sharp-edged, reflected, axial ribs (about 25 on last whorl) which may form thick, corrugated varices randomly. Between ribs and varices are equidistant spiral threads overlain by cancellate sculpture of finer axial and spiral threads. Umbilicus largely filled by rib ends. White. **Habitat:** coral sand offshore and beached. **Distribution:** GO, Mas.

Epitonium replicatum (Sowerby, 1844): **429**
13mm. Moderately high spired with smooth, glossy, pointed protoconch of about 3 whorls. Smooth, polished, disconnected spire whorls bearing well developed, strongly recurved, oblique ribs (about 6 on last whorl) elevated into blunt points at shoulders and joined at sutures; ribs not glossy. Umblicus deep but mostly infilled by rib ends. White. **Habitat:** offshore and beached. **Distribution:** NWG, NG, GO.

Epitonium lyra **425**

Epitonium rissoinaeforme (Melvill & Standen, 1903): **430**
4.2mm. Compactly high spired with smooth, glossy, pointed protoconch of about 3 whorls (its 3rd whorl being conspicuously broader than 1st spire whorl). Gently swollen spire whorls bear low, regular, curved ribs (about 23 on last whorl) succeeded by greatly thickened peristome. Spiral threads between ribs revealed by high magnification. No umbilicus. Chalky white. Could be mistaken for a ribbed rissoid. **Habitat:** deep water. **Distribution:** GO.

Epitonium jomardi **422**

Epitonium malcolmensis **427**

Superfamily: **EPITONIOIDEA**

Epitonium scalare (Linnaeus, 1758): **431**
45mm. High spired with small protoconch (not available for study). Spire whorls, all widely disconnected, bear widely spaced, recurved, slightly oblique ribs (about 9 on last whorl) sometimes connected to each other at sutures. Smooth and glossy between ribs. Umbilicus wide, virtually reaching to protoconch. Pinkish to violet-brown with white ribs and peristome. A small squat form occurs occasionally. Syn: *Scala pretiosa* (Lamarck, 1816) sensu Melvill & Standen, 1901:354; *Epitonium pallasii* (Kiener, 1838) sensu Melvill & Standen, 1903a:345, Bosch & Bosch, 1982:51, and Smythe, 1982:50. **Habitat:** offshore and beached. **Distribution:** NWG, SEG, GO.

Epitonium schepmani (Melvill, 1910): **432**
9mm. Broadly high spired with acutely pointed, smooth, glossy protoconch of about 3 whorls. Gently sloping spire whorls bear increasingly irregular, low, thin ribs (about 36 on last whorl but much more crowded and regular on early whorls) interleaving at sutures. Occasional thicker ribs (or varices). Regular, widely spaced, spiral threads between ribs. No umbilicus. White. **Habitat:** offshore. **Distribution:** NG, SEG.

Epitonium sykesii (Melvill & Standen, 1903): **433**
4mm. Superficially resembles *E. rissoinaeforme* (see above) but outer lip is not greatly thickened and ribs are closer, more sinuous and twice as many (about 45 on last whorl). As each species is known only from the unique haul at 156 fathoms off Muscat, these differences, though apparently significant, may be merely sexual in character. **Habitat:** deep water. **Distribution:** GO.

Epitonium thelcterium (Melvill & Standen, 1903): **434**
4mm. Broadly high spired with smooth, glossy protoconch of about 4 whorls. Spire whorls bear thin, axial, unreflected ribs (about 15 on last whorl) joined at sutures and occasionally thickened. Low, regular threads between ribs. Peristome thickened and slightly recurved. No umbilicus. White. **Habitat:** deep water. **Distribution:** NG, GO.

Epitonium townsendi (Melvill & Standen, 1903): **435**
6mm. Fragile, high spired with smooth, glossy protoconch of about 3 whorls. Spire whorls bear oblique ribs (about 13 on last whorl) joined at sutures, slightly raised and reflected just below; ribs almost straight sided in profile. Irregular, low, spiral threads between ribs. No umbilicus. White. **Habitat:** offshore and beached. **Distribution:** GO, Mas.

Eglisia tricarinata Adams & Reeve, 1850: **436**
15mm. Thin, dull, narrowly high spired, protoconch not seen. Spire whorls have 2-4 strong or weak keels with overriding, fine, oblique lamellae. Base has very fine, crowded, spiral threads crossed by equally fine striae. No umbilicus. Beige sometimes with broad, paler, axial streaks. Syn: *Turritella leptomita* Melvill & Sykes, 1897. **Habitat:** in sand offshore. **Distribution:** NG, GO.

Gyroscala lamellosa (Lamarck, 1822): **437**
30mm. Thin, high spired with pointed protoconch of about 3 smooth, glossy whorls. Spire whorls with strong, evenly rounded and occasionally thickened ribs (8 to 10 on last whorl) connected at sutures and lined up obliquely. A rounded rib encircles base. No umbilicus. Smooth and glossy between dull ribs. Brown to violet-brown, sometimes with darker brown band below sutures and at base and occasional brown spots; ribs white. **Habitat:** offshore and beached. **Distribution:** GO, Mas, SO.

Opalia crassilabrum (Sowerby, 1844): **438**
15mm (but often much smaller). Narrowly high spired with pointed, glossy protoconch of about 4 whorls. Spire whorls bear strong - not lamellate - ribs (about 15 on last whorl) which are highest mid-whorl; usually a strong varix every third whorl. Last whorl has 2 keels. Shallow excavations at sutures. Entire surface of spire has spiral rows of minute, contiguous pits. Umbilicus minute. White. **Habitat:** offshore and beached. **Distribution:** NWG, NG, SEG, GO.

Superfamily: **EPITONIOIDEA, EULIMOIDEA**

Opalia mammosa (Melvill & Standen, 1903): **439**
3mm. Narrowly high spired with cylindrical, smooth, glossy protoconch of about 4.5 whorls. Spire whorls have strong, low, slightly oblique ribs (about 13 on last whorl) meeting irregularly at deep sutures. Obscure spiral threads between ribs. Spiral cord encircles base. Outer lip thick, reflected, with radial rows of punctae on ventral face. White except for amber protoconch. **Habitat:** deep water. **Distribution:** GO.

Family JANTHINIDAE
Fewer than a dozen species belong in this worldwide family. All are pelagic at all stages of their life cycle, a rare circumstance among gastropods. They float at the surface, spire downwards, attached to a raft of mucus bubbles and their fragile, usually violet-coloured shells may drift onto beaches after storms or on-shore winds. Each lacks an operculum and exudes a purple dye for defensive purposes.

Janthina janthina (Linnaeus, 1758): **440**
30mm. Fragile, translucent, trochoidal, early whorls compressed, sutures deep, base keeled. Columella straight, outer lip usually imperfect, lower edge projecting downwards. No umbilicus. Fine, irregular growth ridges crossed by wavy spiral striae. White below sutures, violet above; base dark violet; columella white. **Habitat:** floats passively, sometimes beached. **Distribution:** all.

Janthina prolongata Blainville, 1822: **441**
25mm. Fragile, translucent, globose, low spired with deep sutures. Aperture very large, elongate anteriorly. Columella smooth, sinuous; lip usually imperfect. No umbilicus. Silky surface has faint axial striae forming chevrons at periphery. Violet, paler below suture on last whorl. **Habitat:** floats passively, sometimes beached. **Distribution:** all.

Janthina umbilicata Orbigny, 1840: **442** NOT ILLUSTRATED
5mm. Description as for *J. prolongata* (see above) but smaller, with large, protruding protoconch and an umbilicus not covered by columellar lip. **Habitat:** floats passively, sometimes beached. **Distribution:** Mas.

Recluzia jehennei Petit, 1853: **443**
22mm. Fragile, translucent, glossy, high spired with deep sutures. Columella slightly twisted. Umbilicus a mere chink. Irregular growth striae and faint spiral striae. White or creamy with a thin, amber periostracum. Usually broken when beached. **Habitat:** floats passively, sometimes beached. **Distribution:** GO, Mas.

Superfamily EULIMOIDEA
Species in this superfamily are small or tiny, their shells being mostly elongate, smooth and shiny. Probably all those recorded from Eastern Arabia associate in some way with echinoderms (sea urchins, sea cucumbers, starfishes). The placement here of most species in the family Eulimidae is provisional. Melvill and Standen described as new several species which require critical evaluation and many species collected recently are still unidentified.

Family ACLIDIDAE
Tiny snails with shells which are elongate, many whorled and often needle like. There is a thin, chitinous operculum. Almost nothing is known about their biology but some species are almost certainly carnivorous.

Janthina janthina **440**

Janthina prolongata **441**

Recluzia jehennei **443**

Opalia mammosa **439**

Graphis sp. **444**

Graphis sp: **444**
1.2mm. Fragile, narrowly elongate, sutures impressed; protoconch of 2, bulbous, smooth whorls. Columella straight. Spire whorls have a regular cancellate sculpture. White. **Habitat:** intertidal. **Distribution:** Mas.

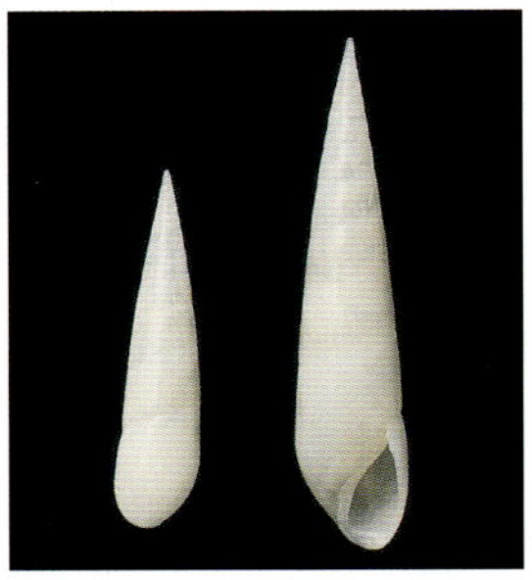

Family EULIMIDAE

The shiny, needle-like or globose shells of eulimids are, for their small size, robust although some are fragile. Most are smooth and shiny and the differences between one species and another are often delicate and subtle, requiring a good microscope for their discrimination. Their mostly colourless shells are unlikely to be seen by the casual observer, except occasionally among shell-sand siftings, because the living molluscs have a close, often parasitic, relationship with various echinoderms. We are almost totally ignorant about the biology of the species recorded from eastern Arabia although some occur on the large sand dollar, *Echinodiscus bisperforatus* (Leske, 1778).

Hypermastus boschorum Warén, 1991: **445**
6mm. Solid, semitransparent, glossy, needle like, spire straight sided with finely incised sutures, very delicate, short plications subsuturally, and occasional, scarcely raised varices; protoconch of about 2 slightly convex whorls. Narrow parietal callus. Outer lip distinctly incurved; curved in profile. White. **Habitat:** intertidal on sand dollars. **Distribution:** Mas.

Hypermastus epiphanes (Melvill, 1897): **446**
12mm. Solid, semitransparent, glossy, needle like, spire straight sided, with finely incised sutures, whorls showing faint, short plications subsuturally; occasional indistinct varices; distinct protoconch of about 1.5 convex whorls. Narrow parietal callus. Outer lip straight, narrowing at base; slightly curved in profile. White. Syn: *H. auritae* Warén, 1991. **Habitat:** intertidal on sand dollars. **Distribution:** NG, NWG, Mas.

Melanella cumingii (A. Adams, 1854): **447**
24mm. Solid, glossy, narrowly elongate, almost straight sided, last whorl rounded (more angular in juveniles), apex tapering evenly to a point. Distinct varical grooves, 1 on each whorl, unevenly distributed. Outer lip gently curved in profile. Milky white. **Habitat:** intertidal and offshore on sea cucumbers. **Distribution:** GO, Mas, SO.

Melanella martinii (A. Adams, 1854): **448**
20mm. Solid, polished, with gradually tapering spire which is often bent towards apex, whorls gently convex, base gently rounded, aperture pear shaped, outer lip curved. In profile varical grooves showing previous positions of lip are obliquely aligned on spire whorls. Shell otherwise smooth. Milky white. **Habitat:** on echinoderms in sandy mud offshore. **Distribution:** NG, GO.

Niso venosa Sowerby, 1895: **449**
15mm. Thin, smooth, glossy, narrowly elongate, straight sided (but base of each whorl slightly broader than its successor), last whorl rounded (more angular in juveniles), apex tapering evenly to a point. Umbilicus narrow but deep. Scarcely raised varices, 1 or more per whorl. Outer lip straight in profile. White to beige with reddish brown, spiral bands (3 on last whorl), varices similarly coloured. **Habitat:** offshore presumably with echinoderms. **Distribution:** all.

Pyramidelloides angusta (Hedley, 1898): **450**
2.7mm. Thin, glossy, translucent, fusiform, aperture about a fourth of total length, outer lip straight sided, sutures wavy; pointed protoconch of 4 smooth whorls. Sinuous axial ribs narrower than their intervals. White. **Habitat:** shallow water. **Distribution:** GO, Mas.

Pyramidelloides miranda (A. Adams, 1861): **451**
3.2mm. Solid, glossy, semitranslucent, elongate-conic; smooth protoconch of about 2.5 whorls. Spire whorls have spiral row of large tubercles at periphery, 4 smooth spiral cords below periphery of last whorl. White. **Habitat:** parasitic upon brittle stars. **Distribution:** GO.

Superfamily MURICOIDEA

An extensive and diverse superfamily containing many families of small to large, carnivorous species, most of them furnished with a thick, chitinous operculum. The shells may be smooth, rough, spiny or ribbed, according to the species; the siphonal canal varies from short to very long. Most of the larger gastropods commonly found on or under intertidal rocks in eastern Arabia belong to this superfamily.

Family MURICIDAE

The small to large gastropods in this extensive family have sturdy shells, usually ornamented with varices and frills or spines, and bear a chitinous operculum with a central or non-central nucleus. They prey upon invertebrates, including other molluscs, and some have a short, stout spine or tooth near the base of the outer lip which

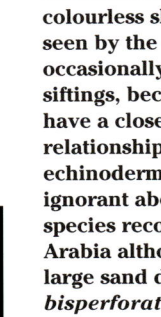

Hypermastus epiphanes 446

Hypermastus boschorum 445

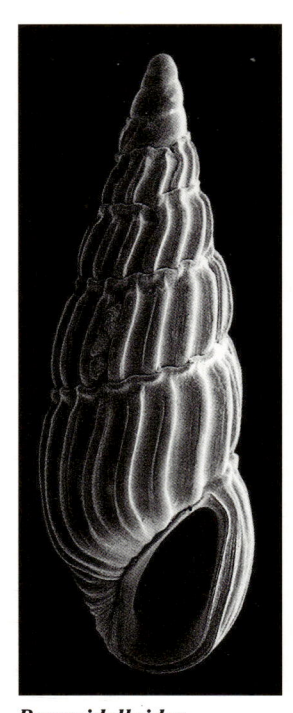

Melanella cumingii 447

Melanella martinii 448

Pyramidelloides angusta 450

Pyramidelloides miranda 451

Niso venosa 449

Superfamily: **MURICOIDEA**

they use to help them force apart the valves of bivalve molluscs. They lay eggs in capsules and occur in all warm waters, especially within the tropics. Although the family is well represented in eastern Arabia, only two of its larger species, *Murex scolopax* and *Hexaplex kuesterianus*, are commonly seen in shallow waters. Women in Oman and elsewhere collect the opercula of these species. The opercula, after crushing, are mixed with other ingredients before being burnt in incense burners, the resultant aroma being highly appreciated locally. If the spines of *M. scolopax* become embedded in the foot it is prudent to remove them immediately to prevent infection.

Subfamily MURICINAE

Aspella producta (Pease, 1861): **452**
16mm. Thick, compressed dorso-ventrally, about 6 whorls; 2 aligned, thin varices per whorl. Each suture has 3 deep pits bordered by an elevated buttress; siphonal canal recurved; 6 or 7 denticles within the aperture. Shell surface has rough spiral threads. White. **Habitat:** offshore and beached. **Distribution:** Mas.

Chicoreus axicornis (Lamarck, 1822): **453**
50mm. Solid, dull, fusiform, high spired. Columellar lip slightly detached. Siphonal canal almost closed. 3 varices with frondose, recurved spines; 1 or 2 nodes between varices. Evenly spaced spiral ribs with fine ridges between. Outer lip wrinkled. Off-white with pale brown ribs and darker brown spines; aperture white. **Habitat:** offshore. **Distribution:** GO.

Chicoreus banksii (Sowerby, 1841): **454**
55mm. Solid, dull, fusiform, high spired. Columellar lip slightly detached. Siphonal canal recurved, almost closed. 3 varices with short, crowded, frondose spines. Coarse spiral ridges encircle whorls. Outer lip wrinkled. Shades of brown, varices and spines darker, aperture white. A single shell from Musandam, found by Martyn Day. **Habitat:** beached. **Distribution:** GO.

Chicoreus brunneus (Link, 1807): **455**
65mm. Thick, heavy, glossy, spire much shorter than last whorl. Siphonal canal almost closed. 3 non-aligned, frondose varices per whorl. Coarse spiral ribs with superimposed riblets. Outer lip wrinkled; aperture and columella smooth. Brown with darker brown ribs; aperture white; columella stained orange. **Habitat:** offshore and beached. **Distribution:** GO.

Chicoreus brunneus **455**

Chicoreus axicornis **453**

Chicoreus banksii **454**

Aspella producta **452**

SEASHELLS OF EASTERN ARABIA 113

Chicoreus ramosus 456

Superfamily: **MURICOIDEA**

Chicoreus ramosus (Linnaeus, 1758): **456**
160mm. Thick, massive, the last whorl much bulkier than spire. Siphonal canal almost closed. 3 non-aligned varices per whorl, each with short, frondose, curved spines (spine at shoulder of last whorl sometimes long and straight). Outer lip coarsely toothed, one tooth much larger than others. White, stained pinkish brown; edge of aperture and columella orange-red. **Habitat:** on intertidal rocks and coral. **Distribution:** SEG, GO, Mas, SO.

Haustellum dolichourus Ponder & Vokes, 1988: **457**
45mm. Solid, high spired, about 7 rounded whorls, sutures indented by varices; protoconch of about 2.5 granulose, rounded whorls. Long, straight, narrowly open siphonal canal lacking spines. 3 rounded varices per whorl with short spines where crossed by spiral cords; axial ridges between varices. Aperture lirate within. Creamy with faint spiral, brown bands at shoulder and base of last whorl; spiral cords topped with reddish brown lines. Illustrated shell is from the Andaman Is. **Habitat:** offshore. **Distribution:** GO.

Haustellum haustellum longicaudus (Baker, 1891): **458**
70mm. Solid, low spired, about 6 swollen, stepped spire whorls (protoconch absent from available material). Long, straight, almost closed, siphonal canal lacking spines. 3 varices per whorl, without spines with 4 prominent axial ribs between them on later whorls. Irregular spiral cords on all whorls, nodular at shoulders. Aperture obliquely circular; inner lip thin and protruding. Creamy with reddish-brown spiral cords and brown blotches in front of varices; aperture white, its margins apricot or violet tinted. **Habitat:** offshore and beached. **Distribution:** GO.

Haustellum malabaricus (Smith, 1894): **459**
90mm. Thick, sturdy, high spired. About 8 whorls with 3 prominent, rounded varices per whorl, axially aligned and each with a short spine; siphonal canal has short spines near aperture. Coarse axial ribs, 4 or 5 between varices on last whorl; crowded, spiral ridges on all whorls, nodular at intersections with axial ribs. Creamy with brown spiral bands, 3 on last whorl; siphonal canal brown tipped; aperture white. **Habitat:** deep water. **Distribution:** SEG, GO, SO.

Haustellum dolichourus **457**

Haustellum malabaricus **459**

Haustellum haustellum longicaudus **458**

Superfamily: **MURICOIDEA**

Hexaplex kuesterianus 460

Hexaplex rileyi 461

Hexaplex kuesterianus
(Tapparone-Canefri, 1875): **460**
80mm. Thick, heavy, short spired, last whorl angled at shoulder, each suture obscured by preceding whorl; protoconch low, about 2 whorls. Last whorl has up to 8 thick, spineless varices (recurved spines present on juveniles). Siphonal canal recurved, narrowly open, with strong spine on apertural side. Coarse spiral cords cover shell; outer lip sharply toothed. Off-white with or without 1 or 2 reddish-brown spiral bands (juveniles uniformly reddish brown); aperture and columella white. Operculum brown. **Habitat:** on and under intertidal rocks. **Distribution:** all.

Hexaplex rileyi D'Attilio & Myers 1984: **461**
40mm. Broadly biconic, short spired, roundly angled at shoulder; protoconch columnar, about 2.5 whorls. Siphonal canal bent to left, recurved and open. About 9 strong varices crossed by strong, lamellose

116 SEASHELLS OF EASTERN ARABIA

Superfamily: **MURICOIDEA**

Murex carbonnieri **463**

Murex forskoehlii **464**

Murex scolopax **465**

Peter Dance holding a giant specimen of *Murex scolopax*.

cords (about 5 on last whorl); usually spiny at shoulder. Umbilicus small. Greyish white with 3 pale brown spiral bands, best seen within the white aperture; early spire whorls reddish brown. Operculum brown. Has different protoconch to *H. kuesterianus* (see above). **Habitat:** offshore on sand. **Distribution:** SEG, GO.

Murex aduncospinosus Sowerby, 1841: **462** NOT ILLUSTRATED
Listed by Smythe, 1979a:67, from UAE (SEG) but probably misidentified.

Murex carbonnieri (Jousseaume, 1881): **463**
85mm. Sturdy, lightweight, globose-ovate. Siphonal canal almost straight. 3 varices per whorl. Nearly all spines curved, including larger ones; markedly nodulous spiral ribs. Outer lip bluntly toothed. Creamy to pale brown; elongated, brown spots conspicuous between paler nodes on stronger spiral cords; spines creamy; aperture white tinged brown. **Habitat:** offshore in sand. **Distribution:** NG, GO.

Murex forskoehlii Röding, 1798: **464**
80mm. Sturdy, lightweight, globose-ovate. Siphonal canal strongly flexed anteriorly. 3 varices per whorl. Most spines slightly curved, larger ones straight; irregular, weakly nodulous spiral ribs. Outer lip lightly toothed. Creamy with pale brown blothes and axial streaks; aperture brown inside, lighter at edge except where spines form. **Habitat:** offshore and beached. **Distribution:** NWG, NG.

Murex scolopax Dillwyn, 1817: **465**
125mm. Sturdy, globose-ovate, spire shorter than last whorl. Siphonal canal open. 3 varices per whorl. All spines slightly curved, longest spine at shoulder of last whorl. 2 or 3 broad, sharp teeth on outer lip. Low spiral cords confluent with spines. Creamy white with cords and tops of spines amber; chocolate inside aperture, its edge white. Operculum concentrically ridged, brown. Specimens up to 208mm long have been collected at Dibba on the east coast of U. A. E. by H. Kauch. **Habitat:** in sand generally. **Distribution:** all.

Superfamily: **MURICOIDEA**

Murex tenuirostrum tenuirostrum Lamarck, 1822: **466**
100mm. Sturdy, globose-ovate, the long siphonal canal open. Pointed protoconch of 3.5 polished whorls. 3 frilled varices per whorl. Spines closed, those on spire whorls slightly curved, shoulder spines longest. Spiral cords connecting spines strongest. Ivory to golden brown, spines darker; reddish brown spots at notches on outer lip; columella and aperture white. Operculum concentrically ridged, reddish brown. **Habitat:** in sand offshore. **Distribution:** GO.

Naquetia cumingii (A. Adams, 1853): **467**
50mm. Solid, high spired, with small, ovate aperture. Siphonal canal slightly recurved, almost closed. 3 axially aligned, frilled varices (revolving viewed from above). Usually 3 low, vertical folds between varices, nodular where crossed by spiral ornament. Outer lip wavy, finely ridged just inside aperture; inner lip raised. Whitish yellow to buff with 2 or 3 pink to brown spiral bands (most evident on last varical frill); aperture white. **Habitat:** offshore and beached. **Distribution:** GO.

Pterynotus albobrunneus Bertsch & D'Attilio, 1980: **468**
50mm. Lightweight, fusiform, spire about half total length. Columella lirate. Siphonal canal narrowly open. 3 varices per whorl (revolving viewed from above; 1 strong axial fold between varices. Spiral ridges cover entire shell; fine, axial scales between ridges. White tinged reddish brown; aperture white. **Habitat:** offshore and beached. **Distribution:** Mas, SO.

Murex tenuirostrum tenuirostrum **466**

Pterynotus albobrunneus 468

Naquetia cumingii **467**

Superfamily: **MURICOIDEA**

Pterynotus elongatus (Lightfoot, 1786): **469**
70mm. Solid, dull, elongate-fusiform, high spired with shouldered whorls and long siphonal canal. Small aperture has raised, smooth inner lip and raised outer lip with crenulated margin. Last whorl has 3 wing-like varices, these becoming much reduced and hooked on spire whorls (but often eroded). Fine spiral and axial threads produce frills on larger varices. White tinged violet in aperture. **Habitat:** offshore among rocks and coral. **Distribution:** GO, Mas.

Subfamily MURICOPSINAE
Favartia cyclostoma (Sowerby, 1841): **470**
20mm. Ovate, short spired, last whorl appearing wider than it is because of well developed, frondose, non-aligned varices (5 to 7 on last whorl). Siphonal canal recurved, almost closed. Rounded spiral cords between varices crossed by fine lamellae. Outer lip corrugated, several weak pairs of teeth within; inner lip has smooth, thin edge. White sometimes with scattered brown spots; aperture white. **Habitat:** under intertidal rocks. **Distribution:** NG, GO, Mas, SO.

Favartia marjoriae (Melvill & Standen, 1903): **471**
25mm. Solid, high spired, sharply pointed apically. Siphonal canal tapering, recurved, almost closed. 6 to 7 thick, elaborately frilled varices; strong spiral cords between varices, about 6 on last whorl. Edge of outer lip wavy, slightly ridged inside. Greyish white; aperture white. **Habitat:** offshore. **Distribution:** NG.

Favartia paulboschi Smythe & Houart, 1984: **472**
18mm. Fusiform, high spired with tabulated whorls. Inner edge of outer lip faintly lirate. Siphonal canal recurved at tip, narrowly open. Strong varices, 6 on last whorl. Strong, rounded, spiral cords, 5 on last whorl, intersected by equally strong axial cords, 1 between each varix; hooked spines develop on cords at shoulders. Yellowish white, aperture yellowish brown. **Habitat:** under muddy stones. **Distribution:** GO, Mas.

Homalocantha anatomica (Perry, 1811): **473**
65mm. Thick, chalky appearance, fusiform, short spired, early spire whorls usually eroded. 5 or 6 frilled varices on later whorls, 4 of those on last whorl bearing spatulate, virtually closed spines. Excavated pits between varices at sutures. Siphonal canal recurved. White or yellowish white. Usually heavily encrusted with coral. Syn: *H. rota* (Mawe, 1823) sensu Melvill & Standen, 1901:398, and Glayzer et al, 1984:321. **Habitat:** intertidal and offshore among loose coral and beached. **Distribution:** NG, NWG, GO.

Pterynotus elongatus **469**

Favartia cyclostoma **470**

Favartia paulboschi **472**

Favartia marjoriae **471**

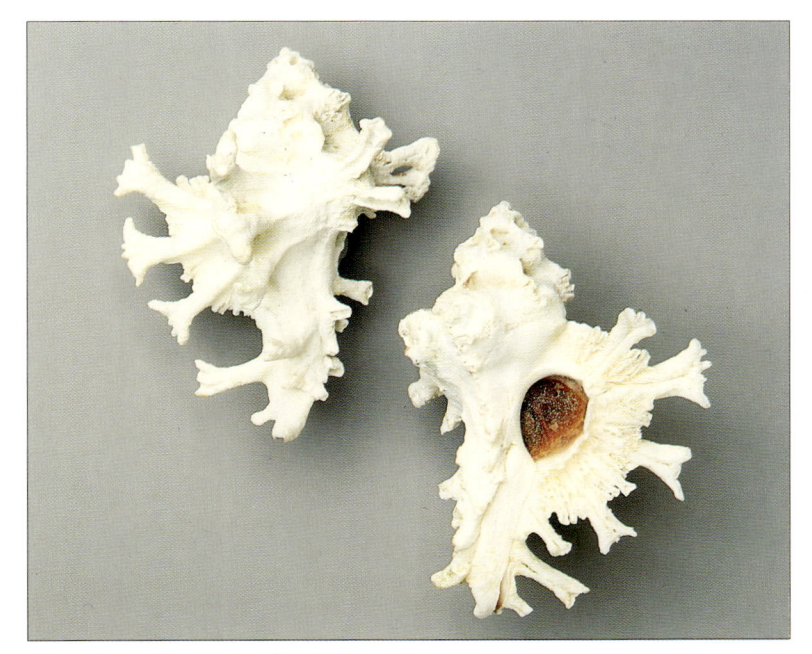

Homalocantha anatomica **473**

Superfamily: **MURICOIDEA**

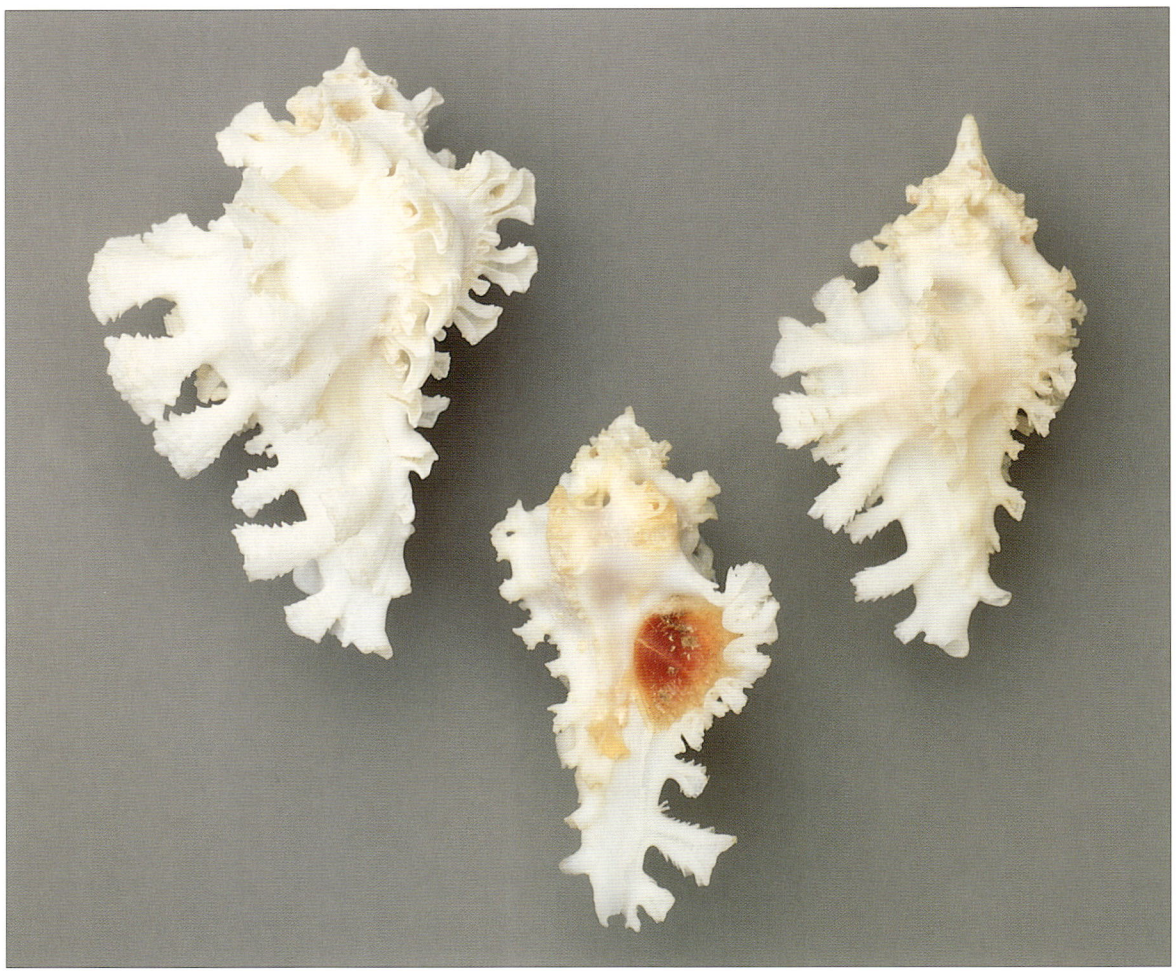

Homalocantha fauroti **474**

Muricopsis bombayanus **476**

Pygmaepterys yemenensis **478**

Muricopsis omanensis **477**

Homalocantha fauroti (Jousseaume, 1888): **474**
55mm. Solid, fusiform, last whorl large but aperture small; early spire whorls usually eroded. 5 varices on later whorls, all those on last whorl bearing 4 spatulate, almost closed, scaly spines, faint spiral cords. Excavated pits at sutures between varices. Siphonal canal recurved. White sometimes tinged violet. Usually heavily encrusted with coral. **Habitat:** intertidal and beached. **Distribution:** GO, Mas.

Homalocantha scorpio (Linnaeus, 1758): **475** NOT ILLUSTRATED
The shells illustrated by Bosch and Bosch, 1982:89, said to be from Oman, cannot now be found. It has not been collected in Eastern Arabia since so confirmation of its presence there is required.

Muricopsis bombayanus (Melvill, 1893): **476**
25mm. Solid, dull, broadly fusiform, spire half total length. Columella smooth. Outer lip has blunt denticles within. Widely spaced, blade-like axial ribs crossed by widely spaced, sharp spiral ribs produce upturned, pointed scales at intersections. Brown or reddish brown. **Habitat:** intertidal rocks. **Distribution:** NG.

Muricopsis omanensis Smythe & Oliver, 1986: **477**
24mm. Solid, fusiform with deep, broad posterior canal. About 5 strong denticles on outer lip. Columella has 3 weak folds towards base. About 7 axial folds crossed by fine spiral ridges, every sixth ridge raised at intersections. Pinkish orange, dark brown on raised ridges; aperture pinkish. Varies greatly in size. **Habitat:** under intertidal rocks and beached. **Distribution:** Mas, SO.

Pygmaepterys yemenensis Houart & Wranik, 1989: **478**
20mm. Solid, high spired, sutures obscured by wavy lamellae. Last whorl has 4 to 6 flanged varices. Aperture has protruding outer lip with wavy edge and 3 elongate folds on lower, inner side. Columellar lip slightly detached, smooth. Coarse spiral cords, strongest at shoulders, crossed by scaly, axial lamellae. Siphonal canal strongly recurved at its end, narrowly open. Greyish white. **Habitat:** under intertidal rocks. **Distribution:** GO, Mas.

Superfamily: **MURICOIDEA**

Cronia cf *konkanensis* **480**

Ocinebrina xuthedra **479**

Subfamily OCINEBRINAE
Ocinebrina xuthedra (Melvill, 1893): **479**
18mm. Elongate-ovate, spire about half total length. Columella has up to 3 slight folds at base. Outer lip has 4 or 5 elongate teeth within. Siphonal canal slightly recurved, wide open. Strong, nodulous, spiral cords, about 6 or 7 on last whorl; up to 4 coarse ridges between cords. Reddish brown with amber cords; aperture orange-white. **Habitat:** offshore. **Distribution:** NG, GO.

Subfamily ERGALATAXINAE
Cronia cf ***konkanensis*** (Melvill, 1893): **480**
26mm. Thick, biconic. Outer lip has 6 blunt teeth. Siphonal canal slightly recurved. Coarse axial folds are nodular where crossed by scaly spiral cords (often eroded). White heavily banded with black, especially on nodules; aperture white. **Habitat:** rocky places exposed at low tide. **Distribution:** SEG, GO, Mas, SO.

Cronia cf ***margariticola*** (Broderip, 1833): **481**
25mm. Biconic, last whorl strongly shouldered, aperture half total length. Columella may have 2 or 3 weak folds at base. Lip has 5 to 7 elongate teeth. Broad axial folds, 7-8 on last whorl, crossed by crowded scaly ridges. Black sometimes with 2 spiral, white bands; aperture purple-brown, teeth paler, may have spiral white lines. Operculum semilunar, dark brown. Shell often bored by sponges. **Habitat:** under intertidal rocks. **Distribution:** SEG, GO, Mas, SO.

Maculotriton serriale (Deshayes in Laborde & Linant, 1834): **482**
16mm. Elongate-fusiform, spire about half total length, with shouldered whorls. Columella has 2 weak folds at base. Outer lip (virtually a varix) has 6 to 8 weak teeth just inside. Up to 16 axial ribs crossed by crowded, nodulous spiral cords. Cream, yellow or greyish with 2 or 3 rows of blackish-brown nodules; aperture whitish. **Habitat:** offshore and beached. **Distribution:** Mas, SO.

Cronia cf *margariticola* **481**

Maculotriton serriale **482**

Subfamily THAIDINAE
Drupella cornus (Röding, 1798): **483**
30mm. Biconic, short or high spired, last whorl always longer than spire. Columella has 2 to 5 low folds at lower end. Outer lip thickened, sharp edged, 5 to 7 blunt teeth just inside. No umbilicus. All whorls nodulous, 4 spiral rows on last whorl. White except for aperture which is bright orange deep inside. Usually encrusted with algae; varies greatly in size. **Habitat:** under intertidal rocks. **Distribution:** GO, Mas.

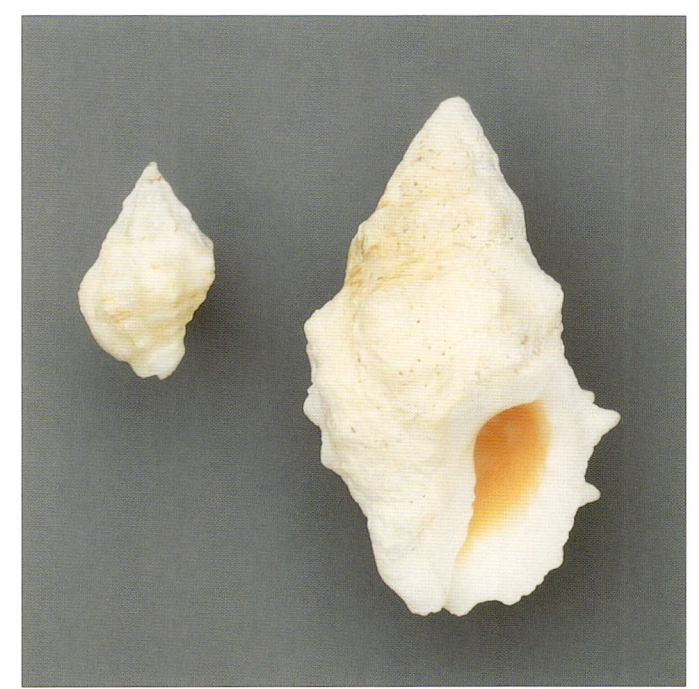

Drupella cornus **483**

Superfamily: **MURICOIDEA**

Morula anaxares (Kiener, 1835): **484**
16mm. Biconic, low spired with narrow, elongate aperture. Columella has 1 or 2 granules at base. Outer lip has 2 widely spaced pairs of teeth just inside. Large, rounded tubercles, 5 axially aligned spiral rows on last whorl. Outer surface covered with fine, scaly, spiral threads. No umbilicus. Brownish black with alternate rows of dull white tubercles; aperture blackish; teeth and columella dull violet. Operculum semilunar, brown. **Habitat:** on rocks. **Distribution:** GO.

Morula chrysostoma (Deshayes, 1844): **485**
25mm. Rotund with short spire and large last whorl. Posterior canal broad and shallow. Columella flexed centrally. Corrugated outer lip has about 6 blunt teeth just inside. Siphonal canal slightly recurved. About 6 strong, axial folds (rotating viewed from above) crossed by 3 strong spiral folds, sharp nodules at intersections; fine, scaly, spiral ridges cover surface. No umbilicus. Orange-white with nodules and base of columella dark brown; aperture and columella bright orange. **Habitat:** under intertidal rocks. **Distribution:** NG, GO, Mas, SO.

Morula granulata (Duclos, 1832): **486**
25mm. Heavy, elongate-ovate or barrel shaped. Columella has 2 to 4 weak folds. Outer lip has 4 or 5 strong teeth. Large, rounded nodules, about 6 spiral rows on last whorl with 2 fine ridges between rows. No umbilicus. Greyish white with dark brown nodules; aperture violet, teeth and columella paler; edge of outer lip dark brown. Operculum kidney shaped, orange brown. Elongate form has smaller, more numerous nodules. **Habitat:** under intertidal rocks, often in unsheltered places. **Distribution:** SEG, GO, Mas, SO.

Nassa situla (Reeve, 1846): **487**
55mm. Ovate with short spire, and finely impressed sutures. Narrow posterior canal. Thin, finely denticulate outer lip. No umbilicus. Spirally striate or smooth. Ivory heavily blotched and spirally lined with dark or pale brown; aperture pinkish brown inside; columella and inner edge of outer lip brown. Operculum oval, dark brown. **Habitat:** under intertidal rocks. **Distribution:** GO, Mas, SO.

Purpura panama Röding, 1798: **488**
60mm. Heavy, ovate, high or low spired. Outer lip has serrated edge. Deep posterior canal with thick callus adjacent. No umbilicus. Spiral rows of nodules, about 5 rows on last whorl, 4 rows of flattened cords between rows; nodules often poorly developed. Brown with beige dashes between nodules; aperture greyish

Purpura panama **488**

Morula anaxares **484**

Morula chrysostoma **485**

Morula granulata **486**

Nassa situla **487**

122 SEASHELLS OF EASTERN ARABIA

Superfamily: **MURICOIDEA**

to white, often with dark spiral ridges, its edge dark brown; columella and callus pinkish. Operculum semilunar, dark brown. **Habitat:** intertidal rocks. **Distribution:** NG, GO, Mas, SO.

Thais bimaculata (Jonas, 1845): **489**
50mm. Heavy, ovate, occasionally globose, low spired with well-defined posterior canal. Broad parietal glaze. Outer lip has saw-tooth edge. Pointed or rounded nodules, 3 spiral rows on last whorl; coarse spiral cords between rows. No umbilicus. Pinkish white with blackish nodules; 1 reddish brown blotch either side of siphonal canal; aperture and columella white, pink-edged outer lip. Operculum semilunar, reddish brown. Spire usually encrusted with algae. **Habitat:** on sheltered side of rocks. **Distribution:** GO, Mas, SO.

Thais bufo (Lamarck, 1822): **490**
55mm. Heavy, ovate, low spired, last whorl occupying most of shell. Narrow, deep posterior canal; thick, broad parietal callus. Columella has slight kink at base. Outer lip toothed, 3 large teeth at its base. Aperture lirate within. Large, widely spaced nodules, 4 or 5 spiral rows on last whorl, prominent at periphery; flat, spiral cords between rows. No umbilicus. Dark or light brown, paler between nodules; sutures, columella and aperture white or pinkish; edge of outer lip brown between teeth. Operculum semilunar. **Habitat:** sides of rocks. **Distribution:** NG, GO.

Thais lacera (Born, 1778): **491**
65mm. Heavy, high spired, last whorl almost scalariform. Outer lip crenulate. Umbilicus very narrow or obscured. Siphonal canal short, deep, slightly recurved. Early spire whorls coronate (but coronations usually absent from adults); last whorl has about 5 spiral rows of nodular cords; flattened spiral cords between rows; massive cord around umbilicus. Cream with reddish-brown blotches and streaks; aperture and columella creamy pink. **Habitat:** subtidal in sand. **Distribution:** all.

Thais rugosa (Born, 1778): **492**
35mm. High spired with shouldered whorls. Aperture almost twice length of spire, spirally ridged inside. Umbilicus a shallow chink. Coarse spiral cords, about 4 on last whorl, the uppermost (occasionally others also) expanded into leaf-like, open scales resembling circular saw viewed from above; spiral rows of fine scales cover surface. Reddish brown; aperture white, edged reddish brown. **Habitat:** under rocks. **Distribution:** Mas.

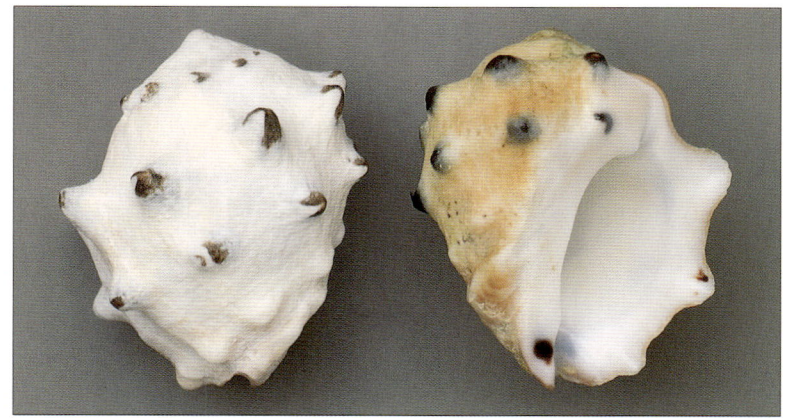

Thais savignyi (Deshayes, 1844): **493**
40mm. Heavy, short or high spired, aperture always taller than spire. Columella may be slightly flexed centrally. Outer lip has weak or strong digitations with paired teeth in their intervals. Pointed nodules on all whorls, 4 spiral rows on last whorl; 1 or 2 strong, spiral cords between rows. No umbilicus. Reddish-brown axial streaks across nodules, greenish white elsewhere; aperture cream, inner edge dark brown; columella white, its base dark brown. Operculum semilunar, dark brown. Upper whorls of mature shells always very eroded. **Habitat:** under intertidal rocks. **Distribution:** all.

Thais tissoti (Petit, 1853): **494**
35mm. High spired, whorls keeled at periphery. Aperture, about twice length of spire, has low spiral ridges inside. Outer lip has low, blunt teeth at ends of apertural ridges. Umbilicus a mere chink. Strong, nodulous, spiral cords, about 5 on last whorl; up to 3 fine, spiral ridges between cords. Surface covered with densely crowded, axial scales. Greyish axially streaked reddish brown across nodules; brown-edged outer lip; apertural ridges may be brown. **Habitat:** under intertidal rocks. **Distribution:** NWG, SEG, GO, Mas, SO.

Thais bimaculata 489

Thais rugosa 492

Thais lacera 491

Thais bufo 490

Thais savignyi 493

Thais tissoti 494

SEASHELLS OF EASTERN ARABIA 123

Superfamily: **MURICOIDEA**

Thais (Mancinella) alouina **495**

Rapana rapiformis **497**

Vexilla vexillum **496**

Rapana venosa **498**

■ *Conspicuous by its barrel-like shape and its brightly tinted aperture, the shell of* Rapana venosa *is unmistakable. Often more than twice as big as the shells illustrated here, it may occur in large numbers. This is bad news for those in the shellfish industry because it is a voracious predator of large bivalves. In the 1940s this native of Chinese and Japanese waters was reported from the Black Sea where it devastated many oyster beds. Since then it has spread farther — in 1992 it was found in the North Sea off the north-eastern coast of England — and may become a widespread pest. Georgina Armour picked up these two shells on a Musandam beach in February 1994. Little did she suspect that they could be the advance troops of a destructive army about to invade the Indian Ocean. The future progress of this unwelcome addition to the molluscan fauna of eastern Arabia may be of interest to others besides shell collectors.*

Thais (Mancinella) alouina (Röding, 1798): **495**
50mm. Heavy, globose, short spired. Thin spiral threads deep inside aperture. Prominent posterior canal. Outer lip crenulate. Spiky or rounded nodules, about 5 spiral rows on last whorl; fine spiral cords between rows. No umbilicus. Greyish orange or pale brown; nodules reddish brown in juveniles, paler in adults; columella and aperture pale orange, apertural threads bright orange. Operculum semilunar, brown. **Habitat:** on exposed rocks. **Distribution:** GO, Mas, SO.

Vexilla vexillum (Gmelin, 1791): **496**
25mm. Lightweight, silky, cylindrically ovate with low spire. Aperture flares towards base. Outer lip sharp edged or slightly thickened, weakly toothed at edge. Covered with fine spiral striae which become stronger on successive whorls. No umbilicus. Off-white to brownish yellow with dark brown spiral bands; aperture and columella whitish to brown; dark brown, silky periostracum. Operculum elongate, dark brown. **Habitat:** under rocks. **Distribution:** GO, Mas, SO.

Subfamily RAPANINAE
Rapana rapiformis (Born, 1778): **497**
70mm. Lightweight or heavy, bulbous with low spire and deeply channelled sutures. Pear-shaped aperture weakly lirate within. Outer lip thin, corrugated. Widely spaced, strong or weak tubercles, 3 or 4 spiral rows on last whorl; prominent spiral grooves between rows. Umbilicus wide and deep, edged by broad, scaly cord. Pale brown blotched and spotted darker brown; aperture and columella white; brown between apertural lirae towards lip edge. Operculum pear shaped, pale brown. **Habitat:** among coral and beached. **Distribution:** all.

Rapana venosa (Valenciennes, 1846): **498**
80mm. Thick and heavy, dull, globose-ovate, spire dwarfed by last whorl and very large aperture, spire whorls flat sided, tabulated and coronate at shoulders. Last whorl has coarse growth ridges which are very pronounced above shoulder and the outer lip develops marginal corrugations; coarse spiral cords are occasionally noduled. Thick ridge encircles deep umbilicus. Yellowish to brown with darker brown spotting on cords; aperture and columella bright reddish orange. 2 shells found at Musandam by Georgina Armour. **Habitat:** offshore. **Distribution:** GO.

Superfamily: **MURICOIDEA**

Babelomurex princeps 499

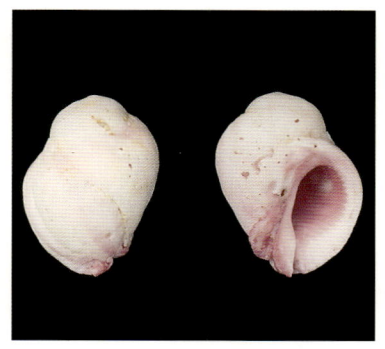

Coralliophila neritoidea 502

Coralliophila radula 504

Coralliophila persica 503

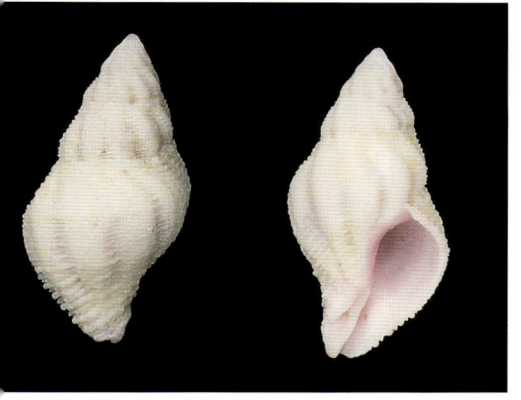

Coralliophila costularis 500

Coralliophila rubrococcinea 505

Coralliophila squamosissima 506

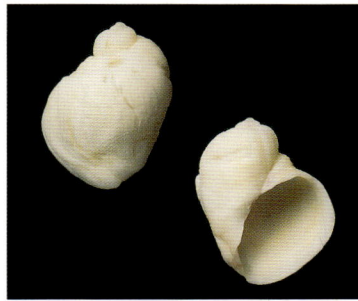

Coralliophila madreporara 501

Family CORALLIOPHILIDAE

Small family notable alike for the beauty of some of its members and the dull, shapeless features of others. Shells from eastern Arabia are mostly colourless but some display a bright violet colour, particularly around the aperture, most are rough externally and one, *Babelomurex princeps,* is spiny. The thin, chitinous operculum is kidney shaped. Predominantly tropical in distribution, little is known about their life history although some are certainly parasitic on corals.

Babelomurex princeps (Melvill, 1912): **499**
35mm. Lightweight, high spired, whorls angulate at periphery, last whorl longer than spire. Inner lip raised. Columella flexed at base. Siphonal canal recurved and open. Spiral row of upturned, broad, open-sided lamellae encircles periphery of each whorl; up to 8 scaly, spiral cords on lower half of last whorl. Umbilicus narrow and deep. White, pale brown or lavender. Usually encrusted with algae. **Habitat:** offshore. **Distribution:** NG, GO.

Coralliophila costularis (Lamarck, 1816): **500**
40mm. Elongate-biconic, whorls gently rounded or almost straight sided. Aperture lirate within. Outer lip corrugated. Strong, aligned axial ribs (revolving viewed from above) crossed by crowded, scaly, spiral cords. Umbilicus open, shallow, surrounded by scaly ridge. Off-white tinted violet; aperture and columella dark violet. Operculum ovate, pale brown. **Habitat:** among coral and beached. **Distribution:** Mas.

Coralliophila madreporara (Sowerby, 1822): **501**
30mm. Solid, ovate, short spired, often irregularly formed. Aperture flattened or concave, usually wide open. Columella shelf like. Outer lip flaring. Siphonal canal a slight embayment. Fine, crowded, scaly spiral cords. No umbilicus. White or grey; columella violet or purple. **Habitat:** in coral and beached. **Distribution:** GO, Mas.

Coralliophila neritoidea (Lamarck, 1816): **502**
30mm. Thick, globose to elongate-ovate, short spired. Aperture strongly lirate within. Callus adjacent to columella. Outer lip sharp edged, bevelled. Tiny umbilicus bordered by thick ridge. Crowded, slightly scaly spiral lines cover surface. Off-white, often tinted violet; aperture and callus purple. Surface often eroded and encrusted. **Habitat:** offshore among coral and beached. **Distribution:** Mas.

Coralliophila persica Melvill, 1897: **503**
17mm. Solid but lightweight, fusiform, spire about half total length. Sutures shallow becoming deeper between later whorls. Aperture strongly and deeply lirate. Outer lip has thin, rough edge. Siphonal canal recurved and open. Covered with fine, regular, scaly, spiral riblets which become obscure on last whorl. No umbilicus. Yellowish brown or whitish. **Habitat:** offshore. **Distribution:** NG, GO.

Coralliophila radula (A. Adams, 1855): **504**
40mm. Heavy, ovate, low spired. Aperture flaring anteriorly, lirate within. Outer lip corrugated. Irregular, obliquely axial ridges crossed by crowded, weak and strong, scaly spiral cords. Umbilicus a mere chink. White; aperture and columella violet to purple. **Habitat:** in coral and beached. **Distribution:** GO, Mas.

Coralliophila rubrococcinea Melvill & Standen, 1901: **505**
25mm. Lightweight, fusiform, spire about half total height, much narrower than last whorl. Sutures deep, wavy. Aperture lirate or smooth within. Columella straight or gently curved. Outer lip thin, rough edged. Coarse, thick, axial folds crossed by closely spaced, scaly, spiral riblets. No umbilicus. White, reddish orange or greyish purple. Very variable in shape and colour. **Habitat:** offshore, presumably on corals or sea fans. **Distribution:** NG, GO.

Coralliophila squamosissima (Smith, 1876): **506**
26mm. Lightweight, ovate with swollen last whorl and deep sutures. Aperture weakly lirate within. Outer lip corrugated. Regular, low, axial ribs crossed by alternately weak and strong, scaly spiral cords. Umbilicus shallow, almost obscured by inner lip and bordered by scaly ridge. White. **Habitat:** in coral and beached. **Distribution:** Mas.

Superfamily: **MURICOIDEA**

Magilus antiquus Montfort, 1810: **507**
160mm. Heavy, resembling solid marble, globular initially, then forming a meandering tube. Coarse spiral ridges on coiled part, irregular growth ridges on tube. White. Tubular extension may reach 300mm and colour may vary according to environment. **Habitat:** insinuated in stony corals and beached. **Distribution:** GO, SO.

Mipus gyratus (Hinds, 1844): **508**
45mm. Fusiform, spire about half total length, all whorls strongly keeled. Aperture weakly lirate within. Outer lip finely corrugated. Siphonal canal gently recurved. Weak vertical folds crossed by crowded, scaly spiral cords. Deep, broad umbilicus encircled by broad, scaly ridge. White sometimes tinted violet or amber. Varies in length, width and development of keels. **Habitat:** offshore and beached. **Distribution:** Mas.

Mipus* cf *rosaceus (Smith, 1903): **509**
35mm. Thick, elongate-globose with swollen last whorl. Aperture lirate within. Outer lip corrugated. Siphonal canal slightly recurved. Strong, widely spaced, obliquely axial ribs crossed by crowded, alternately weak and strong, scaly, spiral cords. Wide, deep umbilicus surrounded by broad, scaly ridge. Off-white; aperture violet to purple. **Habitat:** in coral and beached. **Distribution:** Mas.

Family BUCCINIDAE
Small to large gastropods bearing a robust shell and a chitinous operculum which usually has a nucleus at or close to its margin. Apart from *Babylonia* and some *Cantharus* species the family is represented in eastern Arabia by mostly small species. The siphonal canal is usually short and open, the columella smooth or granulated, the outer lip smooth or toothed. The surface ornament, when present, is weak. Occasionally the colour pattern is striking but is usually subdued and may be hidden under a thick periostracum. Most species are carnivorous, feeding on dead or moribund animal matter, and lay their eggs in capsules attached to other objects. They are plentiful intertidally among rocks.

Subfamily BUCCININAE
Babylonia areolata (Link, 1807): **510** NOT ILLUSTRATED
A worn shell found by L Hubers at Salalah was probably adventitious. Unrecorded otherwise from eastern Arabia.

Babylonia pintado Kilburn, 1971: **511**
55mm. Thick, high spired, whorls almost straight sided; protoconch domed. Sutures narrowly and deeply channelled. Thick parietal callus; posterior canal well developed. Columella gently curved, truncate. Outer lip flaring. Siphonal canal broad and shallow. Smooth with fine growth lines. Umbilicus a narrow, deep slot bordered by strong, rounded ridge. Cream with 2 spiral rows of orange-brown, oblong blotches and a broad zone of regularly spaced, slightly paler spots between rows; early whorls fawn. Known mostly from beached shells. **Habitat:** offshore and beached. **Distribution:** Mas.

Babylonia spirata (Linnaeus, 1758): **512**
65mm. Heavy, ovate, last whorl very inflated, protoconch pointed; sutures deeply channelled. Thick parietal callus. Deep, narrow posterior canal. Outer lip sharp and smooth, strongly flexed at top. Fine growth striae crossed by fine spiral striae. Broad, deep umbilicus usually obscured by parietal callus; umbilical area bordered by thick ridge. Greyish white with reddish brown or orange crescents, oblique streaks and spots; early whorls bright purple. Operculum semilunar, brown. Attains a large size at Masirah. Syn: *Eburna valentiana* Swainson, 1822. **Habitat:** in sand. **Distribution:** NG, GO, Mas, SO.

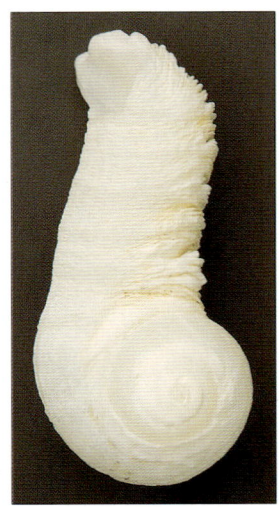

Magilus antiquus 507

Babylonia pintado 511

Mipus gyratus 508

Mipus cf *rosaceus* 509

Babylonia spirata 512

Superfamily: **MURICOIDEA**

Nassaria acuminata 513

Nassaria pusilla 514

Phos roseatus 515

Cantharus mollis 517

Cantharus fumosus 516

Cantharus spiralis 519

Cantharus sowerbyanus 518

Subfamily PHOTINAE
Nassaria acuminata (Reeve, 1844): **513**
25mm. High spired with smooth, glossy protoconch, last whorl about half total height. Aperture lirate within, crenulate at raised edge. Prominent posterior canal. Columella plicate. Outer lip greatly thickened. Siphonal canal strongly recurved. Strong, regular, axial ribs and occasional varices crossed by crowded, smooth or scaly spiral cords. No umbilicus. White or yellowish with interrupted, brown spiral band; aperture white. Syn: *N. bitubercularis* (A. Adams, 1851), *N. suturalis* A. Adams, 1853, *N. recurva* Sowerby, 1859. **Habitat:** rock crevices and beached. **Distribution:** NG, GO, Mas.

Nassaria pusilla (Röding, 1798): **514**
28mm. Solid, high spired, bulbous apically, last whorl half total height. Aperture strongly lirate within. Columella lirate. Outer lip finely corrugated. Siphonal canal obliquely recurved. Strong axial ribs and occasional varices crossed by strong spiral cords, nodulous at intersections. No umbilicus. Off-white with brown banding; aperture white. Syn: *N. nivea* (Gmelin, 1791). **Habitat:** offshore and beached. **Distribution:** NG, GO.

Phos roseatus (Hinds, 1844): **515**
35mm. Solid, elongate-fusiform, with glossy protoconch. Aperture sharply angled at top, lirate within. Columella has 2 to 4 folds. Outer lip thickened, corrugated. Prominent axial ribs and occasional varices crossed by strong and weak spiral cords. Yellowish white with 2 interrupted brown or orange spiral bands; aperture white or lavender, columella white. Operculum paddle shaped, brown. Syn: *P. gladysiae* Melvill & Standen, 1901. **Habitat:** among shells and algae in mud offshore. **Distribution:** NWG, SEG, GO.

Subfamily PISANIINAE
Cantharus fumosus (Dillwyn, 1817): **516**
25mm. Thick, biconic, aperture half total length. Fold within posterior canal. Columella flexed at base, with irregular folds. Thickened outer lip lirate within. Strong, rounded, axial ribs crossed by regular spiral ridges. No umbilicus. Yellowish with dark brown ribs, sometimes encircled by white bands; aperture white; outer lip edge orange-yellow. **Habitat:** intertidal rocks. **Distribution:** GO.

Cantharus mollis (Gould, 1860): **517**
22mm. Solid, biconic, with short spire. Posterior canal distinct. Columella bluntly toothed. Outer lip lirate within. Weak axial ribs crossed by narrower spiral ridges, nodulous at intersections. No umbilicus. Brown with white aperture and columella. Covered with velvety, greenish-brown periostracum when fresh. **Habitat:** among intertidal rocks and stones. **Distribution:** SO.

Cantharus sowerbyanus (Melvill & Standen, 1903): **518**
34mm. Solid, almost biconic, spire about as long as aperture. Blunt, glossy protoconch of about 3.5 whorls. Columella strongly flexed at base. Posterior canal very shallow. Outer lip thin, wavy edged. Aperture lirate within. Strong, rounded, axial ribs (about 10 on last whorl) crossed by regular spiral cords. White with upper half of spire whorls pale brown; aperture and columella white. Covered with straw-coloured periostracum when fresh. Operculum claw shaped, pale brown. Like *C. spiralis* (see below) but less shouldered, paler and with different operculum. **Habitat:** deep water. **Distribution:** GO.

Cantharus spiralis (Gray, 1846): **519**
40mm. Heavy, dull, almost biconic with strongly shouldered whorls, spire much narrower than last whorl. Columella flexed at base. Outer lip wavy edged. Aperture lirate within. Strong, rounded, axial ribs (about 10 on last whorl) crossed by strong and weak spiral cords. No umbilicus. Brown blotched and streaked dark brown; aperture white or purplish, edged orange or brown. Covered with greenish-brown periostracum when fresh. Operculum oval, amber. **Habitat:** under rocks and in sand. **Distribution:** Mas, SO.

Superfamily: **MURICOIDEA**

Cantharus undosus (Linnaeus, 1758): **520**
35mm. Heavy, biconic with short, straight-sided spire. Sutures weakly channelled. Aperture lirate within. Posterior canal toothed each side. Columella irregularly plicate. Outer lip greatly thickened, its edge corrugated. Keeled, evenly spaced spiral ribs. No umbilicus. Ribs dark purple, their intervals mostly white or orange; aperture white, its edge and columellar lip orange. Covered with thick, brown periostracum when fresh. **Habitat:** under rocks. **Distribution:** GO, Mas, SO.

Cantharus wagneri (Anton, 1838): **521**
25mm. Solid, ovate, with deep sutures. Columella strongly flexed and folded at base. Outer lip not thickened. Obliquely axial, rounded cords (about 9 on last whorl) crossed by thin, slightly raised, spiral ridges and threads. Brownish yellow with amber spiral ornament; columella and aperture white. **Habitat:** offshore on sand. **Distribution:** GO.

Engina mendicaria (Linnaeus, 1758): **522**
11mm. Thick, biconic, aperture longer than spire, whorls roundly shouldered. Swollen lower half of columella bears row of connected nodules. Thickened outer lip has large, lumpy teeth centrally. Coarse growth lines cover surface. No umbilicus. Yellow with broad, black, spiral bands; aperture white; columella and outer lip dark brown. **Habitat:** under intertidal rocks. **Distribution:** GO, Mas, SO.

Engina natalensis Melvill, 1895: **523**
12mm. Thick, biconic, spire almost straight sided, shorter than aperture. Swollen columella toothed towards base, plicate above. Outer lip swollen centrally on inner edge, with about 5 blunt teeth. Low axial ribs, nodulous spiral cords, fine striae between cords. No umbilicus. White with spiral rows of blackish nodules; aperture lilac; columella and edge of outer lip dotted reddish brown. Operculum paddle shaped. Syn: has been confused with ***E. zonalis*** (Lamarck, 1822). **Habitat:** under intertidal rocks. **Distribution:** Mas, SO.

Engina rawsoni (Melvill, 1897): **524**
15mm. Solid, dull, fusiform, whorls gently rounded. Short, broad, open siphonal canal. About 11 denticles on outer lip. Columella has about 6 indistinct folds. Many axial folds on all whorls (revolving viewed from above) crossed by strong, regular, rounded, spiral cords. White with interrupted, pale brown, spiral bands; denticles orange. **Habitat:** Offshore among shell rubble. **Distribution:** NG, GO.

Engina zea Melvill, 1893: **525**
18mm. Thick, biconic, spire straight sided, aperture half total length. Deep-seated parietal fold adjacent to posterior canal. Columella plicate along entire length. Outer lip wavy edged, 5 blunt teeth just inside. Axial ribs crossed by strong spiral cords, nodulous at intersections. No umbilicus. White with ochre and brownish-black streaks on ribs; aperture and columella pale lilac; edge of outer lip dotted reddish brown. **Habitat:** intertidal muddy rocks. **Distribution:** NG, Mas.

Pisania ignea (Gmelin, 1791): **526**
38mm. Solid, lightweight, aperture half total length. Posterior canal bordered by tooth each side. Columella flexed at base. Outer lip thin, scalloped. Fine growth striae give silky appearance; weak spiral cords at base of last whorl. No umbilicus. Yellowish brown with dark brown streaks and blotches; aperture bluish white. Operculum elongate with basal nucleus, pale brown. Early whorls usually absent from adults. **Habitat:** in branched corals and beached. **Distribution:** GO, Mas, SO.

Pisania tritonoides (Reeve, 1846): **527**
27mm. Solid, lightweight, aperture slightly more than half total length. Weak tooth each side of posterior canal. Columella flexed and twisted at base. Outer lip has many elongated teeth inside. Shallow spiral grooves at base and on upper spire whorls. No umbilicus. White heavily blotched and lined with reddish brown; aperture and columella white. **Habitat:** offshore and beached. **Distribution:** Mas.

Family COLUMBELLIDAE
Small to tiny gastropods with strong, elongated or inflated shells, the columbellids (or dove shells as they are often called) form a large family, well represented in warm waters around the globe. The *Anachis* group has axially ribbed shells. Other shells are mostly smooth and often glossy. Many have striking colour patterns which vary greatly, even within a single population. The narrow aperture often appears even narrower because the outer lip may be thickened and toothed centrally and the columella may have one or two rows of denticles. There may be a shallow posterior canal but never an umbilicus. Most species have a thin, chitinous operculum which is oval or sickle shaped with

Engina mendicaria 522

Engina natalensis 523

Cantharus undosus 520

Cantharus wagneri 521

Engina rawsoni 524

Engina zea 525

Pisania ignea 526

Pisania tritonoides 527

Superfamily: **MURICOIDEA**

Anachis fauroti **529**

Anachis raysutana **530**

Anachis donnae **531**

Euplica varians **532**

Columbella aspersa **528**

Mitrella agatha **533**

an apical nucleus. Dove shells live on and under rocks, on and in sand or among algae, often close to coral reefs. They are omnivorous but seem to be carnivorous by preference. There is no satisfactory classification of species and genera in this family; some authorities dispute even the family name, preferring the alternative name Pyrenidae. As species vary considerably in size, sculpture, and colour pattern they are difficult to identify correctly.

Subfamily COLUMBELLINAE
Columbella aspersa Sowerby, 1844: **528**
24mm. Solid, ovate-fusiform with slightly channelled sutures. Basal portion of columella has groove deep within aperture; a fused group of teeth within central portion of outer lip. Smooth except for spiral grooves around base. White with streaks and netted lines arranged in spiral bands; aperture white.
Habitat: offshore and beached.
Distribution: GO, Mas, SO.

Subfamily PYRENINAE
Anachis fauroti (Jousseaume, 1888): **529**
12mm. Solid, fusiform, aperture less than half total length, spire whorls appear slightly telescoped; sutures impressed, wavy; protoconch of 1.5 smooth, translucent whorls. Strong axial ribs (14-15 on last whorl). Straight columella has few weak plications; centrally thickened outer lip has a few weak denticles within. White with black axial or wavy streaks and blotches on ribs; white columella and aperture have occasional black or brown blotches. Syn: *A. rassierensis* Smythe, 1985.
Habitat: among intertidal rocks.
Distribution: NG, SEG, GO, Mas, SO.

Anachis raysutana Smythe, 1985: **530**
13mm. Thick, ovate-biconic with deep, wavy sutures; protoconch of 2.5 whorls, usually eroded. Strong axial ribs (9-10 on last whorl) much narrower than their intervals, absent from lower half of last whorl; spiral grooves at base. Columella straight, smooth; outer lip thickened centrally, plicate along its length. White variegated with yellow to reddish brown producing tent-like markings, blotches and lines.
Habitat: under rocks or in sand.
Distribution: Mas, SO.

Anachis donnae Moolenbeek & Dance, 1994: **531**
20mm. Thick, moderately glossy, elongate-fusiform, aperture less than half total length, spire whorls almost straight sided, with deep sutures; protoconch of about 1.5 smooth whorls. Low, broad, axial ribs (14-16 on penultimate whorl, about 11 on last whorl); 3 or 4 spiral grooves at base. Columella smooth; flared outer lip has 4-7 weak teeth within. Early spire whorls brown, later whorls white with brown blotches and a network of brown lines, 2 brownish spiral bands on last whorl; aperture white. **Habitat:** intertidal.
Distribution: Mas.

Euplica varians (Sowerby, 1832): **532**
10mm. Ovate-fusiform, short spired with slightly shouldered last whorl. Broad axial ribs (not always present) on spire whorls and upper half of last whorl; up to 9 nodules on columella; about 8 teeth on thickened outer lip. Pale amber with white flecks and 2 or 3 spiral bands of dark brown flecks; outer lip and columella violet. **Habitat:** intertidal rocks and beached. **Distribution:** NWG, SEG, GO, Mas, SO.

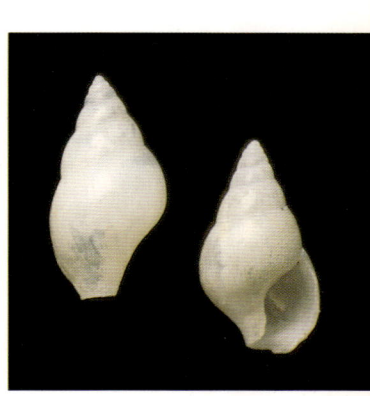

Mitrella agatha (Melvill, 1904): **533**
7mm. Thin, translucent, glossy, ovate-fusiform, aperture half total length; spire whorls rounded, sutures appressed; lop-sided protoconch of about 1.5 whorls. Siphonal canal short, broad, slightly recurved. Thin, broadly expanded outer lip, thickened behind, lacks denticles within. Smooth externally. White. **Habitat:** deep water.
Distribution: GO.

SEASHELLS OF EASTERN ARABIA 129

Superfamily: **MURICOIDEA**

Mitrella albina (Kiener, 1841): **534**
15mm. Solid, biconic to elongate-fusiform, aperture half total length; spire whorls shouldered, sutures slightly channelled; protoconch pointed. Smooth or slightly coronated at shoulders; spiral grooves at base. Columella thickened and plicate; outer lip sinuous, thickened and toothed centrally. White with dark brown spiral bands or blotches; columella and outer lip violet; sometimes all white or all brown. **Habitat:** offshore and among intertidal stones. **Distribution:** SEG, GO, Mas.

Mitrella alizonae (Melvill & Standen, 1901): **535**
11mm. Thick, dull, ovate-fusiform, aperture half total length, last whorl swollen; sutures deep; protoconch of 3.5 glossy whorls. Lower half of columella has 4 or 5 folds; about 5 elongate teeth inside thickened outer lip. Smooth except for strong spiral grooves around base. White with linked beige blotches; aperture white. **Habitat:** offshore and beached. **Distribution:** NG, SEG, GO, SO.

Mitrella blanda (Sowerby, 1844): **536**
17mm. Elongate-fusiform, very glossy, with almost straight sides, sometimes elevated subsuturally; protoconch acutely pointed, often missing. Smooth except for weak spiral grooves at base. Greyish or whitish with wavy and zigzag, axial, brown streaks which vary in width and disposition and may be entirely absent. Very variable colour pattern; occasionally represented by the white form ***candidans*** Melvill & Standen, 1901. Syn: *M. doriae* (Issel, 1865). **Habitat:** in sand. **Distribution:** NWG, NG, SEG, GO, Mas.

Mitrella cartwrighti (Melvill, 1897): **537**
7mm. Solid, glossy, narrowly fusiform, with straight sides and finely incised sutures; protoconch pointed. Smooth except for spiral grooves at base. Siphonal canal short, narrow, recurved backwards. Columella smooth; thickened outer lip toothed along its length. Whitish or yellowish with axial chestnut streaks and blotches, or yellowish chestnut with round white spots, reddish brown between denticles. Syn: *Columbella agnesiana* Melvill & Standen, 1901. **Habitat:** offshore in muddy sand. **Distribution:** NG, NWG, SEG, GO, Mas.

Parviterebra thyraea (Melvill & Standen, 1897): **538**
12mm. Thin, translucent, elongate-fusiform, whorls gently rounded, sutures impressed and wavy; smooth, glassy protoconch of 2.5 well rounded whorls. Strong axial ribs crossed by widely spaced, almost microscopic striae. White with a spiral row of brown spots between ribs just below sutures of later whorls. Resembles a small ***Terebra***. **Habitat:** intertidal in sand. **Distribution:** GO, Mas, SO.

Pyrene flava (Bruguière, 1789): **539**
24mm. Solid, broadly biconic, aperture more than half total length; spire whorls slightly telescoped; bulbous protoconch of 2 whorls. Very fine axial and spiral striae and weak spiral grooves. Columella smooth or with 4-5 denticles at base; outer lip has 8-10 denticles. There are 2 distinct colour patterns: a) white mottled, streaked and netted light or dark brown with brown flames below sutures; b) white blotched and netted dark or pale brown, sometimes all brown; protoconch bright pink. Very variable. Has been confused with ***P. testudinaria*** (Link, 1807). **Habitat:** among intertidal rocks. **Distribution:** Mas, SO.

Mitrella cartwrighti 537

Parviterebra thyraea 538

Mitrella albina 534

Mitrella alizonae 535

Pyrene flava 539

Mitrella blanda 536

SEASHELLS OF EASTERN ARABIA

Superfamily: **MURICOIDEA**

Pyrene nomadica (Melvill & Standen, 1901): **540**
12mm. Elongate-fusiform, straight sided, sutures distinct, aperture less than half total length. Spiral grooves at base, rest of shell smooth, silky. Columella smooth; thickened outer lip has 6 or 7 teeth, not easily seen. Thick, netted, dark brown lines enclose circular or irregular greenish-white areas; coloration may occur on columella and in aperture. **Habitat:** under intertidal rocks. **Distribution:** GO, Mas, SO.

Pyrene propinqua (Smith, 1891): **541**
17mm. Thick, moderately glossy, broadly biconic, aperture slightly more than half total length, roundly shouldered, sutures deep, outer lip greatly thickened and slightly incurved centrally. Strong teeth inside outer lip. Smooth except for spiral grooves on last whorl which are most pronounced on thickened outer lip. White clouded grey-brown, spiral grooves orange-brown; aperture violet, teeth white. Has yellow periostracum when fresh. **Habitat:** offshore and beached. **Distribution:** NG, GO.

Zafra comistea (Melvill, 1906): **542**
3mm. Solid, glossy, pear shaped with attenuated base, short spire, deep sutures; protoconch well rounded. Aperture very narrow, made narrower by thickening inside outer lip, inner lip raised. Strong axial ribs on later whorls, grooves around base. Greyish white (but available shells probably long dead when dredged). **Habitat:** deep water. **Distribution:** GO.

Zafra phaula (Melvill & Standen, 1901): **543**
2mm. Sturdy, translucent, glossy, elongate-fusiform, aperture less than half total length. Strong, slightly oblique, rounded, axial ribs, as wide as their intervals; strong spiral grooves at base. Columella and outer lip parallel, upper half of outer lip thickened. Dark brown initially becoming amber on last whorl, with bold white patches on ribs just below sutures. **Habitat:** offshore and beached. **Distribution:** NG, NWG, SEG, GO.

Zafra selasphora (Melvill & Standen, 1901): **544**
4mm. Solid, ovate, aperture half total length, with protoconch of about 3.5 smooth whorls; sutures well defined. Columella straight, smooth; central portion of outer lip has 4 or 5 teeth. Rounded axial ribs (about 12 on last whorl), wider than their intervals. White with amber dashes on alternate ribs and axial streaks descending from them on last whorl. **Habitat:** offshore and beached. **Distribution:** all.

Family NASSARIIDAE

A large family characterised by a barrel-shaped or elongate-ovate shell containing an animal with long, slender tentacles and a long siphon. Shells are smooth or ribbed and several have a parietal callus which may be very thick. The siphonal canal is deep but short, the outer lip often lirate within and the columella lacks well developed ornament. Shells of *Bullia*, common objects on beaches in eastern Arabia, have high spires and lack columellar folds. The chitinous operculum, associated with members of the family, often has a serrated edge. Most nassariids (or mud snails as they are sometimes called) are carnivorous scavengers; some are active predators. At low tide they may be seen, sometimes abundantly, crawling over or burrowing obliquely into muddy sand, their siphons erect. The species recorded include some of the smallest in the family and one or two of the larger. The subgenera adopted here are taken from Cernohorsky (1984).

Pyrene nomadica 540

Zafra selasphora 544

Zafra phaula 543

Zafra comistea 542

Pyrene propinqua 541

SEASHELLS OF EASTERN ARABIA

Superfamily: **MURICOIDEA**

Subfamily NASSARIINAE

Nassarius (Aciculina) ischnus (Melvill, 1899): **545**
9mm. Solid, dull, narrowly elongate-ovate, tall spired, sutures wavy and impressed; protoconch of about 3 smooth whorls. Outer lip greatly thickened, slightly notched towards base, bearing about 6 elongate teeth within. Thick, widely spaced axial ribs crossed by widely spaced spiral threads; short, recurved siphonal canal. White with 2 reddish brown, spiral bands. **Habitat:** offshore. **Distribution:** GO.

Nassarius (Alectrion) glans glans (Linnaeus, 1758): **546**
35mm. Lightweight, elongate-ovate, spire whorls well rounded; protoconch of 1.5 glassy whorls. Axial riblets crossed by wavy spiral grooves becoming obsolete on penultimate and last whorls. Columella has basal fold; tooth by posterior canal; spiny denticles at edge of outer lip. White with brown streaks and blotches, later whorls with thin, orange-brown, spiral lines; columella and lip white. **Habitat:** offshore and beached. **Distribution:** NWG, SEG, GO, Mas, SO.

Nassarius (Hima) pauperus (Gould, 1850): **547**
12mm. Solid, elongate-ovate with gently rounded spire whorls; protoconch of about 3 glassy whorls. Strong, rounded axial ribs crossed by thin spiral cords with very fine spiral threads between. Columella has weak folds; thickened outer lip has elongate teeth. Cream to brown, occasionally with 1 or 2 brown spiral bands on last whorl. **Habitat:** in sand offshore. **Distribution:** GO, Mas.

Nassarius (Nassarius) arcularia plicatus (Röding, 1798): **548**
22mm. Thick, broadly ovate, spire whorls tabulated at sutures; protoconch of about 3 smooth whorls. Crowded, widely spaced, wavy, axial ribs crossed by regular, deep grooves (giving saw-tooth profile to ribs on last whorl); large nodules just below sutures; broad callus extends above penultimate whorl; columella has weak folds; aperture lirate. Greenish grey, cream or brown, sometimes banded with darker colour; subsutural nodes always pale; callus and aperture white. **Habitat:** in sand intertidally. **Distribution:** NWG, GO, Mas.

Nassarius (Nassarius) coronatus (Bruguière, 1789): **549**
30mm. Thick, broadly ovate, stepped spire much shorter than last whorl; protoconch of about 3 smooth whorls. Early spire whorls have flattened axial ribs crossed by widely spaced spiral grooves; later whorls have coronations, sometimes extending as smooth ribs. Columella has weak nodules; aperture weakly lirate, edge of outer lip sometimes spiny. Brown, greenish or yellowish, dark brown between coronations; aperture has reddish bands; columella, callus and lip white. **Habitat:** on sand intertidally and offshore. **Distribution:** GO, Mas.

Nassarius emilyae emilyae Moolenbeek & Dekker, 1994: **550**
16mm. Thick, ovate, aperture a third total length, spire whorls almost straight sided, periphery of last whorl; protoconch not seen. Coarse axial ribs; 5-6 denticles on inner side of outer lip. Cream with broad, greyish brown, spiral zone (more intense dorsally) bearing, on last whorl, 5 spiral bands of reddish brown dots; above the zone are irregular, blackish, triangular markings, outer lip with reddish brown markings. **Habitat:** intertidal pools. **Distribution:** GO, Mas, SO.

Nassarius (Niotha) albescens gemmuliferus (A. Adams, 1852): **551**
10mm. Thick, elongate-ovate, spire whorls tabulated at sutures; smooth protoconch of about 4 whorls. Crowded, slightly oblique, axial ribs crossed by irregular spiral grooves give rough surface; large nodules just below sutures; callus on parietal wall. Strong fold at base of columella; aperture weakly lirate. White blotched, spotted and streaked dark and pale brown; aperture and callus white. **Habitat:** under intertidal and offshore rocks. **Distribution:** NWG, SEG, GO, Mas.

Nassarius (Niotha) conoidalis (Deshayes, 1832): **552**
25mm. Thick, ovate with acuminate spire shorter than last whorl, deep sutures; protoconch of about 3.5 smooth whorls. Broad axial ribs (about 15-30 on last whorl) crossed by deep spiral grooves producing elongate nodules. Columella has many irregular nodules; aperture lirate, edge of outer lip sometimes

Nassarius (Aciculina) ischnus 545

Nassarius (Hima) pauperus 547

Nassarius (Niotha) albescens gemmuliferus 551

Nassarius (Nassarius) arcularia plicatus 548

Nassarius (Niotha) conoidalis 552

Nassarius emilyae emilyae 550

Nassarius emilyae 550

Nassarius (Alectrion) glans glans 546

Nassarius (Nassarius) coronatus 549

Superfamily: **MURICOIDEA**

spiny. Yellowish to pale brown; aperture and columellar callus white. **Habitat:** sand offshore. **Distribution:** NWG, GO, Mas.

Nassarius (Niotha) deshayesianus (Issel, 1866): **553**
16mm. Thick, elongate-ovate, almost straight-sided spire about half total length; pointed protoconch of about 3.5 smooth whorls. Strong, widely spaced axial ribs; growth striae between ribs. Columella has strong folds or denticles; size of aperture reduced by callus thickening; a few denticles within outer lip. Cream to grey, occasionally orange; brown spiral band at centre of last whorl; edge of aperture white to orange. **Habitat:** intertidal under rocks and in sand. **Distribution:** SEG, GO, Mas.

Nassarius (Niotha) himeroessa (Melvill & Standen, 1903): **554**
6mm. Thick, translucent, ovate, with distinctly keeled, smooth protoconch of 3-4.75 whorls. Rounded, slightly oblique, axial ribs (about 13 on last whorl), nodulous at their upper ends, with fine spiral threads between; spiral grooves at base. Columella has 3 or 4 denticles at base; thickened outer lip denticulate. White with pale reddish brown spiral bands (3 on last whorl). **Habitat:** offshore. **Distribution:** GO.

Nassarius (Niotha) jactabundus (Melvill, 1906): **555**
18mm. Solid, dull, elongate-ovate, spire slightly tapering, much narrower than last whorl, sutures deep; protoconch of about 3.5 keeled whorls. Outer lip thickened, bearing about 9 teeth within, columella wrinkled. Wide, crescentic axial ribs crossed by flat spiral cords. White to yellowish, sometimes with 2 broad, pale brown spiral bands on last whorl. **Habitat:** shallow water. **Distribution:** NG.

Nassarius (Niotha) nodulosus (Marrat, 1873): **556**
15mm. Solid, ovate, short spired; protoconch of about 2 whorls. Spiral rows of rounded nodules (2 rows on penultimate whorl and 4 or 5 rows on last whorl) which are also axially aligned; encircled with spiral striae and rows of minute pits. Columella has about 6 denticles; thickened outer lip has about 6 elongate teeth. Orange, sometimes greyish orange with reddish-brown nodules; aperture white. **Habitat:** beached. **Distribution:** Mas, SO.

Nassarius (Niotha) obesus (Nevill & Nevill, 1875): **557** NOT ILLUSTRATED
A junior synonym is *Nassa (Alectryon) eranea* Melvill & Standen, 1901, under which name it was recorded from "Persian Gulf (no precise locality)". No more information is available.

Nassarius (Niotha) jactabundus **555**

Nassarius (Niotha) nodulosus **556**

Nassarius (Niotha) splendidulus **558**

Nassarius (Niotha) splendidulus (Dunker, 1846): **558**
15mm. Lightweight, glossy, elongate-ovate, whorls convex; protoconch of 3.5 glassy whorls. Surface covered with quadrate nodules which align axially and spirally. Whorls stepped at sutures. Columellar callus narrow or spreading over last whorl. White, cream or beige, weakly or prominently banded and streaked brown; aperture and callus white. **Habitat:** offshore and intertidal. **Distribution:** SEG, GO, Mas.

Nassarius (Plicarcularia) fissilabris (A. Adams, 1852): **559**
18mm. Solid, elongate-ovate; protoconch of 1.5 smooth, glassy whorls. Spire whorls stepped and coronate; strong axial ribs become obsolete on lower half of last whorl, crossed by regular, deep spiral grooves. Nodules at base of columella; aperture coarsely lirate. Columellar callus reaches penultimate whorl; thick callus above deep posterior canal. White to

Nassarius (Plicarcularia) fissilabris **559**

Nassarius (Niotha) deshayesianus **553**

yellowish, usually with complete or interrupted, reddish brown spiral bands; callus yellowish; protoconch golden. **Habitat:** intertidal in sand. **Distribution:** NG, SEG, GO, Mas.

Nassarius (Plicarcularia) persicus (Martens, 1874): **560**
18mm. Lightweight, elongate-ovate; protoconch of 3 glassy whorls. Stepped spire whorls have strong axial ribs, constricted at sutures to form coronations but becoming obsolete on lower half of last whorl where spiral grooves are present. Columella has basal plication. Columellar callus reaches penultimate whorl. White to pale grey with 1 or 2 spiral dark brown bands (visible in aperture); callus and lip white. **Habitat:** intertidal muddy sand. **Distribution:** NWG, GO, Mas.

Nassarius (Plicarcularia) persicus **560**

Nassarius (Niotha) himeroessa **554**

Superfamily: **MURICOIDEA**

Nassarius (Telasco) marmoreus (A. Adams, 1852): **561**
25mm. Solid, elongate-ovate, spire whorls gently rounded; protoconch of about 1.5 smooth whorls. Early spire whorls axially ribbed, later ones smooth except for spiral cords at base. Deep posterior canal. Callused columella has 2 or more weak folds; aperture has short spiral lirae near lip. White heavily blotched, spotted and netted dark brown; callus by posterior canal brown; columella paler. Covered by reddish-brown periostracum when fresh. **Habitat:** under intertidal rocks and offshore. **Distribution:** GO, Mas.

Nassarius (Zeuxis) castus (Gould, 1850): **562**
23mm. Thick, elongate-ovate, short pointed spire, deep or tabulated sutures; protoconch of about 3 smooth, glassy whorls. Narrow or broad, rounded, axial ribs, widely spaced on last whorl with large, rounded nodules at tops and crossed by widely spaced, spiral grooves. Columella has raised rim and weak folds; aperture lirate. Cream to shades of brown, usually banded; aperture dark brown; columella and outer lip glazed white. **Habitat:** sandy mud offshore. **Distribution:** GO.

Nassarius (Zeuxis) concinnus (Powys, 1835): **563**
12mm. Solid, elongate-ovate, almost straight sided, last whorl gently rounded; protoconch of about 3 glassy whorls, the last gently carinate. Flattened, slightly curved axial ribs; fine spiral striae, sometimes restricted to intervals between ribs; subsutural groove on all spire whorls. Up to 13 denticles on columella; up to 10 denticles in aperture. White to pale brown with darker spiral bands (3 on last whorl). **Habitat:** in sand offshore. **Distribution:** GO.

Nassarius (Zeuxis) frederici (Melvill & Standen, 1901): **564**
13mm. Lightweight, moderately glossy, elongate-ovate, spire whorls rounded and stepped, sometimes channelled; protoconch of about 3 glassy, keeled whorls. Crescentic axial ribs as wide as their intervals, crossed by widely spaced spiral grooves. Thickened outer lip corrugated at edge and notched near base. Whitish with 2 pale brown, spiral bands. Syn: *N. townsendi* (Melvill, 1897). **Habitat:** offshore to deep water. **Distribution:** NG.

Nassarius (Zeuxis) mammilliferus (Melvill, 1897): **565**
10mm. Thick, elongate-ovate, spire whorls stepped and almost straight sided; domed protoconch of about 2.5 frosted whorls. Narrow axial ribs crossed by flattened spiral cords, noduled at intersections; occasionally a varix. Columella has 3 or 4 denticles basally; about 8 denticles within thickened outer lip. Off-white sometimes spirally banded brown (about 3 bands on last whorl). **Habitat:** muddy sand offshore. **Distribution:** NG, GO.

Nassarius (Zeuxis) pseudoconcinnus (Smith, 1895): **566**
15mm. Lightweight, dull, elongate-ovate with gently rounded spire whorls; protoconch of 3.5 glassy whorls. Slender, close-set, crescentically axial ribs with regular, spiral grooves between them, a deeper groove cutting the ribs below sutures. Outer lip thickened, corrugated below, bearing 9-11 denticles within. White to fawn. Syn: *N. sturtiana* (Melvill & Standen, 1901) and possibly *N. idyllius* (Melvill & Standen, 1901). **Habitat:** offshore to deep water. **Distribution:** GO.

Subfamily DORSANIINAE

Bullia (Bullia) mauritiana Gray, 1839: **567**
44mm. Solid, glossy, high spired with flat-sided whorls and sloping platform below each suture; pointed protoconch. Callus ridge above sutures and 2-3 spiral grooves on each spire whorl, more on last whorl (grooves sometimes obsolete). Fasciolar ridge may be sharp. Pale brown to cream often with chocolate stain on fasciolar ridge and by posterior canal; aperture shades of brown. **Habitat:** intertidal and offshore in sand. **Distribution:** SEG, GO, Mas, SO.

Bullia (Bullia) melanoides Deshayes in Belanger, 1832: **568**
24mm. Solid, dull, narrowly elongate, spire whorls becoming rounder with growth, well-defined sutures; large, pointed protoconch of about 3 smooth whorls; 2nd post-nuclear whorl broader than 3rd. Strong axial ribs become obsolete on all but tops of later whorls; close-set spiral threads on all whorls

Superfamily: **MURICOIDEA**

Bullia (Bullia) persica 570

becoming cords at base. Brown to purple-brown with pale subsutural band; apical whorls and base yellowish white; aperture reddish brown. Syn: *B. cumingiana* Dunker, 1856, *B. strenaria* Melvill, 1904. **Habitat:** in sand. **Distribution:** NG, GO.

Bullia (Bullia) otaheitensis (Bruguière, 1789): **569**
40mm. Solid, high spired with rounded whorls; protoconch not seen. Crowded, regular, flattened spiral ribs crossed by fine, irregular axial threads which develop points on ribs below sutures. Strong fasciolar ridge; outer lip corrugated. Greyish violet with occasional brown streaks; ribs paler; aperture dark brown. **Habitat:** in sand. Syn: *B. tahitensis* (Gmelin, 1791). **Distribution:** NEG, Mas, SO.

Bullia (Bullia) persica Smith, 1878: **570**
25mm. Thin, dull, inflated last whorl, spire same length as aperture; protoconch of 2 dull, nipple-like whorls. Smooth callus cord at sutures, continuous with parietal callus. Spiral grooves on spire whorls, becoming partially obsolete on last whorl, crossed by coarse axial grooves; sharp fasciolar ridge. Beige, yellow, purple or white; apex purple. **Habitat:** intertidal in sand. **Distribution:** NG.

Bullia (Bullia) rogersi Smythe in Smythe & Chatfield, 1981: **571**
20mm. Lightweight, glossy, with rounded whorls, spire slightly longer than aperture; mammillate protoconch. Thick, smooth callus cord at sutures. Spiral grooves on early whorls become mostly obsolete on later whorls; up to 6 fasciolar cords; outer lip smooth. Ivory with brown axial lines and subsutural streaks; columella white; aperture white to pale brown. **Habitat:** in sand offshore and beached. **Distribution:** Mas.

Bullia (Bullia) rogersi 571

Bullia (Bullia) semiplicata Gray in Griffith & Pidgeon, 1834: **572**
30mm. Lightweight, glossy, with straight-sided, stepped whorls, the early whorls needle like; protoconch blunt topped. Thick, lumpy callus cord at sutures. Strong axial ribs on spire whorls become obsolete on last whorl. Low fasciolar ridge; outer lip smooth. White, yellowish or shades of brown inside and out, occasionally spirally banded, columella paler. **Habitat:** in sand. **Distribution:** Mas, SO.

Bullia (Bullia) smytheae Moolenbeek & Dekker, 1994: **573**
14mm. Lightweight, glossy, elongate-ovate, about twice as long as wide; protoconch mammillate. Smooth callus at sutures channelled below. Fine, widely separated spiral grooves on all whorls. About 7 fasciolar cords; outer lip smooth. Cream or white with pale brown zigzag markings; aperture and sutural callus white. **Habitat:** in sand offshore and beached. **Distribution:** Mas.

Bullia (Bullia) otaheitensis 569

Bullia (Bullia) semiplicata 572

Bullia (Bullia) smytheae 573

Superfamily: **MURICOIDEA**

Bullia (Bullia) tranquebarica (Röding, 1798): **574**
18mm. Lightweight, elongate-ovate, about twice as long as wide; pointed protoconch. Irregular spiral grooves on all whorls. About 5 fasciolar cords; outer lip smooth. Smooth callus cord at sutures, channelled below. Yellowish to pale brown with darker brown, wavy, axial streaks; callus cord and columella white. Axial brown streaks sometimes lacking and larger shells may be paler and have more grooves. **Habitat:** in sand. **Distribution:** NWG, SEG, GO

Bullia (Bullia) tranquebarica 574

Family MELONGENIDAE
The crown conchs inhabit tropical or sub-tropical mud flats or sandy estuaries where they prey on other molluscs. Some members of this small family have large, heavy shells with a prominent siphonal canal and the animal usually has a thick, chitinous operculum. Only two species recorded.

Taphon striatum (Sowerby, 1833): **575**
50mm. Lightweight, short spired, with inflated and roundly tabulated later whorls; protoconch raised but usually missing. Columella smooth, straight; siphonal canal straight or curved. Regular spiral cords with fine spiral threads between; aperture lirate. Amber with brown, axial streaks, the brown tint often limited to spiral cords; aperture and columella white; sometimes uniformly white. Systematic position uncertain. **Habitat:** offshore and beached. **Distribution:** Mas, SO.

Taphon striatum 575

Volema paradisiaca nodosa (Lamarck, 1822): **576**
50mm. Thick, pear shaped, with wavy and deep sutures, silky gloss; protoconch raised but usually encrusted. Last whorl lightly or heavily coronate at shoulder, otherwise smooth except for low spiral cords at base. Columella smooth and slightly sinuous; outer lip smooth and sharp edged. Spirally banded chocolate, orange and greyish white, with occasional white axial streaks; columella white. **Habitat:** intertidal on sand. **Distribution:** SO.

Family FASCIOLARIIDAE
The genera placed in this family differ from each other by their very different shells but the animal inhabitants are all active predators. The larger species occur in sandy places while many of the smaller ones live among coral and rocks. The single species of *Fasciolaria* in eastern Arabia is the only one with a large, thick, heavy shell. The genus *Colubraria*, with heavily variced shells, is poorly represented, presumably because its members live offshore. Some of the species of *Latirus* and *Peristernia* are peculiar to the area. One species, belonging to the genus *Sinistralia*, has a sinistral shell. The claw-like, chitinous operculum has an apical nucleus.

Volema paradisiaca nodosa 576

Subfamily FASCIOLARIINAE
Pleuroploca trapezium (Linnaeus, 1758): **577**
200mm. Very thick and heavy, high spired with pointed protoconch (usually encrusted). Spire whorls encircled by large or small, rounded tubercles; coarse growth lines cover surface. Edge of outer lip often has prickles. Smooth columella has 3 folds at base; siphonal canal long and broad. Orange-brown often with darker brown spiral lines; aperture may have faint lirae; columella purple-brown or orange. May grow to 300mm. **Habitat:** intertidal on sand. **Distribution:** GO, Mas, SO.

Subfamily FUSININAE
Fusinus arabicus (Melvill, 1898): **578**
90mm. Lightweight, narrowly spindle shaped with rounded or slightly angulate spire whorls; bulbous protoconch of 1.5 whorls. Axial ribs on spire whorls crossed by thin spiral ridges which are raised on ribs at periphery. Columella smooth; inner lip raised. Brown with paler spiral ridges; columella and aperture white. Variable in shape and colour pattern. Syn: possibly a form of ***Fusinus colus*** (Linnaeus, 1758). **Habitat:** offshore and intertidal. **Distribution:** NWG, SEG, GO, Mas.

Fusinus arabicus 578

Superfamily: **MURICOIDEA**

Pleuroploca trapezium 577

SEASHELLS OF EASTERN ARABIA

Superfamily: **MURICOIDEA**

■ *All members of the genus Sinistralia display a peculiarity unusual among marine gastropods: they are sinistral, their shells coiling in an anti-clockwise direction with the aperture opening on the left when the protoconch is uppermost. Most gastropods have dextral shells, the aperture opening on the right when similarly oriented. About a hundred normally dextral species are known to have produced sinistral monstrosities, notably among the marginellids.* Sinistralia gallagheri *is one of the much rarer instances of a normally sinistral species producing a dextral monstrosity. Indeed, the first specimen found may have been dextral! Not until sinistral examples turned up were specialists convinced that here was a species new to science. Many sinistrally coiled examples have been found since the species was described in 1981 — but the dextral monstrosity remains unique.*

Sinistralia gallagheri 581

Fusinus townsendi 580

Fusinus forceps (Perry, 1811): **579**
140mm. Solid, spindle shaped with well-rounded spire whorls. Broad, rounded axial ribs crossed by sharp spiral ridges. Columella smooth, inner lip not raised; aperture strongly lirate. White, covered by fibrous, pale brown periostracum when fresh. **Habitat:** amongst loose rocks and coral sand offshore and beached. **Distribution:** NG, GO.

Fusinus townsendi (Melvill, 1899): **580**
70mm. Thick, spindle shaped with angulate whorls; protoconch of 2 bulbous whorls. Axial ribs well developed mid-whorl becoming almost plate-like at periphery where crossed by sharp spiral ridges. Columella smooth, inner lip slightly raised; aperture lirate. Pale brown to purplish brown and covered by a thin, fibrous, pale brown periostracum when fresh. Often encrusted by *Lepralia*, a polyzoan. **Habitat:** offshore. **Distribution:** NWG, NG, SEG.

Sinistralia gallagheri (Smythe & Chatfield, 1981): **581**
18mm. Solid, sinistral, biconic, with impressed sutures; protoconch flat topped. First 2 spire whorls axially ribbed, later whorls with sinuous axial ribs crossed by coarse spiral cords, noduled at periphery of whorls. Columella smooth, sinuous; outer lip thin edged. Purplish brown with whitish and reddish brown ribs; columella white. A single dextral example known (illustrated at right of photo). **Habitat:** under rocks or in reef crevices at extreme low tide level. **Distribution:** Mas.

Superfamily: **MURICOIDEA**

Fusinus forceps 579

Superfamily: MURICOIDEA

Subfamily PERISTERNIINAE
Latirus bonnieae Smythe, 1985: **582**
50mm. Thick, fusiform, twice as long as broad; sutures shallow; protoconch usually eroded. Nodular axial ribs crossed by up to 4 spiral ridges, uppermost ridge very prominent; coarse growth lines between nodes. Base of columella has 2 or 3 plications; aperture lirate; sometimes with false umbilicus. Reddish brown with cream ridges; aperture apricot; apical whorls white. Usually found worn. **Habitat:** offshore and beached. **Distribution:** GO, Mas.

Latirus pagodaeformis Melvill, 1899: **583**
30mm. Solid, dull, elongate-fusiform, 3 times as long as broad with long, recurved siphonal canal; sutures deep; protoconch of 3.5 glossy whorls. Nodular axial ribs crossed by alternately thick and thin spiral ridges. Base of columella has 4 weak plications. Beige with brown blotches on ribs and spiral band on last whorl; aperture pale violet. **Habitat:** deep water. **Distribution:** GO.

Latirus pulchellus (Reeve, 1847): **584**
28mm. Solid, dull, elongate-fusiform, twice as long as broad, with slightly recurved siphonal canal; sutures deep; large protoconch of 4.5 glossy whorls. Nodular axial ribs crossed by regular, scaly spiral ridges. Columella smooth. Beige or whitish with orange-brown spiral band on spire whorls (2 on last whorl), ribs mostly white; aperture and columella crimson fading to pink. **Habitat:** offshore. **Distribution:** GO.

Peristernia nassatula forskalii (Tapparone-Canefri, 1875): **585**
28mm. Thick, biconic, aperture half total length, sutures impressed. Axial ribs broadest at periphery, crossed by sharp spiral ridges. Columella has 2 or 3 weak folds at base; aperture lirate; outer lip thin, corrugated. Siphonal canal slightly recurved. White with broad, orange-brown, spiral bands; aperture purple. **Habitat:** under intertidal rocks. **Distribution:** GO, Mas, SO.

Peristernia cf *rhodostoma* (A. Adams, 1855): **586**
38mm. Thick, biconic, aperture shorter than spire; blunt-topped protoconch of 3 smooth whorls. Widely spaced, rounded, axial ribs crossed by crowded, scaly ridges. Blunt tooth at base of smooth columella; aperture weakly lirate; outer lip corrugated; false umbilicus wide and deep. Cream with broad, orange-brown, spiral bands; aperture violet; protoconch orange brown. Possibly an undescribed species. **Habitat:** offshore and beached. **Distribution:** Mas.

Peristernia tayloriana (Reeve, 1847): **587**
23mm. Solid, fusiform with angulate whorls, aperture less than half total length; pointed protoconch of 3 axially ribbed whorls. Axial ribs crossed by strong, scaly, spiral ridges form sharp nodes at periphery. Columella smooth and sinuous; outer lip corrugated. False umbilicus shallow. Brown with cream nodes; aperture purplish or violet; protoconch reddish brown. **Habitat:** intertidal under stones. **Distribution:** Mas.

Subfamily COLUBRARIINAE
Colubraria antiquata (Hinds, 1844): **588**
22mm. Thin, fusiform, whorls rounded but shell superficially straight sided; protoconch usually missing. Each whorl has 2 thin varices; gently curved axial ribs crossed by coarse spiral riblets. Elongate aperture with weak denticles on outer lip; small, recurved siphonal canal. Whitish or yellow with brown bar at centre of each varix. **Habitat:** offshore on sand. **Distribution:** GO.

Peristernia tayloriana 587

Peristernia cf *rhodostoma* 586

Latirus pulchellus 584

Peristernia nassatula forskalii 585

Colubraria antiquata 588

Latirus bonnieae 582

Latirus pagodaeformis 583

Superfamily: **MURICOIDEA**

Colubraria soverbii **590**

Colubraria ceylonensis **589**

Acamptochetus daphnelloides **591**

Lyria lesliehoschae **592**

Colubraria ceylonensis (Sowerby, 1833): **589**
40mm. Thin, fusiform, almost straight sided, whorls coiling irregularly, sutures shallow; pointed protoconch of 4.5 smooth, glossy whorls. Each whorl has 2 broad varices; gently curved axial ribs crossed by spiral grooves produce beaded ornament; varices weakly beaded. Columella smooth; outer lip has a rounded notch towards base; outer lip weakly lirate. Greyish brown with orange-brown spiral bands, darker brown on varices; aperture and protoconch creamy brown. Syn: *C. concinnata* (Melvill, 1904). **Habitat:** offshore among rocks and coral and beached. **Distribution:** NG, GO, Mas.

Colubraria soverbii (Reeve, 1844): **590**
75mm. Thick, fusiform, aperture less than half total length, whorls coiling irregularly, sutures shallow; protoconch not seen. Each whorl has 2 broad, flattened varices; low axial ribs crossed by well incised grooves produce coarse, prickly ornament; varices smoother. Columella weakly pimpled; outer lip smooth edged, lirate within. Cream mottled and spotted pale and darker brown with grooves pencilled dark brown; aperture pinkish orange. **Habitat:** caves and crevices offshore and beached. **Distribution:** GO, Mas.

Acamptochetus daphnelloides (Melvill & Standen, 1903): **591**
15mm. Solid, silky, elongate-fusiform, aperture half total length, whorls gently convex, sutures deep, siphonal canal broad; large protoconch of 4 or 5 whorls, the first with 1 keel, the others with 2 keels. Spire whorls have sinuous, close-set, axial ribs with spiral threads between them; prominent subsutural cord. Greyish white with beige spiral bands (2 on last whorl). Distantly resembles a turrid. **Habitat:** offshore. **Distribution:** NG.

Family VOLUTIDAE

The volutes comprise a large family of carnivorous gastropods, mostly found in warm waters. Their conspicuously attractive shells have made them popular with collectors and several books have been written about them. The animals, too, are often strikingly marked in bold colours. As they thrive best in nutrient-rich environments, such as extensive coral reefs within the tropics, it is not surprising that only two species have been recorded from eastern Arabia. The fusiform to oblong shell has a short siphonal canal and the columella usually has strong plications. A small, chitinous operculum is usually present.

Subfamily LYRIINAE

Lyria lesliehoschae Emerson & Sage, 1986: **592**
110mm. Moderately thick, elongate-fusiform, last whorl more than half total length, sutures indented; protoconch of 2.5 smooth whorls. Axial ribs weakly developed on last whorl, about 21 on penultimate whorl, more prominent on earlier whorls; 6 spiral lirae below sutures. Columella has 3 basal folds; outer lip thin, smooth; posterior canal long, narrow. Cream with interrupted reddish brown spiral bands and spiral lines on ribs; aperture and columella cream. Known only from a few beached shells, it may live in moderately deep water. **Habitat:** beached. **Distribution:** Mas.

■ *Of all the seashells recorded from eastern Arabia* Lyria lesliehoschae *may be the one that knowledgable collectors covet most and are least likely to own. For serendipity plays a large part in successful shell collecting. The observant collector who is in the right place at the right time may still overlook the treasure lying at his feet, because he lacks the happy faculty of making unexpected chance finds. Donald Bosch has never lacked that faculty. On what proved to be a memorable day in January 1985 he was shelling at Haql on the Indian Ocean side of Masirah Island after heavy surf had been pounding the beach. Among the shells he picked up were four volutes resembling* Festilyria festiva, *a species often stranded on Masirah's beaches. They were different enough, however, for him to send them to a specialist in New York. A year later* Lyria lesliehoschae *was described as new to science. In June 1994 Martyn Day found an attractive specimen cast up on a beach near Haql, a few kilometres from the original site. Again, serendipity was at work.*

Festilyria festiva 593

Superfamily: **MURICOIDEA**

Subfamily FULGORARIINAE
Festilyria festiva (Lamarck, 1811): **593**
85mm. Thick, moderately glossy, spire tall or short, last whorl more than half total length, sutures impressed; bulbous protoconch of about 2 smooth whorls. Strong axial ribs, lacking above periphery of last whorl and sometimes absent entirely, with growth lines between. Columella has plications for most of its length; outer lip slightly thickened, smooth. Ivory blotched reddish brown and encircled by about 6 double, spiral lines pencilled in brown; base of columella chocolate; aperture pinkish orange; protoconch ivory stained reddish brown. Operculum present. The slender form has been named *F. deceptrix* Palazzi, 1981.
Habitat: offshore and beached.
Distribution: Mas.

Family HARPIDAE
Carnivorous or scavenging sand dwellers, the harps are distinguished for the beauty of their shells. They are almost exclusively tropical and although dwelling intertidally are not easily detected in the sand in which they may bury themselves almost completely. The animal, which lacks an operculum, has a large, crescent-shaped foot which it may detach when molested, leaving the detached portion wriggling independently. Each of the few species has a thick, glossy, usually strongly ribbed and highly coloured shell with a large aperture, a smooth, glazed columella and no umbilicus. Only two species have been recorded from eastern Arabia.

Harpa amouretta Röding, 1798: **594**
50mm. Thick, ovate, aperture narrow, spire short; protoconch of 4 smooth whorls. Flattened axial ribs (about 10-12 on last whorl), pointed below sutures, last whorl with fine axial threads between. Pale and dark brown bars and lines on ribs with bars, zigzags and chevrons between; large brown blotch on columella; aperture bluish white; early whorls violet. **Habitat:** intertidal in sand.
Distribution: Mas.

Harpa ventricosa Lamarck, 1801: **595**
65mm. Thick, globose-ovate, aperture wide open, spire short; protoconch of 4 smooth whorls. Flattened, recurved axial ribs (10-19 on last whorl), pointed below sutures, last whorl with coarse axial threads between. Reddish and violet-brown bars on ribs with bars, zigzags and chevrons between; large brown blotch on columella, another on parietal wall, cream area between; aperture cream; early whorls violet. **Habitat:** intertidal in sand. **Distribution:** Mas, SO.

Family VASIDAE
Small family of carnivorous gastropods with a characteristically thick, heavy shell with strong folds on the columella, the animal bearing a chitinous, claw-like operculum. Widespread in sandy and stony places in the tropical Pacific, the family has one species in eastern Arabia.

Vasum turbinellus (Linnaeus, 1758): **596**
40mm. Thick, heavy, aperture twice as long as spire; protoconch always eroded. Coarse axial ribs with short or long, bluntly pointed or upturned spines are well developed at shoulder of last whorl. Columella has 4 or 5 folds. No umbilicus. Cream or white interrupted, spiral, chocolate bands; aperture and columella orange-yellow. Record based on a beached shell from Masirah; species is usually adult at about 75mm. **Habitat:** Intertidal and beached. **Distribution:** Mas.

Harpa amouretta 594

Harpa ventricosa 595

Vasum turbinellus 596

SEASHELLS OF EASTERN ARABIA

Superfamily: **MURICOIDEA**

Oliva bulbosa 597

Ancilla (Chilotygma) exigua exigua 599

Family OLIVIDAE
A very large family of carnivorous and scavenging gastropods abounding in sand or mud in many tropical and sub-tropical areas. The foot is large and may envelop the shell entirely. An operculum is present in some genera (but not in the genus *Oliva*). All species have a smooth and usually very glossy shell with an aperture which may be narrow (as in *Oliva*) or wide (as in *Ancilla*). Shell shape varies from obese to slender but is constant within each species. Conversely, the colour pattern may vary greatly within a species. Protoconch and columellar plications are useful characters for identification, as they are constant in form and disposition within each species. There is no umbilicus. A feature of the shell in this family, conspicuous in *Ancilla*, is the "fasciolar band" (with associated grooves) winding around its lower portion. Typical of *Ancilla*, too, is the glazing of the spire and the consequent absence of sutures. In species of *Oliva* the shell is probably adult if the outer lip is thickened. At least one eastern Arabian *Ancilla* is endemic.

Subfamily OLIVINAE
Oliva bulbosa (Röding, 1798): **597**
45mm. Thick, heavy, ovate, swollen centrally in older shells, short spired with deeply grooved sutures and smoothly corrugated callus between them. Columella plicate in juveniles, heavily callused in older shells with 2 prominent ridges; outer lip greatly thickened in adults. Colour and pattern very variable, white, cream, greyish tan, greenish brown, brown or black, with dots, blotches, zigzags, spiral bands, streaks; columella white tinged brown on ridges. Juvenile shells of this protean species may resemble other species. **Habitat:** intertidal and offshore in sand. **Distribution:** SEG, GO, Mas, SO.

Oliva tremulina Lamarck, 1811: **598**
70mm. Thick, elongate-ovate, short spired with deeply grooved sutures and smoothly corrugated callus between them. Columella plicate throughout its length, plications prominent around its base. Creamy yellow with 2 spiral bands of reddish or greyish brown and irregular, paler blotches and tent-like markings; spire whorls have axial brownish streaks; aperture and columella creamy. **Habitat:** offshore in sand. **Distribution:** GO

Subfamily ANCILLINAE
Ancilla (Chilotygma) exigua exigua (Sowerby, 1830): **599**
8mm. Thick, oblong-ovate, spire very low, blunt topped, siphonal canal very shallow and broad. Columella has 2-3 strong plications; axial groove in parietal wall just within aperture with callus below. Whitish, pale orange, or brownish, spire greyish. **Habitat:** intertidal in sand. **Distribution:** Mas.

Ancilla (Chilotygma) testudae Kilburn, 1977: **600** NOT ILLUSTRATED
12mm. Characterised by a transversely ridged columella. Known from eastern Arabia by a single shell from the F. W. Townsend collection in the Natural History Museum London, labelled merely "Persian Gulf". Confirmation of its presence in eastern Arabia is required.

Superfamily: **MURICOIDEA**

Ancilla (Sparella) boschi Kilburn, 1980: **601**
25mm. Lightweight, elongate-biconic, aperture twice as long as spire and wider anteriorly. Upper fasciolar groove ends in small denticle by broad siphonal canal. Columella has 2-3 plications anteriorly. Orange-yellow to pale chocolate with white subsutural band and upper fasciolar groove; columella white, protoconch and aperture pale brown or white. **Habitat:** intertidal in sand. **Distribution:** Mas, SO.

Ancilla (Sparella) castanea (Sowerby, 1830): **602**
30mm. Solid, oblong-ovate, aperture more than twice as long as spire and almost straight sided centrally. Upper fasciolar groove (visible as ridge internally) ends in conspicuous denticle by very broad siphonal canal. Columella has up to 6 strong or weak plications. Dark brown to yellowish, protoconch and upper fasciolar groove paler, columella white or yellowish; sometimes a whitish callus by posterior canal. Entirely white forms occur rarely. **Habitat:** intertidal in sand. **Distribution:** NWG, SEG, GO, Mas, SO.

Ancilla (Sparella) djiboutina (Jousseaume, 1894): **603**
25mm. Solid, bullet shaped, almost straight sided, spire short and pointed. Upper fasciolar groove ends in blunt denticle by broad siphonal canal. Columella has 3-5 plications. Pale brown often with darker axial streaks, spire banded pale brown; upper fasciolar groove and columella white, aperture pale brown. **Habitat:** offshore in sand. **Distribution:** GO, Mas.

Ancilla (Sparella) farsiana Kilburn, 1981: **604**
15mm. Solid, cylindrical, short spired, aperture usually twice length of spire, almost straight sided centrally. Upper fasciolar groove ends in sharp denticle. Columella has 3-8 plications; axial groove in parietal wall just within aperture. White to orange-yellow. **Habitat:** intertidal and offshore in sand. **Distribution:** NG, NWG, SEG, GO, Mas.

Ancilla (Sparella) castanea **602**

Ancilla (Sparella) castanea **ploughing through sand and leaving trails.**

Ancilla (Sparella) djiboutina **603**

Ancilla (Sparella) farsiana **604**

Ancilla (Sparella) boschi **601**

Oliva tremulina **598**

SEASHELLS OF EASTERN ARABIA

Superfamily: **MURICOIDEA**

Canalispira replicata **609**

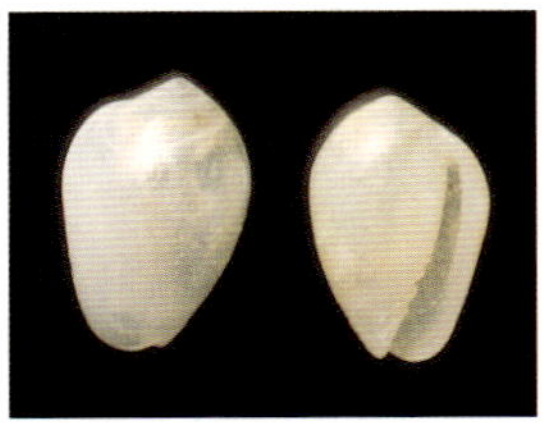

Ancilla (Sparella) inornata (Smith, 1879): **605** NOT ILLUSTRATED
Differs from **A. djiboutina** (see above) by its smaller protoconch, shorter columella, posteriorly narrower aperture and longer, more pointed spire. Known from eastern Arabia by a single shell from Muscat in the Natal Museum, ex W. Falcon collection.

Ancilla (Sparella) scaphella (Sowerby, 1859): **606**
28mm. Thin, lightweight, cylindrical-ovate, spire very low and nipple-like, outer lip widening towards base, siphonal canal shallow and broad. Upper fasciolar groove shallow, no terminal denticle. Columella has 3-5 weak plications. White often tinged orange-brown on spire. **Habitat:** intertidal in sand. **Distribution:** SEG, GO.

Ancilla (Sparellina) ovalis (Sowerby, 1859): **607**
15mm. Solid, oblong-ovate, spire low, blunt topped, siphonal canal very shallow and broad. Upper fasciolar groove ending in prominent denticle (often eroded). Columella has 2-4 plications. Greyish white to yellowish with pale brown, wavy, axial lines and pale brown spiral lines; fasciolar band entirely or intermittently streaked brown; upper fasciolar groove whitish; columella white or brown; aperture orange-brown; protoconch pale brown. **Habitat:** intertidal and offshore in sand. **Distribution:** all.

Family MARGINELLIDAE
Widely distributed in tropical and subtropical waters, the margin shells are well represented in eastern Arabia, although most species are small or tiny. Some are known to be active predators which feed upon other molluscs. Many burrow just below the sand, some live under rocks. Most have smooth, glossy shells, slippery to the touch, with a narrow aperture, a thickened outer lip and a columella with oblique folds. The spire may be raised or sunken, according to the species. The collector appreciates the vivid colour pattern adorning the shells of many species and two or three of those occurring off the coast of Oman are strikingly marked. Most, however, are small, featureless and colourless. The identity of some of the smaller eastern Arabian species is still uncertain.

Subfamily CYSTISCINAE
Granulina oodes (Melvill, 1898): **608**
2mm. Solid, globular, distantly resembling a tiny cowry, aperture the length of shell and of equal width throughout, outer lip thickened and weakly toothed, its upper end covering spire; 4 columellar teeth. White. Syn: evidence for this being a junior synonym of **G. isseli** (G. & H. Nevill, 1875) is not convincing. **Habitat:** in sand intertidal and below. **Distribution:** NG, NWG, SEG, GO, Mas.

Subfamily MARGINELLINAE
Canalispira replicata Melvill, 1912: **609**
5.25mm. Described by Melvill, 1912:250, from 48 fathoms in the Arabian Gulf (NG).

Dentimargo alchymista (Melvill & Standen, 1903): **610**
5.5mm. Solid, translucent, fusiform, whorls well rounded, sutures distinct; aperture about twice length of spire, outer lip constricted by a prominent tooth; 4 columellar teeth. White or golden brown. Melvill and Standen included the varieties *chrysalchyma* and *leucalchyma* in their original description, both merely colour forms. **Habitat:** offshore and deep water. **Distribution:** GO.

Gibberula charbarensis (Melvill, 1897): **611**
8mm. Thick, cylindrical, spire very short, sides almost parallel, outer lip thickened and slightly inturned. Columella with up to 5 weak folds. White with grey zone around each spire whorl. **Habitat:** intertidal and offshore in muddy sand. **Distribution:** GO, Mas.

Gibberula* cf *mazagonica (Melvill, 1892): **612**
3mm. Thin, translucent, ovate, spire nipple-like, aperture equal width

Gibberula cf *mazagonica* **612**

Granulina oodes **608**

Ancilla (Sparellina) ovalis **607**

Dentimargo alchymista **610**

Ancilla (Sparella) scaphella **606**

Gibberula charbarensis **611**

throughout, outer lip sharp edged, smooth or minutely toothed within and thickened behind. About 6 columellar folds. Milky white. **Habitat:** in sand intertidal and below. **Distribution:** NG, NWG, SEG, GO, Mas.

Gibberula sueziensis (Issel, 1869): **613** NOT ILLUSTRATED
Listed by Glayzer et al, 1984:322, from Kuwait (NWG) and by Smythe, 1979a:68, from U. A. E. (SEG).

Marginella obtusa Sowerby, 1846: **614**
24mm. Thick, broadly biconical, spire a third total length, margined outer lip weakly toothed, thickest centrally. Elongate coronations on spire whorls and shoulder of body whorl. Yellowish white with zigzag, violet-brown markings overlain by spiral rows of tiny, dark brown dots; outer lip white or pale violet, brown spotted; aperture and columellar teeth white. Syn: *M. guillaini* (Petit, 1851). **Habitat:** in sand at low tide. **Distribution:** GO, Mas, SO.

Persicula brinkae Bozetti, 1993: **615**
12mm. Solid, cylindrical-ovate, spire intorted and glazed, outer lip thickened, slightly inturned centrally. Columella has oblique folds, outer lip seemingly toothed within but teeth are actually ends of apertural lirae. Yellow-orange with 3 darker, spiral zones and fine, wavy, reddish-brown lines from apex to base, apex reddish brown; aperture, teeth and columellar folds white. **Habitat:** beached. **Distribution:** Mas, SO.

Persicula masirana Roth & Petit, 1972: **616**
9mm. Solid, globose-ovate, spire intorted and glazed, outer lip thickened, straight centrally. Central and lowest columellar folds enlarge greatly outside of aperture. Beige mottled whitish, overlain by encircling lines of elongate or rounded, white blotches alternating with smaller, orange-brown spots; spire glaze reddish brown; lip, columella and folds white. **Habitat:** intertidal in sand. **Distribution:** Mas, GO.

Prunum terverianum (Petit, 1851): **617**
10mm. Thick, broadly ovate, spire heavily callused, apex rounded, aperture widening towards base, outer lip greatly thickened when adult, its inner edge tending to obstruct aperture centrally. About 6 columellar teeth. Yellowish white. Syn: may be a junior synonym of *Gibberula monilis* (Linnaeus, 1758). **Habitat:** intertidal in sand. **Distribution:** GO, Mas.

Volvarina amydrozona (Melvill, 1906): **618**
8mm. Thin, translucent, cylindrical with slightly convex sides, spire totally immersed in callus at top of outer lip; aperture the length of shell and narrowest centrally, outer lip only slightly thickened, gently curved in profile; 3 columellar teeth. Straw coloured or greenish white with opaque, white, axial lines showing through, aperture and outer lip white. **Habitat:** offshore. **Distribution:** GO, Mas.

Persicula brinkae **615**

Marginella obtusa **614**

Persicula masirana **616**

Volvarina amydrozona **618**

Prunum terverianum **617**

Superfamily: **MURICOIDEA**

Volvarina eumorpha Melvill, 1906: **619**
9mm. Solid, cylindrical, spire very short, outer lip thickened and slightly concave. Columella with 4 strong folds. Yellowish to pale violet-orange with 3 obscure, darker, spiral zones. **Habitat:** offshore. **Distribution:** GO.

Volvarina cf *obscura* (Reeve, 1865): **620**
15mm. Solid, oblong-ovate to cylindrical, spire nipple-like, margined outer lip straight for most of its length, aperture wide basally. Pale yellowish brown spirally banded reddish brown, sometimes glazed blue-grey; columellar folds whitish, margin creamy conspicuously banded reddish brown. **Habitat:** under rocks in sand low tide and below. **Distribution:** GO, Mas, SO.

Volvarina pergrandis Clover, 1974: **621**
18mm. Thick, oblong-ovate, spire nipple-like, margined outer lip thickest below mid-whorl level, aperture wide basally. Pale greyish violet with faint spiral bands of greyish rose, overlain by whitish flecks and streaks; spire and aperture rosy; margin creamy with 2 broad, violet-rose bands. **Habitat:** in sand low tide and below. **Distribution:** Mas.

Family MITRIDAE
Together with the Costellariidae, with which they were formerly united by systematists, members of this large family are known as mitres, alluding to the fancied resemblance of some of them to episcopal headgear. Shells have 3 to 10 columellar folds, according to the species and, with few exceptions, have spiral ribs and grooves. Colour patterns provide useful clues to identification but sculptural features are equally important. Few are smooth externally but the aperture in mature shells is always smooth inside (not lirate as in the Costellariidae). An operculum is never present. Sand and reef dwellers, most species appear to be carnivorous. Mitres abound in tropical environments and form a significant part of the molluscan fauna of eastern Arabia. We do not use subgeneric names here, preferring to raise subgenera to the status of genera, a preference having the twin merits of convenience and simplicity.

Subfamily MITRINAE
Mitra aurantia aurantia (Gmelin, 1791): **622**
30mm. Thick, dull gloss, broadly fusiform, sutures deep; protoconch not seen. Aperture more than half total length. Low, rounded, spiral ridges on all whorls. Yellowish to whitish brown, lower half of last whorl violet brown; aperture violet. **Habitat:** beached. **Distribution:** GO, Mas.

Volvarina cf *obscura* **620**

Volvarina eumorpha **619**

Volvarina pergrandis **621**

Mitra aurantia aurantia **622**

148   SEASHELLS OF EASTERN ARABIA

Superfamily: **MURICOIDEA**

Mitra aurantia subruppeli Finlay, 1927: **623**
22mm. Description as for ***M. aurantia aurantia*** (see above) but more slender, lower half of last whorl not noticeably darker, the whole shell being yellowish brown often with whitish spots arranged in roughly axially aligned groups; sometimes a pale brown subsutural zone. **Habitat:** offshore and beached. **Distribution:** Mas, SO.

Mitra bovei Kiener, 1838: **624**
35mm. Thick, moderately glossy, elongate-fusiform, spire whorls slightly convex or straight sided, aperture longer than spire, sutures incised. Regular spiral grooves and irregular, subsutural coronations sometimes extending as weak axial ribs. White with 2 greenish brown, spiral bands bearing whitish spots; spiral grooves often orange-brown; aperture brownish. **Habitat:** offshore and beached. **Distribution:** NWG, GO, Mas.

Mitra fasciolaris Deshayes in Laborde & Linant, 1834: **625**
22mm. Solid, narrowly elongate-ovate, aperture about half total length, spire whorls almost flat sided, last whorl gently rounded, sutures slightly channelled. Aperture flares slightly towards base. Regular, punctate, spiral grooves on all whorls. Yellowish, lower half of last whorl brown with faint yellowish spots. **Habitat:** beached. **Distribution:** GO.

Mitra floccata Reeve, 1844: **626**
50mm. Thick, moderately glossy, elongate-ovate, last whorl slightly shouldered, outer lip slightly concave, aperture longer than spire, sutures deep, apex always worn. All whorls wrinkled at sutures; earlier whorls weakly, spirally punctate. Orange-brown mottled and spotted greyish white, encircled by unequally spaced red lines; columella and aperture orange-brown, lip edge white. Syn: *M. vaticinator* Melvill, 1918 may be the same. **Habitat:** offshore and beached. **Distribution:** Mas.

Mitra mitra (Linnaeus, 1758): **627**
100mm. Heavy, silky, elongate-ovate, last whorl straight sided, aperture half total length, apex eroded. Thickened outer lip has serrated edge; columellar lip thin and raised basally. Early whorls have finely punctate spiral grooves; later whorls smooth. Cream with mostly rectangular, spirally arranged reddish or orange blotches; aperture apricot. **Habitat:** intertidal and offshore in sand. Largest eastern Arabian mitre. **Distribution:** GO, Mas, SO.

Mitra aurantia subruppeli 623

Mitra fasciolaris 625

Mitra floccata 626

Mitra bovei 624

Mitra mitra 627

Superfamily: **MURICOIDEA**

Nebularia fulvescens (Broderip, 1836): **628**
25mm. Solid, elongate-ovate, aperture half total length, spire almost straight sided, sutures deep. Spiral rows of weak, punctate grooves on all whorls, prominent or obsolete over much of last whorl. Dark or pale brown with narrow, pale spiral zone encircling whorls and ending just below last suture. Known from a shell found at Khor Fakkan by H. Kauch. **Habitat:** beached. **Distribution:** GO.

Nebularia luctuosa (A. Adams, 1853), **forma *rutila*** A. Adams, 1853: **629**
25mm. Solid, elongate-ovate, whorls gently convex, sutures deep, tops of whorls minutely crenulate. Spiral, punctate grooves on all whorls, visible even on worn shells. Dark to pale brown with paler subsutural zone; aperture white. **Habitat:** intertidal and offshore under stones. **Distribution:** GO, Mas, SO.

Nebularia punctostriata (A. Adams, 1855): **630**
40mm. Thick, dull, fusiform, aperture about same length as spire, whorls well rounded, sutures deep, apex eroded. Outer lip thickened below centre point, weakly crenulate. Whorls sometimes weakly coronate below sutures; evenly spaced, spiral, punctate grooves overall. Greenish brown with yellowish white flecks below sutures, grooves darker brown, columella white, aperture violet-brown. **Habitat:** intertidal in sand under rocks. **Distribution:** Mas, SO.

Nebularia sacerdotalis (A. Adams, 1853): **631** NOT ILLUSTRATED
A shell collected by Don Bosch at Masirah has been illustrated and identified as this by Turner, 1992; His illustration shows a worn shell similar to a large *N. punctostriata* (A. Adams, 1853) (see above).

Nebularia sanguinolenta (Lamarck, 1811): **632**
26mm. Solid, moderately glossy, elongate-ovate, aperture more than half total length, sutures incised, apex usually worn. Whorls slightly coronated and encircled by punctate, spiral grooves. White with reddish brown lines in grooves and broader axial streaks; aperture cream to yellow. **Habitat:** offshore and beached. **Distribution:** Mas.

Nebularia townsendi (Melvill, 1904): **633**
30mm. The chestnut-coloured, minutely punctate shell closely resembles that of ***N. luctuosa*** (A. Adams, 1853) which may be conspecific. Described from Muscat (GO).

Strigatella litterata (Lamarck, 1811): **634**
23mm. Thick, dull, broadly ovate, aperture twice as long as spire, sutures finely incised, apex eroded. Outer lip thickened centrally; spiral rows of minute punctures, obsolete on middle portion of last whorl. Cream or white with brown blotches, streaks and zigzags; aperture and columellar teeth white. Amber periostracum. **Habitat:** intertidal under rocks and coral. **Distribution:** GO, Mas, SO.

Strigatella nebrias (Melvill, 1895): **635**
27mm. Thick, dull, elongate-ovate, shoulders of later whorls slightly shouldered, sutures finely incised, aperture more than half total length. Outer lip thickened; columellar lip slightly raised. Shallow spiral grooves overall. White with brownish blotches, brown lines in grooves; aperture violet. Green periostracum. **Habitat:** intertidal under rocks. **Distribution:** Mas, SO.

Nebularia townsendi 633

Nebularia fulvescens 628

Nebularia sanguinolenta 632

Nebularia luctuosa 629 forma *rutila*

Strigatella nebrias 635

Nebularia punctostriata 630

Strigatella litterata 634

SEASHELLS OF EASTERN ARABIA

Superfamily: **MURICOIDEA**

Subfamily CYLINDROMITRINAE

Pterygia crenulata (Gmelin, 1791): **636**
26mm. Thin, dull, elongate-ovate, spire very short, last whorl gently convex, sutures incised. Punctate spiral grooves and weak subsutural crenulations on all whorls. White heavily blotched and streaked pale brown; aperture white. **Habitat:** offshore and beached. **Distribution:** GO.

Subfamily IMBRICARIINAE

Domiporta carnicolor (Reeve, 1844): **637** NOT ILLUSTRATED
Listed by Cernohorsky, 1991:93, from Oman but its presence there requires confirmation.

Domiporta filaris (Linnaeus, 1771): **638**
25mm. Thick, dull, elongate-ovate, aperture less than half total length, spire whorls convex and shouldered at sutures. Strong spiral cords (about 11 on last whorl) with 1 or 2 lesser cords between them, crossed by regular axial grooves which give beaded appearance to stronger spiral cords. Yellowish or whitish with orange-brown on stronger spiral cords; aperture white. A more slender form occurs sparingly. **Habitat:** offshore. **Distribution:** SEG, GO.

Domiporta granatina (Lamarck, 1811): **639**
60mm. Solid, dull, elongate-fusiform, aperture half total length, whorls shouldered, sutures shallow. Spiral cords with lesser cords between, cut by close-set, axial grooves; outer lip corrugated. Cream with broad, spiral, reddish brown bands (2 on last whorl) and finer spiral lines of similar colour elsewhere; aperture white. **Habitat:** offshore in sand. **Distribution:** GO.

Neocancilla circula (Kiener, 1838): **640**
35mm. Solid, lightweight, elongate-fusiform, whorls shouldered, sutures inconspicuous, aperture half total length. Widely spaced, sharp-edged, spiral ribs with up to 5 lesser spiral ribs between them; growth ridges between ribs give coarsely cancellate appearance. Outer lip smoothly corrugated. Dirty white or yellowish, with ribs occasionally orange tinted; aperture brown or white. **Habitat:** offshore in sandy mud. **Distribution:** NG, GO, SEG.

Neocancilla clathrus (Gmelin, 1791): **641**
35mm. Solid, dull, elongate-fusiform, whorls gently convex, aperture more than half total length, pointed protoconch of 2.5 straight-sided whorls. Spiral cords crossed by close-set, axial grooves; outer lip corrugated. Creamy white with 1 or 2 reddish-brown, spiral bands on last whorl and blotches on spire whorls; occasionally also pink spiral bands; aperture white or pink. **Habitat:** offshore in sand. **Distribution:** GO.

Scabricola desetangsii (Kiener, 1838): **642**
28mm. Solid, elongate-ovate, whorls gently convex, sutures incised. Regular spiral grooves crossed by axial grooves which become obsolete on last whorl. White blotched brown (fading to orange). Known from a faded shell found at Khor Fakkan by H. Kauch. **Habitat:** beached. **Distribution:** GO.

Pterygia crenulata **636**

Domiporta filaris **638**

Domiporta granatina **639**

Neocancilla circula **640**

Neocancilla clathrus **641**

Scabricola desetangsii **642**

Superfamily: **MURICOIDEA**

Costellaria acuminata **648**

"Ziba" dupilirata **645**

Swainsonia fissurata **644**

Costellaria acupicta **649**

Scabricola potensis **643**

"Ziba" pretiosa **647**

"Ziba" flammea **646**

Scabricola potensis (Montrouzier, 1858): **643**
23mm. Solid, lightweight, broadly ovate, straight-sided spire a third of total length, sutures finely incised, apex pointed. Regular, punctate spiral grooves on all whorls. Cream with greyish axial streaks and mottlings, brown spots below sutures; aperture brown. Syn: *S. nux* (Sowerby 1874) sensu Melvill & Standen, 1901:424. **Habitat:** offshore in broken coral and shingle. **Distribution:** NG, SEG.

Swainsonia fissurata (Lamarck, 1811): **644**
40mm. Solid, glossy (and slippery to touch), narrowly elongate, aperture longer than spire, sutures shallow, apex sharply pointed. Widely spaced, punctate, spiral grooves on all whorls, weak or obsolete centrally on last whorl. Creamy beige, darker on lower half of last whorl and above sutures, a netted white pattern overall; aperture beige-orange. **Habitat:** offshore and beached. **Distribution:** GO.

"Ziba" dupilirata (Reeve, 1845): **645**
14mm. Thick, narrowly elongate, whorls slightly convex, sutures deep, siphonal canal attenuated; pointed protoconch of 3 smooth, glassy whorls. Ledge-like, beaded, spiral ridges, 4 to each spire whorl, stronger and unbeaded around base, rows of pits between ridges; outer lip crenulate. White or straw coloured. Syn: *M. lalage,* Melvill & Standen, 1901. **Habitat:** offshore in mud. **Distribution:** GO.

"Ziba" flammea (Quoy & Gaimard, 1833): **646**
25mm. Solid, dull, elongate-fusiform, sutures imperceptible, aperture more than half total length, pointed protoconch of 2.5 whorls. Widely spaced, sharp-edged, spiral ribs (occasionally a lesser, beaded, spiral ridge) with close-set, axial ridges between them; outer lip (of mature shells) thickened and smoothly corrugated. White with brown blotches, spots and dashes; aperture white. The name given to this species here is provisional; it is merely the earliest of several names which could be applied. **Habitat:** offshore in sand. **Distribution:** SEG, GO, Mas.

"Ziba" pretiosa (Reeve, 1844): **647**
25mm. Solid, dull, elongate-fusiform, aperture half total length, pointed protoconch of 3.5 smooth whorls. Sharp spiral ridges, crossed by axial grooves. White with reddish brown spiral band and streaks on last whorl, brown spots on more prominent spiral ridges; aperture orange brown. Syn: *Z. antoniae* (H. Adams, 1871). **Habitat:** offshore in sand. **Distribution:** NG, NWG, SEG, GO.

Family COSTELLARIIDAE
The description of the previous family, the Mitridae, applies equally to this, except that shells are often ornamented with axial ribs, are spirally striate and the aperture is always lirate inside. Specialists differ in their use of generic and subgeneric names, sometimes placing species in subgenera within the genus *Vexillum*. As with the Mitridae we use generic names throughout, although we are aware that some workers prefer to place each subgenus in *Vexillum*.

Costellaria acuminata (Gmelin, 1791): **648**
30mm. Solid, dull, elongate-fusiform, aperture less than half total length, slightly shouldered below deep sutures. Close-set axial ribs, sinuous initially with regular spiral striae between them. Pinkish brown to greenish brown with a thin, paler, spiral band on later whorls (fades to dull pink and violet). **Habitat:** intertidal and offshore. **Distribution:** SEG, GO.

Costellaria acupicta (Reeve, 1844): **649**
30mm. Solid, elongate-fusiform, shouldered below sutures, aperture less than half total length, pointed protoconch. Close-set, sometimes sinuous, axial ribs cut by irregular, spiral striae. Cream with pale or dark brown bands (1 or 2 on last whorl) and isolated blotches, spots and dashes; aperture cream. **Habitat:** offshore in coral sand. **Distribution:** NG, SEG, Mas.

Superfamily: **MURICOIDEA**

Pusia aethiopica (Jickeli, 1874): **650**
4mm. Solid, fusiform, aperture half total length, spire whorls convex, sutures deep. Coarse axial ridges crossed by regular spiral grooves produce cancellate pattern on later whorls. Dark brown, apex paler. **Habitat:** beached. **Distribution:** SO.

Costellaria alauda (Sowerby & Sowerby, 1874): **651**
21mm. Solid, glossy, elongate-fusiform, aperture less than half total length, shouldered at sutures, siphonal canal slightly curved. Widely spaced, rounded, axial ribs with fine growth lines between them, weak spiral grooves on lower half of last whorl and on early whorls. Columellar folds flattened. Bluish white with spiral rows of brown spots and dashes on ribs. **Habitat:** offshore. **Distribution:** GO.

Costellaria albatum (Cernohorsky, 1988): **652** NOT ILLUSTRATED
This small, white species has been dredged off As Sib (GO).

Costellaria caliendrum (Melvill & Standen, 1901): **653**
22mm. Solid, dull, fusiform, sutures deep, aperture less than half total length. Sinuously axial ribs as broad as their intervals, with flattened spiral ridges between them. Columella has 4 folds. Cream or white with brown spiral bands (1 narrow, 1 broad on last whorl). **Habitat:** unknown. **Distribution:** Arabian Gulf, without precise locality.

Costellaria daedala (Reeve, 1845): **654**
15m. Thick, moderately glossy, elongate-ovate, aperture less than half total length. Broad, axial ribs; base with 5 or 6 nodulose, spiral cords, the stronger continuous with columellar folds. Brown fading to orange (columella paler) with white spiral band on each whorl (fine, continuous or interrupted line within band). **Habitat:** offshore and beached. **Distribution:** NG, GO, Mas.

Costellaria diaconalis (Melvill & Standen, 1903): **655**
13mm. Solid, moderately glossy, fusiform, aperture less than half total length, almost straight sided, sutures deep, aperture long and narrow. Axial ribs about same width as their intervals, spiral grooves between ribs. Whitish with scattered greyish-brown blotches. **Habitat:** offshore in coral sand. **Distribution:** NG, SEG.

Costellaria exasperata (Gmelin, 1791): **656**
17mm. Solid, dull, ovate-fusiform, aperture half total length, early spire whorls rounded, later ones strongly shouldered. Sharp, widely spaced, axial ridges, projecting at shoulders; fine or coarse spiral grooves crossed by similar axial grooves producing reticulated sculpture overall. White with reddish-brown ribs. **Habitat:** beached. **Distribution:** GO.

Costellaria helena (Bartsch, 1915): **657**
10mm. Thick, fusiform, spire whorls gently rounded, sutures well indented; protoconch not seen. Broad axial ribs with spiral threads between them. Off-white blotched brown, often tinged violet, encircled by interrupted, fine, brown, spiral lines; columella violet. **Habitat:** intertidal and beached. **Distribution:** Mas.

Costellaria infausta (Reeve, 1845): **658**
15mm. Thick, broadly fusiform, spire whorls almost stepped at sutures; protoconch not seen. Broad axial ribs with very fine, crowded, axial threads between, crossed by evenly spaced spiral grooves which are deeper towards base. Off-white with reddish-brown spiral band at sutures and at centre of last whorl; faint brown where grooves cross ribs; columella reddish brown. Syn: *Mitra elizae* Melvill, 1899. **Habitat:** coral sand offshore. **Distribution:** NG.

Costellaria caliendrum **653**

Costellaria alauda **651**

Pusia aethiopica **650**

Costellaria diaconalis **655**

Costellaria daedala **654**

Costellaria exasperata **656**

Costellaria helena **657**

Costellaria infausta **658**

Superfamily: **MURICOIDEA**

Vexillum plicarium 663

Costellaria obeliscus 660

Costellaria polygona 664

Costellaria pacifica 661

Costellaria
malcolmensis **659**

Costellaria pasithea **662**

Costellaria sculptilis
666

Costellaria malcolmensis (Melvill & Standen, 1901): **659**
10mm. Solid, glossy, fusiform, deep sutures; pointed protoconch of 3.5 smooth whorls. Strong, crescentic, axial ribs, nodulous at their tops and narrower than their intervals, with crowded, flattened, spiral ridges between them. Columella has 4 strong folds. White with occasional, amber, spiral bands. **Habitat:** offshore in mud. **Distribution:** NG, GO.

Costellaria obeliscus (Reeve, 1844): **660**
22mm. Solid, dull, elongate-fusiform, sutures incised, aperture less than half total length. Thin, sinuously axial ribs with flat-topped, spiral ribs between them producing a cancellate appearance. Columellar folds grooved along their length. Orange to brown with a spiral white band on later whorls (sometimes a thin coloured line within band). **Habitat:** offshore. **Distribution:** SEG, GO.

Costellaria pacifica (Reeve, 1845): **661**
20mm. Thick, dull, elongate-ovate, whorls shouldered, aperture half total length. Sinuously axial ribs, beaded at shoulders and crossed by shallow spiral grooves; sinuous outer lip lightly corrugated. White tinged brown (seen as fine axial lines resembling coffee stains under magnification); aperture white. Syn: *Mitra revelata* Melvill, 1899. **Habitat:** offshore in shingle and dead coral and beached. **Distribution:** NG, SEG, GO, Mas.

Costellaria pasithea (Melvill & Standen, 1901): **662**
10mm. White, with obliquely axial ribs, it was described from Muscat (GO).

Vexillum plicarium (Linnaeus, 1758): **663**
45mm. Thick, dull, elongate-fusiform, sutures incised, later whorls shouldered, aperture half total length. Axial ribs, nodulose at shoulder on last whorl, obsolete towards base; grooves encircling all whorls strongest on shoulders and lower half of last whorl. White with broad, spiral, brown bands; aperture white, chocolate-edged outer lip. Single beached shell found by Mrs Eli Morrison. **Habitat:** intertidal and beached. **Distribution:** GO.

Costellaria polygona (Gmelin, 1791): **664**
28mm. Thick, glossy, elongate-fusiform, aperture half total length, spire whorls gently rounded with sloping shoulders, siphonal canal slightly recurved; protoconch not seen. Sharp, sinuous, axial ribs with widely spaced spiral threads between them. Shades of brown with paler spiral zone sometimes brown spotted, visible within aperture. **Habitat:** beached. **Distribution:** GO.

Costellaria scitula (A. Adams, 1853): **665** NOT ILLUSTRATED
Listed by Melvill & Standen, 1901:423, from Muscat (GO) and the Arabian Gulf.

Costellaria sculptilis (Reeve, 1845): **666**
23mm. The regularly disposed cancellations, characteristic of this species, may or may not cross the glossy axial riblets. Syn: *Mitra iteina* Melvill, 1918. **Habitat:** offshore. **Distribution:** NG.

154 SEASHELLS OF EASTERN ARABIA

Superfamily: **MURICOIDEA**

Costellaria stephanucha **667**

Pusia osiridis **673**

Pusia pardalis **674**

Pusia microzonias depexa **672**

Pusia blanfordi **670**

Pusia geoffreyana **671**

Pusia amabile **668**

Pusia aureolata **669**

Costellaria stephanucha (Melvill, 1897): **667**
38mm. Solid, moderately glossy, elongate-fusiform, whorls shouldered, sutures incised. Later whorls have sinuously axial ribs, smooth or sharply noduled at shoulders; spiral grooves, strongest at base, between ribs. Greyish red with paler blotches and spiral bands; aperture orange-red. The var. *astephana* (Melvill, 1904) is merely a smooth form. **Habitat:** offshore in sandy mud and beached. **Distribution:** NWG, SEG, GO, Mas.

Pusia amabile (Reeve, 1845): **668**
10mm. Solid, dull, biconic, sutures deep, aperture less than half total length. Low, rounded, axial ribs crossed by shallow spiral grooves (weak at centre of last whorl); columellar folds continuous with cords between grooves; outer lip corrugated. White with greyish-brown spiral bands, darker in grooves; aperture brown. **Habitat:** offshore and beached. Seldom found undamaged. **Distribution:** GO, Mas.

Pusia aureolata (Reeve, 1844): **669**
14mm. Solid, dull, elongate-ovate, sutures well incised, aperture half total length. Rounded axial ribs with spiral ridges between them (giving shell pitted appearance when worn); strong spiral cords towards base. White with broad or narrow, orange-brown, spiral bands. Illustrated are 2 faded, worn specimens from Muscat, ex Townsend collection, in Manchester Museum. **Habitat:** offshore. **Distribution:** GO.

Pusia blanfordi (Melvill & Standen, 1901): **670**
5mm. Closely resembles and may be mistaken for ***P. geoffreyana*** (see below). **Habitat:** offshore. **Distribution:** GO.

Pusia geoffreyana (Melvill, 1910): **671**
5mm. Solid, oblong-fusiform, spire less than half total length, whorls slightly stepped, sutures deep, apex blunt. Aligned axial ribs with regular spiral ridges between them, strong spiral folds on lower half of last whorl. Columellar callus forms straight, diagonal line; 4 folds on columella. Creamy blotched reddish brown. Resembles a columbellid. **Habitat:** offshore. **Distribution:** NG, GO.

Pusia microzonias depexa (Deshayes, 1834): **672**
15mm. Thick, moderately glossy, broadly ovate, whorls well rounded and slightly telescoped, sutures deep; protoconch not seen. Close-set, thin axial ridges which are occasionally thickened, fine spiral grooves between ridges; thick ridges encircle base. Purple-brown with broad, wavy-edged and brown-blotched, yellowish-white zone around each whorl; aperture violet to purple or whitish. **Habitat:** beached. **Distribution:** "Oman".

Pusia osiridis (Issel, 1869): **673**
30mm. Thick, dull, elongate-ovate, whorls shouldered, aperture less than half total length. Widely spaced axial ribs, pointed at shoulders; weak spiral grooves; base with nodulous spiral cords. White with broad, reddish brown, spiral band on last whorl, continuing narrowly above sutures and visible in aperture. **Habitat:** offshore in coral sand and beached. **Distribution:** NG, NWG, SEG, GO, Mas.

Pusia pardalis (Küster, 1840): **674**
15mm. Solid, ovate, spire less than half total length, whorls distinctly stepped, sutures deep, apex pointed. Coarse axial ribs become obsolete on lower half of later whorls; 3 or 4 spiral rows of beads towards base. Spire whorls yellow becoming reddish brown on last whorl, blotched white (blotches forming spiral band on last whorl); beads whitish; aperture orange. **Habitat:** beached. **Distribution:** GO.

SEASHELLS OF EASTERN ARABIA

Superfamily: **CANCELLARIOIDEA**

Pusia unifasciale 675

Merica melanostoma 678

Cancellaria agalma 676

Nipponaphera paucicostata 679

Scalptia cf *fusca* 682

Scalptia articularis 680

Scalptia hystrix 683

Scalptia contabulata 681

Pusia unifasciale (Lamarck, 1811): **675**
20mm. Thick, dull, biconic, twice as long as wide, whorls shouldered, sutures deep. Strong axial ribs as wide as their intervals, crossed by deep, spiral grooves; nodulous spiral cords towards base. Reddish brown to yellowish orange, often with thin, white and/or dark brown, spiral lines. **Habitat:** offshore and beached. **Distribution:** GO.

Superfamily CANCELLARIOIDEA
See description under next family.

Family CANCELLARIIDAE
Little is known about the biology of the nutmegs, as collectors call shells of this family, although they are known to occur in sand and mud and on rocks or stones, according to the species. In eastern Arabia their shells are usually found washed up, empty, on beaches. Most are thick and globular with strong axial ribs and some have strongly stepped whorls. The aperture is often triangular and lirate, the columella thickened and bearing up to 4 folds. An umbilicus may be present but there is never an operculum. Most species live offshore but some are beached occasionally.

Cancellaria agalma Melvill & Standen, 1901: **676**
7.5mm. Solid, dull, oblong-fusiform, aperture a third total length, sutures very deep. Spire whorls and aperture trigonal in outline; narrow umbilicus. Strong, obliquely axial ribs (11 on last whorl) crossed by alternately thick and thin, spiral threads; outer lip thickened; columella with 2 folds. Brown. **Habitat:** offshore in sand and mud. **Distribution:** GO.

Merica elegans (Sowerby, 1822): **677** NOT ILLUSTRATED
Listed by Melvill & Standen, 1901:450, from GO, but probably misidentified.

Merica melanostoma (Sowerby, 1849): **678**
25mm. Thick, dull, globose-ovate, aperture twice as long as spire; blunt-topped protoconch of 3 smooth whorls. Whorls well rounded; no umbilicus. Columella has 3 large folds; broad parietal callus; aperture lirate. Coarse axial ridges crossed by spiral grooves produce cancellate sculpture. Creamy yellow with reddish brown, interrupted, spiral bands, protoconch darker; aperture and callus creamy yellow. **Habitat:** offshore and beached. **Distribution:** NWG, SEG, GO.

Nipponaphera paucicostata (Sowerby, 1895): **679**
20mm. Solid, lightweight, dull, biconic, aperture more than half total length, deep sutures; pointed protoconch of 2 whorls. Whorls angulated at periphery; no umbilicus. Columella has 3 folds (lower 2 fused); aperture smooth. Thick, widely spaced, axial folds crossed by crowded, coarse, spiral ridges. Pinkish orange; aperture paler. **Habitat:** offshore on other shells. **Distribution:** NG, SEG, GO.

Scalptia articularis (Sowerby, 1832): **680**
18mm. Solid, dull, globose-ovate, aperture about half total length, deeply channelled sutures; blunt protoconch of 2.5 smooth whorls. Spire whorls rounded, aperture flaring; small, deep umbilicus. Sharp-edged, widely spaced, obliquely axial ridges (12 on last whorl) extend into points above; fine, irregular, spiral threads between ridges; outer lip lirate inside; columella with 3 folds. Orange-brown to white. Syn: *Cancellaria crenifera* Sowerby, 1833, and var. *serrata* Reeve, 1856, sensu Melvill & Standen, 1901:451. **Habitat:** offshore. **Distribution:** NG, SEG, GO.

Scalptia contabulata (Sowerby, 1833): **681**
20mm. Thick, dull, globose-ovate, aperture more than half total length, spire whorls broadly tabulated, sutures shallow; blunt protoconch. Whorls almost straight sided, base rounded, aperture flaring; small, deep umbilicus. Close-set axial ridges (sharp at shoulder) crossed by regular, finer, spiral ridges; outer lip ridged inside; columella with 3 folds. Yellowish white. **Habitat:** offshore and beached. **Distribution:** SEG, GO.

Scalptia cf *fusca* (Sowerby, 1889): **682**
10mm. Solid, glossy, oblong-fusiform, aperture half total length, whorls tabulated, sutures deep; blunt-topped protoconch of 2.5 smooth whorls. Whorls gently rounded, narrowing sharply towards base; small, deep umbilicus. Columella has 3 folds; outer lip ridged inside. Widely spaced, obliquely axial ribs, pointed above; faint spiral threads between ribs but becoming nodular on them. Chocolate with white nodules on ribs. **Habitat:** offshore. **Distribution:** GO.

Scalptia hystrix (Reeve, 1856): **683**
25mm. Solid, dull, globose-ovate, aperture more than half total length, sutures very deep; blunt-topped protoconch of 2 smooth whorls. Whorls well rounded; umbilicus obscured by columellar callus. Columella has 3 folds; outer lip

ridged inside. Sinuous, reflected, lamellate axial ribs, roundly pointed above; thin threads between ribs, thicker on ribs. Cream with 2 brown spiral bands on last whorl. **Habitat:** offshore and beached. **Distribution:** NG, GO.

Scalptia nassa (Gmelin, 1791): **684**
18mm. Solid, dull, globose-ovate, aperture about half total length; blunt-topped protoconch of 2 smooth whorls. Spire whorls straight sided, last whorl well rounded; large, deep umbilicus. Columella has 3 strong folds; aperture lirate. Obliquely axial, lamellated (often multi-lamellated) ribs, pointed above and comb-like in profile; weak threads between ribs. White tinged brown between ribs below sutures; aperture white. Syn: *Cancellaria lamellosa* Hinds, 1843, sensu Melvill & Standen, 1901:451. **Habitat:** offshore. **Distribution:** NG.

Scalptia obliquata (Lamarck, 1822): **685**
15mm. Solid, glossy, globose-ovate, aperture about half total length; bulbous protoconch of 2 whorls. Whorls well rounded; large, deep umbilicus. Columella has 3 folds; aperture lirate. Obliquely axial, sharp-edged ribs, pointed above; lirate between ribs. Creamy yellow tinged pink; usually a narrow, brown, spiral band below periphery of last whorl and brown spots on ribs; aperture white. **Habitat:** offshore. **Distribution:** SEG.

Scalptia cf *scalarina* (Lamarck, 1822): **686**
22mm. Thick, moderately glossy, ovate-biconic, aperture half total length, deeply channelled sutures; protoconch of 2.5 glossy whorls. Spire whorls gently rounded; broad, deep umbilicus. Columella has 3 folds; triangular aperture constricted towards base, lirate within. Axial ribs, pointed above, prickly where crossed by spiral threads. Cream to beige with 1 or 2 thin, brown, spiral bands; aperture white. Syn: *Trigonostoma costifera* (Sowerby, 1833) sensu Bosch & Bosch, 1982:118. **Habitat:** **Distribution:** NG, SEG, GO.

Trigonostoma antiquata (Hinds, 1843): **687**
20mm. Thin, dull, oblong-fusiform, aperture a third total length, spire scalariform, sutures deep; blunt protoconch of about 2.5 smooth whorls. Spire whorls and aperture trigonal in outline; broad umbilicus open to apex. Thin, widely spaced, axial ribs extend into points above; surface finely cancellate; outer lip ridged inside; columella with 2 folds. Off-white. **Habitat:** offshore. **Distribution:** GO.

Superfamily CONOIDEA

Although each of the three families included in this extensive superfamily has its own distinctive shell type all the contained species prey on other organisms. In many species the radular teeth are reduced in number while others, notably the cones, have detachable teeth associated with a poison apparatus.

Family CONIDAE

Cones have long attracted the shell collector and the biologist, not least because there are many different species and varieties. Instantly recognisable by their thick, conical and mostly unornamented shells, most species have a narrow aperture and all have a smooth columella. When fresh the shell usually has a thick or thin, sometimes fibrous periostracum. The animal has a small, narrowly oblong, chitinous operculum. Some are primarily reef dwellers but others are usually found under rocks, in rock crevices, and in clean or muddy sand. They feed on worms, other molluscs or even fish, using their detachable teeth, harpoon-like, to sting and paralyse prey. Some species are capable of inflicting painful stings on humans and there are well authenticated cases of humans dying painfully after being stung (none, mercifully, authenticated for eastern Arabia) so it is necessary to handle cones carefully when picking them up, the larger ones especially. The proboscis, through which the animal delivers its sting, issues from the narrower anterior end; for this reason always pick up a cone shell by the larger end. Because many species have similarly shaped shells and vary considerably in size, colour and pattern their correct identification is sometimes a job for experts. Apart from the Arabian Gulf, where species are few, they are well represented in eastern Arabia. Almost certainly, however, other species await discovery. One or two of those described and illustrated here, for instance, were first recognised as new to science during the preparation of this book.

Conus achatinus Gmelin, 1791: **688**
45mm. Thick, with short, slightly stepped spire, pointed apex; almost twice as long as wide, it has a rounded shoulder and slightly convex sides. Pale blue, greyish blue, greenish, or shades of brown with irregular blotches and more regular, spirally arranged dots and dashes; aperture white or bluish white. Very variable colour pattern. **Habitat:** intertidal under stones. **Distribution:** GO, Mas, SO.

Scalptia cf *scalarina* **686**

Trigonostoma antiquata **687**

Scalptia nassa **684**

Scalptia obliquata **685**

Conus achatinus **688**

Superfamily: **CONOIDEA**

Conus ardisiaceus **689**

Conus betulinus **691**

Conus arenatus **690**

Conus biraghii omanensis **693**

Conus biliosus **692**

Conus ardisiaceus Kiener, 1845: **689**
35mm. Sturdy, lightweight, dull, rounded or slightly keeled shoulder and well-defined sutures, short spired. Last whorl has spiral ridges towards base and coarse growth lines; spire whorls have strong spiral ridges. Pale brown to greyish white with brown blotches, zigzags, streaks and bands; also covered with small brown spots; aperture violet, white edged. **Habitat:** under rocks and in sand beneath rock ledges. **Distribution:** Mas, SO.

Conus arenatus Hwass, 1792: **690**
50mm. Heavy, dull or glossy, broadly conical, convex sided, noticeably narrowed towards base, shoulder rounded and coronated, short spired with slightly elevated early whorls. Columella twisted at base, a swollen ridge above; low spiral ridges at base; later spire whorls coronated and spirally striate. White or cream covered by zigzag patterns of tiny brown spots; aperture white, orange-pink deep within. **Habitat:** among coral and beached. **Distribution:** Mas.

Conus betulinus Linnaeus, 1758: **691**
80mm. Heavy, broadly conical, almost flat topped, sometimes only the apical whorls elevated, shoulder rounded, sutures incised. Low spiral ridges towards base, otherwise almost smooth. Creamy yellow with spiral rows of large and small spots; aperture creamy white. May become large and ponderous. **Habitat:** offshore and beached. **Distribution:** SEG, GO.

Conus biliosus (Röding, 1798): **692**
50mm. Heavy, almost straight sided, spire moderately elevated and slightly coronated, shoulder keeled. Wavy spiral ridges and coarse growth lines; columella slightly twisted at base. Orange-violet with paler spiral band centrally, short brown flecks on spire whorls and spiral rows of brown spots on last whorl; aperture white with violet edge. **Habitat:** offshore and beached. **Distribution:** Mas.

Conus biraghii omanensis
Moolenbeek & Coomans, 1993: **693**
7.5mm. Solid, biconic, slender. First spire whorl has strong spiral groove, a 2nd and 3rd groove developing subsequently; last whorl smooth except for groove just below shoulder. White with greyish upper spiral band bearing 5 white spiral lines; greyish band at base; between bands are 3 whitish, spotted spiral lines. **Habitat:** subtidal and beached. Smallest eastern Arabian cone. **Distribution:** Mas, SO.

Superfamily: **CONOIDEA**

Conus boschorum **694**

Conus chaldeus **695**

Conus coronatus **696**

Conus flavidus **699**

Conus ebraeus **697**

Conus elegans **698**

Conus boschorum Moolenbeek & Coomans, 1993: **694**
10mm. Thin, short spired with exserted whorls, last whorl with convex sides and keeled shoulder, spiral grooves towards base. 2 or 3 channels around spire whorls. White with reddish black spiral band below shoulder, irregular blotches below; pale central band with milky spots. **Habitat:** subtidal and beached. **Distribution:** Mas.

Conus chaldeus Hwass, 1792: **695**
30mm. Heavy, dull, broadly conical with convex sides and low coronations, short spired; apex usually eroded. Last whorl has rounded, sometimes granular ridges towards base, spiral grooves on spire whorls. White to pinkish, last whorl with axial, dark brown strips separated by a thin white band at shoulder and another mid-whorl, the latter spotted dark brown; early whorls pinkish; aperture bluish white. **Habitat:** offshore and beached. **Distribution:** NWG, Mas.

Conus coronatus Gmelin, 1791: **696**
30mm. Heavy, dull to glossy, globose-ovate, widest just below lightly coronated shoulder, short spired; blunt apex usually eroded. Spiral ridges on lower half of last whorl. Pinkish to grey-blue with 2 broad bands of bluish-brown blotches and paler spiral rows of dots and dashes superimposed; aperture bluish brown, pale edged. Greenish periostracum. **Habitat:** intertidal rocks. **Distribution:** GO, Mas, SO.

Conus ebraeus Linnaeus, 1758: **697**
35mm. Heavy, dull, broadly conical, short spired with weak or strong coronations; blunt apex usually eroded. Spiral ridges on spire whorls. Spiral ridges, sometimes nodular, on lower half of last whorl. White to pinkish with large, dark brown, more or less rectangular blotches (3 spiral rows on last whorl); aperture white or pinkish. **Habitat:** offshore and beached. **Distribution:** GO, Mas.

Conus elegans Sowerby, 1895: **698**
30mm. Thin, glossy, fusiform, sides of spire slightly concave, each spire whorl has pointed nodules just above suture; pointed, glossy protoconch of 1.5 whorls. Regular, punctate, spiral grooves on lower half only of last whorl and between rows of nodules on spire. White with spiral rows of reddish-brown dots and occasional blotches and dashes. **Habitat:** offshore and beached. **Distribution:** NG, SEG, GO.

Conus flavidus Lamarck, 1810: **699**
45mm. Heavy, dull, straight sided, short, domed spire; early whorls always eroded. Last whorl has irregular, weak or granulose, spiral ridges crossed by growth ridges; 3-4 weak, spiral ridges on spire whorls. Orange, yellow or yellow-grey with 2 whitish, spiral bands on last whorl; base bright purple; aperture pale purple. Thick, brown periostracum. **Habitat:** intertidal among rocks. **Distribution:** SEG, GO, Mas, SO.

Superfamily: **CONOIDEA**

Conus geographus **701**

Conus generalis maldivus **700**

Conus inscriptus **702**

Conus lischkeanus tropicensis **703**

Conus lividus **704**

■ *Few shells are collected more enthusiastically than cones and a varied selection of them may be obtained without much difficulty in shallow water around the coasts of eastern Arabia. Most species are as innocuous to the collector as they look, but one or two are potentially dangerous to human beings. Conus geographus, in particular, should be treated with great respect. Like other cones it has a poison apparatus, including replaceable, harpoon-like teeth, which it uses to paralyse small fish and other invertebrates before ingesting them. Occasionally it may inject its powerful poison into the anatomy of an unwary collector, usually a finger which has strayed within the reach of its extensible proboscis. So it is important to pick up cones in such a way that the hand is not close to the aperture. Also make sure that a cone cannot use its proboscis to pierce through the fabric of a collecting bag attached to the person. Undoubtedly the safest way to collect cones is to pick up those without animals.*

Conus generalis maldivus Hwass, 1792: **700**
55mm. Thick, moderately glossy, elongate-conic with short spire, early whorls of unworn shells elevated and sharply pointed. Spire whorls shallowly channelled; last whorl smooth or with growth flaws. Dark brown spiral lines and dashes partly covered by paler brown or orange blotches and bands; brown crescents on later spire whorls; base violet; aperture white. **Habitat:** offshore and beached. **Distribution:** GO, Mas, SO.

Conus geographus Linnaeus, 1758: **701**
85mm. Thin, lightweight, moderately glossy, last whorl ovately cylindrical and bearing widely spaced coronations, spire very short, whorls coronated; pointed protoconch. Last whorl smooth; above coronated shoulder finely, spirally ridged. Pinkish, cream or bluish white with 2 broad, interrupted, brown bands and fine brown reticulations from base to shoulder; spire radially blotched brown; early whorls bright pink; aperture bluish white. Venomous so handle carefully! **Habitat:** intertidal in sand. **Distribution:** GO, Mas, SO.

Conus inscriptus Reeve, 1843: **702**
55mm. Solid, moderately glossy, elongate-conic, keeled at shoulder otherwise straight sided; short, concave-sided spire; pointed protoconch. Upper third of last whorl smooth, lower two thirds spirally grooved, the grooves progressively deeper towards base; above shoulder finely, spirally ridged. Creamy white covered with pale or darker brown, squarish spots and axial streaks; spire radially streaked. **Habitat:** offshore and beached. **Distribution:** GO, Mas.

Conus lischkeanus tropicensis Coomans & Filmer, 1985: **703**
45mm. Heavy, broadly conic, keeled at shoulder otherwise straight sided; short spire, apex usually worn. Shallow spiral grooves at base of last whorl; spire whorls spirally striate. White suffused with tan, often a white spiral band mid-whorl; brown streaks at shoulder and on spire whorls; aperture white with 2 broad, violet zones. Syn: *C. kermadecensis* Iredale, 1912, sensu Bosch & Bosch, 1982:128. **Habitat:** intertidal in sand under rocks. **Distribution:** Mas.

Conus lividus Hwass, 1792: **704**
45mm. Heavy, elongate-conic, coronate at shoulder otherwise straight sided; short, coronated spire, apex usually worn. Low spiral ridges at base, less evident on rest of last whorl; ridges encircle spire whorls. Olive-tan or bluish orange with paler zone mid-whorl, white at shoulder and on spire whorls. Base

160 SEASHELLS OF EASTERN ARABIA

Superfamily: **CONOIDEA**

Conus melvilli **707**

Conus milesi **709**

Conus cf *luctificus* **706**

Conus miles **708**

purple; aperture purple with deep orange lip. **Habitat:** intertidal under rocks. **Distribution:** Mas, SO.

Conus longurionis Kiener, 1845: **705** NOT ILLUSTRATED
Listed by Melvill & Standen, 1901:431, from Arabian Gulf and Muscat (GO), and by Melvill, 1928:110, from Musandam (GO), but identity of specimens uncertain. They may have been smooth-shelled *C. milesi* Smith.

Conus* cf *luctificus Reeve, 1848: **706**
65mm. Solid, broadly conic, keeled at shoulder, edges of spire whorls slightly stepped; apex worn. Weak, ripple-like ridges at base; fine ridges encircle spire whorls. Brown or pinkish brown with white spiral band mid-whorl and darker brown spiral lines of dots and dashes; reddish brown, crescentic strips on spire whorls; aperture pink or bluish white. **Habitat:** beached. **Distribution:** Mas.

Conus melvilli Sowerby, 1879: **707**
25mm. Thick, broadly conical, sides convex, short spire topped by pimple-like apex. Ivory or faintly bluish white netted with dark brown lines, brown blotches forming a band mid-whorl and towards base; spire whorls irregularly streaked brown, apex white; aperture purplish. Syn: *C. boschi* Clover, 1972. **Habitat:** intertidal muddy sand. **Distribution:** GO, Mas.

Conus miles Linnaeus, 1758: **708**
65mm. Heavy, silky, broadly conical, sides convex, keeled shoulder, short spired with blunt apex. Widely spaced, low, spiral ridges on lower half of last whorl. Fine, wavy, orange-brown, axial lines cover shell; lines eclipsed on lower third of last whorl by opaque chocolate brown; a paler brown, spiral band about mid-whorl allowing lines to show through; aperture white with 2 brown zones. **Habitat:** near coral and beached. **Distribution:** Mas.

Conus milesi Smith, 1887: **709**
35mm. Thin, glossy, fusiform, sides of spire slightly concave, each spire whorl beaded just above suture; glossy, pointed protoconch of 2-3 whorls. Regular, punctate, spiral grooves on last whorl and between rows of beads on spire. White with reddish-brown blotches and axial streaks; occasionally spirally banded. Syn: *C. dictator* Melvill, 1898. **Habitat:** offshore. **Distribution:** NG, NWG, GO, Mas.

Superfamily: **CONOIDEA**

Conus milneedwardsi 710

Conus milneedwardsi Jousseaume, 1894: **710**
100mm. Heavy, glossy, elongate-biconic, straight sided or slightly concave centrally, shoulder of last whorl weakly keeled, slightly concave above, sometimes with shallow spiral grooves; outer lip curved forwards; spire whorls stepped, concave above and spirally ridged. White netted with large, orange-brown to darker brown tents and irregular areas and 3 irregular, spiral bands of reddish-brown blotches; spire whorls with similar blotches; aperture white to cream. Syn: *C. clytospira* Melvill & Standen, 1899. **Habitat:** offshore and beached. **Distribution:** GO.

Conus monile Hwass, 1792: **711**
NOT ILLUSTRATED
Listed by Melvill & Standen, 1901:432, from Muscat (GO) and from Sheyk Shoeyb Island (NG). Possibly misidentifications for **C. generalis maldivus**.

Conus namocanus badius Kiener, 1845: **712**
55mm. Heavy, moderately glossy, broadly conical, shoulder roundly keeled. Last whorl weakly, spirally ridged; weak spiral ridges on spire whorls. Yellowish or bluish green with white spiral bands mid-whorl and at shoulder, faint, brown spiral lines cover last whorl; spire whorls and shoulder streaked brown; aperture purple with orange-yellow lip edge. **Habitat:** intertidal. **Distribution:** GO, Mas.

Conus nigropunctatus Sowerby, 1857: **713**
30mm. Lightweight, low conical, deep sutures, sloping shoulder and gently convex sides. Low spiral ridges at base, 2 or 3 spiral ridges on each spire whorl. White to bluish white with irregular pale brown blotches on last whorl, darker brown blotches on spire whorls; last whorl encircled by fine lines of alternately brown and white dashes; aperture bluish white, brown blotches at lip edge. **Habitat:** intertidal. **Distribution:** GO, Mas.

Conus nussatella Linnaeus, 1758: **714**
50mm. Solid, glossy, elongate-cylindrical, straight sided with steeply sloping shoulder, spire whorls slightly telescoped, apex pointed. Regular, low spiral ridges on last whorl, ridges finer and crescentic above shoulder. White mottled reddish brown or orange, dark brown flecks at tops of whorls, dark brown dots on ridges; aperture white. **Habitat:** beached. **Distribution:** GO, Mas.

■ *It is unfortunate, perhaps, that the shell upon which Jousseaume based the description of his* Conus milneedwardsi *in 1894 was, with a length of only 46 millimetres, a paltry example of a species which may attain a length — and an enhanced elegance — of about 175 millimetres; doubly unfortunate that he christened it so unimaginatively. Five years later, in December 1899, Melvill and Standen published the description of* Conus clytospira, *based on the two specimens illustrated here. These had been hauled up from a depth of 45 fathoms in the Arabian Sea, attached to a submarine cable, and had been pocketed by their discoverer, F. W. Townsend. In a letter Townsend wrote to Melvill he said that another, much longer specimen was knocked off the cable and lost overboard. "The dredging of this remarkable textile cone", said Melvill and Standen, "undeniably constitutes one of the most important discoveries of the kind during the nineteenth century. It will rank amongst the most select of a genus unusually distinguished in both form, texture, and coloration." It is still so ranked. Since receiving the very apt name bestowed upon them by Melvill and Standen these two shells have been segregated, the smaller one ending up in the Natural History Museum, London, the larger in the Manchester Museum. Appropriately they have been brought together again, a century later, to pose for the camera.*

Superfamily: **CONOIDEA**

Conus obscurus Sowerby, 1833: **715**
34mm. Thin, lightweight, elongate-ovate, low spired, tops of spire whorls concave and striate; last whorl angled at shoulder, its sides slightly convex or almost straight, smooth; aperture broad basally. Purplish orange mottled white and violet and encircled by fine, interrupted, brown lines. Sometimes mistaken for a juvenile *C. geographus*. **Habitat:** intertidal. **Distribution:** Mas.

Conus orbignyi Audouin, 1831: **716**
NOT ILLUSTRATED
Listed by Melvill & Standen, 1901:431, and by Melvill, 1928:110, from the Arabian Gulf.

Conus parvatus sharmiensis Wils, 1986: **717**
24mm. Thick, broadly conical, sides slightly convex, shoulder weakly coronated, spire almost flat topped and usually eroded. Widely spaced, low, spiral ridges at base. White or bluish white with faint bluish, spiral band mid-whorl, purple at base; spiral rows of squarish brown spots on last whorl, thin brown streaks at shoulder; aperture purple. Syn: *C. musicus* Hwass, 1792, sensu Bosch & Bosch, 1982:125. **Habitat:** intertidal among stones. **Distribution:** GO, Mas, SO.

Conus pennaceus quasimagnificus Da Motta, 1982: **718**
50mm. Heavy, glossy, low conical, sides slightly convex, shoulder roundly keeled; apex pointed. Faint spiral threads all over. White to pinkish covered by brown-edged, overlapping tents; also 2 spiral rows of large, reddish-brown patches in which are fine, brown-and-white, spiral lines; aperture white or pinkish; apex pink. Syn: *C. omaria* Hwass, 1792, sensu Smythe, 1979a:68. **Habitat:** intertidal under stones and beached. **Distribution:** SEG, GO, Mas, SO.

Conus namocanus badius **712**

Conus nigropunctatus **713**

Conus obscurus **715**

Conus nussatella **714**

Conus pennaceus quasimagnificus **718**

Conus parvatus sharmiensis **717**

SEASHELLS OF EASTERN ARABIA

Superfamily: **CONOIDEA**

Conus quercinus Lightfoot, 1786: **719**
70mm. Heavy, dull, broadly conical, sides nearly straight, shoulder keeled or rounded, flat topped to short spired with elevated early whorls; aperture slightly flared at base. Shades of yellow, occasionally a darker spiral band mid-whorl, covered with fine, amber, spiral lines; aperture and columella white.
Habitat: offshore and beached.
Distribution: GO, Mas, SO.

Conus rattus Hwass, 1792: **720**
50mm. Heavy, moderately glossy, broadly conical, sides gently convex, shoulder keeled, short spired; protoconch pimple-like. Low spiral ridges towards base and on tops of spire whorls. White to bluish white, 2 broad spiral bands of brown to violet brown extending flame-like over shoulder and onto spire; white flecks scattered over last whorl; early whorls yellow; aperture purple.
Habitat: offshore and beached.
Distribution: GO, Mas, SO.

Conus saecularis Melvill, 1898: **721**
30mm. Solid, fusiform, stepped spire whorls concave sided in profile, last whorl acutely angled at shoulder, concave sided and attenuate basally. Tops of spire whorls cancellated below sutures; last whorl encircled by regular, axially striated grooves. White spotted and streaked pale orange-brown, especially at angled periphery. Lip usually broken.
Habitat: in mud offshore.
Distribution: GO.

Conus stocki Coomans & Moolenbeek, 1990: **722**
32mm. Solid, elongate-conical, outer lip slightly concave, shoulder sharply keeled; spire short, concave sided, pointed protoconch of 1.5 smooth whorls. Lower half of last whorl has regular, spiral grooves; spire whorls stepped, channelled above. Bluish white blotched brown, encircled by brown and white dashes; spire radially striped brown. Syn: *C. lemniscatus* Reeve, 1849, sensu Bosch & Bosch, 1982:127.
Habitat: beached. **Distribution:** Mas.

Conus striatellus Link, 1807: **723**
50mm. Heavy, dull, elongate-conical, slightly convex sided narrowing noticeably towards base, shoulder keeled; short spired with blunt apex; tops of whorls channelled. Weak spiral wrinkles at base. White with 2 broad, interrupted, orange to brown, spiral bands; spire radially striped brown; apex and aperture white. Syn: *C. pulchrelineatus* Hopwood, 1921. **Habitat:** offshore. **Distribution:** GO.

Conus striatus Linnaeus, 1758: **724**
75mm. Heavy, silky, cylindrical, widest well below shoulder with corresponding curve to outer lip; short spired with pointed apex; tops of all whorls channelled. Fine spiral ridges cover shell. White, pink, reddish or bluish, mottled brown, purplish or blackish (spirally lined under magnification), spire radially striped similar colour; apex pink; aperture white. Venomous, so handle carefully! **Habitat:** beached. **Distribution:** SEG, GO, Mas, SO.

Conus taeniatus Hwass, 1792: **725**
35mm. Thick, broadly conical, convex sided, shoulder slightly coronated, short spired, apex usually eroded. Widely spaced, low ridges towards base. Bluish grey with whitish spiral band mid-whorl and at shoulder, spire radially striped same colour; last whorl encircled by regular blackish and white dashes; apex white; aperture purplish with central white band. Commonest cone in eastern Arabia.
Habitat: intertidal among rocks.
Distribution: NG, GO, Mas, SO.

Conus striatus 724

Conus quercinus 719

Conus striatellus 723

Conus rattus 720

Conus saecularis 721

Conus stocki 722

Superfamily: **CONOIDEA**

Conus tessulatus **727**

Conus vexillum sumatrensis **729**

Conus taeniatus **725**

Conus terebra thomasi **726**

Conus textile **728**

Conus virgo **730**

Conus terebra thomasi Sowerby, 1881: **726**
70mm. Heavy, dull, elongate-cylindrical, sides slightly concave, outer lip almost straight edged, shoulder and tops of spire whorls rounded, deep sutures; apex usually eroded. Fine, widely spaced, spiral ridges cover last whorl; finely, spirally striate above shoulder. White to pale yellow or pale violet, 2 darker, broad, spiral bands mid-whorl, base violet; aperture white to violet. **Habitat:** intertidal and beached. **Distribution:** GO, Mas, SO.

Conus tessulatus Born, 1778: **727**
55mm. Heavy, glossy, broadly conical, upper half slightly convex, shoulder roundly keeled; only early whorls elevated. Widely spaced, low, spiral ridges at base; tops of whorls spirally ridged. White with violet base, last whorl covered with spiral rows of reddish-brown or orange rectangles and squares, fusing into 2 bands either side of mid-whorl; spire radially streaked same colour; apex white; aperture white edged violet. **Habitat:** beached. **Distribution:** NG, GO, Mas.

Conus textile Linnaeus, 1758: **728**
70mm. Heavy, glossy, ovate-conical to broadly conical, convex-sided, shoulder rounded; spire concave sided, tops of later whorls channelled, apex pointed. Weak spiral ridges at base. White to bluish white covered by brown-edged, overlapping, open, large and small tents; 2 broad, reddish-brown, interrupted, spiral bands incorporating axial and spiral brown lines; similar pattern on shoulder and spire; early whorls pinkish; aperture white or pinkish. Venomous, so handle carefully! **Habitat:** intertidal under rocks. **Distribution:** GO, Mas, SO.

Conus vexillum sumatrensis Hwass, 1792: **729**
65mm. Heavy, glossy, broadly conical, convex below keeled shoulder then straight sided; short spired or almost flat topped. Weak spiral ridges towards base, tops of spire whorls spirally ridged. White with 2 broad, pale brown, spiral bands above and below mid-whorl; dark brown axial streaks continue onto spire; early whorls yellow; aperture white. Juveniles greenish yellow (described as *C. sulphuratus* Kiener, 1845). **Habitat:** intertidal. **Distribution:** SEG, GO, Mas, SO.

Conus virgo Linnaeus, 1758: **730**
80mm. Heavy, dull, low-conical, sharply keeled shoulder, almost straight sided; short spired or flat topped. Widely spaced, weak spiral ridges towards base; fine spiral threads on spire whorls. White to pale yellow, early whorls paler, base purple; aperture white except for purple base. **Habitat:** beached. **Distribution:** GO, Mas.

SEASHELLS OF EASTERN ARABIA

Superfamily: **CONOIDEA**

Conus zeylanicus Gmelin, 1791: **731**
50mm. Thick, heavy, moderately glossy, broadly conical, shoulder lightly coronated, convex sided, short spired. Columella twisted at base, a swollen spiral ridge above. Low spiral ridges on lower half of last whorl; tops of spire whorls lightly coronated. White to pinkish with pinkish brown zigzags encompassing darker brown dots and dashes; 2 spiral bands of dark brown blotches, continuing on spire; base violet; aperture white or pink. **Habitat:** beached. **Distribution:** Mas, SO.

Etrema spurca **732**

Conus zeylanicus **731**

Clavatula navarchus **734**

Family TURRIDAE

The largest family of gastropods in world seas today, the turrids are carnivorous, feeding on polychaete worms, or scavengers, consuming dead animal matter. Some of the more advanced species have harpoon-like teeth associated with a poison apparatus and thus show an affinity with the cones. Turrid classification is confusing at best, incomprehensible at worst. The importance of the protoconch as an aid to species identification has long been recognised but that feature is often damaged or lacking and often its details may not be revealed without the aid of a powerful microscope or scanning-electron micrographs. The diagnostic significance of protoconch details at the generic level has also been emphasised but they seem to be of questionable value at this level. The shape of the posterior sinus as an aid to distinguishing genera has also received some attention from specialists and this feature is easily examined (although it, too, may vary in configuration within a species). So the following descriptions focus attention more upon this feature than the protoconch. Ultimately anatomical features, especially the radula, may provide the best clues to turrid relationships. The comprehensive review by Melvill (1917a) deals with 184 so-called turrid species, many of them from deep water. Most of these occur in eastern Arabia. Eventually systematists may reduce that number significantly. A comprehensive survey of eastern Arabian turrids is beyond the scope of this book and so only a few species are described and illustrated here.

Subfamily CLATHURELLINAE

Etrema spurca (Hinds, 1844): **732**
11mm. Thick, fusiform, sutures distinct, siphonal canal slightly recurved. Outer lip edge curved in profile, greatly thickened behind and supporting about 6 weakly defined teeth; posterior sinus very deep and rounded. Broad, non-aligned, axial ribs crossed by strong, evenly spaced, spiral ridges (prominent on ribs). Amber, but ribs and columella paler. **Habitat:** offshore. **Distribution:** NG, NWG, GO.

Lienardia soror (Smith, 1882): **733**
13mm. Thick, glossy, fusiform, sutures appressed, siphonal canal slightly recurved. Outer lip edge almost straight in profile, greatly thickened behind and supporting about 6 blunt teeth; posterior sinus broad, rounded. Broad, usually non-aligned, axial ribs crossed by sharp, widely spaced, spiral ridges (prominent on ribs). Amber with orange bands or blotches; aperture orange banded. **Habitat:** offshore. **Distribution:** NG, GO.

Subfamily CLAVATULINAE

Clavatula navarchus (Melvill & Standen, 1903): **734**
60mm. Thick, elongate-fusiform, last whorl about same length as spire, siphonal canal long, sutures deep. Last whorl spirally and obliquely grooved; each spire whorl encircled by a beaded cord below suture, a broad, obliquely grooved cord occupying central portion. Pale brown with darker brown streaks and beads. **Habitat:** deep water. **Distribution:** GO.

Subfamily CONORBINAE

Conorbis coromandelicus (Smith, 1894): **735**
38mm. Solid, dull, almost biconic, aperture about twice as long as spire, gently convex below angled shoulder of last whorl, spire whorls slightly keeled, sutures incised. Last whorl and part of penultimate whorl encircled by strong, regular, sometimes nodulous ribs; early whorls have beaded periphery, intermediate whorls almost smooth. Pale brown, ribs darker, aperture white. **Habitat:** offshore and deeper water. **Distribution:** GO.

Lienardia soror **733**

Conorbis coromandelicus **735**

Superfamily: **CONOIDEA**

Funa variabilis **737**

Funa tayloriana **736**

Inquisitor flavidulus **740**

Inquisitor philotima **741**

Daphnella thia **738**

Hemidaphne axis **739**

Inquisitor sinensis **742**

Splendrillia lucida **743**

Subfamily CRASSISPIRINAE

Funa tayloriana (Reeve, 1846): **736**
35mm. Solid, elongate-fusiform, sutures shallow, aperture moderately wide. Outer lip strongly convex in profile; deep posterior sinus opening slightly upwards; callus on columellar wall visible through sinus. Strong, widely spaced, axial ribs crossed by weak spiral threads. Yellowish white with spiral rows of brown dots and occasional blotches. Syn: *Drillia theoreta* (Melvill, 1899), *D. topaza* (Melvill & Standen, 1901). **Habitat:** offshore. **Distribution:** NG, GO, Mas.

Funa variabilis (Smith, 1877): **737**
40mm. Thick, fusiform, sutures shallow, aperture wide. Outer lip strongly convex in profile; deep posterior sinus; thick callus on columellar wall joins outer lip to form short spout. Axial ribs form pointed nodules at periphery of later whorls. Thin, smooth, spiral ridges encircle all whorls; prominent subsutural cord. Off-white with spiral zones of brown blotches and spots; callus also tinged brown. **Habitat:** mud and coral sand offshore and beached. **Distribution:** SEG, GO, Mas.

Subfamily DAPHNELLINAE

Daphnella thia Melvill & Standen, 1903: **738**
10mm. Thin, semitranslucent, elongate-fusiform with rounded whorls, deep sutures, narrow aperture and extended siphonal canal; protoconch of about 3.5 whorls bearing regular cancellate ornament. Siphonal canal narrowly U-shaped, deep, upturned, its inner edge continuous with inner lip. All whorls have bold cancellate pattern, spiral cords towards base. White or straw coloured. **Habitat:** offshore to deep water. **Distribution:** NG, GO.

Hemidaphne axis (Reeve, 1846): **739**
20mm. Thin, dull, elongate-fusiform, spire more than half total length. Aperture narrow, outer lip thin, posterior sinus roundly U-shaped. Early whorls have cancellate pattern of neat, sharp-edged squares, later whorls having sharp spiral ridges with fine axial striae between them. White with orange-brown blotches and streaks, a white spiral band at periphery of last whorl. **Habitat:** offshore on coral sand. **Distribution:** NG, GO.

Subfamily DRILLIINAE

Inquisitor flavidulus (Lamarck, 1822): **740**
50mm. Thick, elongate-fusiform, last whorl half total length, sutures fine. Aperture flaring, outer lip thickened, posterior sinus deep, callus on upper columellar wall. Axial ribs prominent at periphery of whorls; unequal ridges encircle whorls. Amber, yellow or whitish sometimes with occasional reddish-brown blotches and spiral lines; aperture violet-orange. **Habitat:** offshore and beached. **Distribution:** NG, NWG, GO, Mas.

Inquisitor philotima (Melvill & Standen, 1903): **741**
25mm. Thick, dull, narrowly fusiform, shallow sutures; protoconch of about 3 smooth, glossy whorls. Outer lip thin, strongly curved in profile, greatly thickened behind; posterior sinus broadly V-shaped, joined to callus on columellar wall; columella straight. Coarse axial ribs with almost smooth interspaces initially becoming spirally ribbed later. Cream with spiral bands and blotches of pale brown; aperture white. **Habitat:** offshore. **Distribution:** NWG.

Inquisitor sinensis (Hinds, 1843): **742**
15mm. Solid, glossy, elongate-fusiform, sutures deep; protoconch not seen. Siphonal canal recurved towards apertural side. Outer lip strongly curved in profile; posterior canal deeply U-shaped, slightly upturned and joining callus on columellar wall. Nodulous, obliquely axial ribs, strong pre-sutural cord, close-set, flat-topped spiral ribs. Off-white or cream with faint, axial, orange-brown streaks. Varies in sculpture and colour. **Habitat:** offshore. **Distribution:** GO.

Splendrillia lucida (G. & H. Nevill, 1875): **743**
10mm. Thick, glossy, elongate-fusiform, sutures shallow; protoconch smooth, bulbous. Outer lip thickened behind, posterior sinus broad and deep, partly filled by thick callus projecting from columellar wall. Smooth axial ribs interrupted by broad, spiral groove towards their upper ends give noduled appearance to earlier whorls. Few spiral cords at base. Distinct growth lines. White. **Habitat:** intertidal and offshore in muddy sand. **Distribution:** NG, GO.

Superfamily: **CONOIDEA**

Splendrillia persica (Smith, 1888): **744**
13mm. Thick, glossy, elongate-fusiform, sutures incised; protoconch of 2.5 whorls, bulbous intitially, becoming keeled. Outer lip strongly curved in profile; posterior sinus broad and deep, its upper edge joined to callus on columellar wall. Broad, smooth, axial ribs restricted at their tops by broad subsutural channel. Weak spiral ridges at base. Fine, sinuously axial striae. Translucent-white, sometimes chalky. **Habitat:** offshore. **Distribution:** NG, GO.

Splendrillia resplendens (Melvill, 1898): **745**
19mm. Thick, glossy, elongate-fusiform, sutures incised, sinuous; protoconch bulbous. Outer lip strongly curved in profile; posterior sinus broad and deep, its upper edge joined to callus on columellar wall. Broad, smooth, axial ribs restricted at their tops by smooth subsutural cord. Few spiral cords at base. Ribs straw coloured, their intervals bright red (often faded); aperture white. **Habitat:** intertidal and offshore. **Distribution:** GO, Mas, SO.

Tomopleura pouloensis (Jousseaume, 1883): **746**
25mm. Solid, fusiform, sutures incised. Outer lip deeply corrugated in profile; posterior sinus moderately deep. Strong spiral ridges on all whorls (widely spaced over all of the last) with well-defined axial threads between them (weak or obsolete at sinus level). Obscure spiral bands of amber and greyish brown, ridges white; aperture pale brown. **Habitat:** intertidal and offshore among sand and stones. **Distribution:** NG, GO, Mas.

Tomopleura vertebrata (Smith, 1875): **747**
20mm. Solid, elongate-fusiform, sutures incised. Outer lip corrugated in profile; posterior sinus moderately deep, V-shaped. Strong spiral ridges on all whorls (widely spaced on middle portion of the last) with well defined axial threads between them. Yellowish white with zigzag, chestnut streaks down length of shell; aperture white to violet. **Habitat:** offshore and beached. **Distribution:** GO.

Tylotiella sacra (Reeve, 1845): **748**
22mm. Thick, fusiform, aperture almost half total length, sutures shallow. Outer lip strongly curved in profile; posterior sinus deep and rounded, its upper edge joined to plate-like callus on columellar wall. Widely spaced axial ribs swollen at periphery of whorls; fine spiral threads. White above periphery of whorls and around base, violet-brown below and spotted violet-brown subsuturally. **Habitat:** offshore and beached. **Distribution:** GO, SO.

Subfamily MANGELIINAE
Citharomangelia townsendi (Sowerby, 1895): **749**
13mm. Thin, shiny, elongate-fusiform, sutures well defined, aperture narrow. Edge of outer lip smooth, scarcely curved in profile; shallow posterior sinus. Thick, widely spaced, axial ribs crossed by crowded, fine, spiral threads. Translucent white or greenish grey, a line of brown or grey spots between ribs above periphery of each whorl. **Habitat:** intertidal and offshore. **Distribution:** NWG, SEG, GO.

Subfamily TURRICULINAE
Paradrillia inconstans prunulum (Melvill & Standen, 1901): **750**
10mm. Thick, moderately glossy, elongate-fusiform, whorls appear keeled at periphery because of sculpture, sutures deep; protoconch of 2.5 smooth, keeled whorls. Posterior canal deeply U-shaped; columella slightly curved. Strong axial ribs at periphery of whorls, a beaded cord above sutures, and spiral threads overall. Cream, pale brown to purplish. **Habitat:** offshore. **Distribution:** GO.

Turricula catena (Reeve, 1843): **751**
50mm. Solid, silky gloss, elongate-fusiform, spire almost half total length, long siphonal canal, sutures appressed. Aperture narrow, outer lip thin, posterior sinus broadly V-shaped. Obliquely axial ribs at periphery of earlier whorls, obsolete on later ones; subsutural cord. Spiral grooves above periphery and around siphonal canal. White with dark and pale brown, flame-like, axial streaks. **Habitat:** offshore. **Distribution:** GO, Mas, SO.

Turricula nelliae spuria (Hedley, 1922): **752**
28mm. Solid, ovate-fusiform with long siphonal canal, sutures appressed. Outer lip thin; posterior sinus broadly V-shaped. Nodulous keel at periphery of whorls and

Splendrillia persica **744**

Splendrillia resplendens **745**

Citharomangelia townsendi **749**

Tomopleura pouloensis **746**

Paradrillia inconstans prunulum **750**

Tomopleura vertebrata **747**

Turricula catena **751**

Tylotiella sacra **748**

Superfamily: **CONOIDEA**

nodulous subsutural cord; fine spiral threads on upper part of whorls and beaded spiral ridges below periphery of last whorl. Cream to orange-brown, nodules paler. **Habitat:** offshore. **Distribution:** GO.

Subfamily TURRINAE
Lophiotoma acuta (Perry, 1811): **753**
45mm. Solid, dull, fusiform, spire more than half total length, long siphonal canal, sutures appressed; pointed protoconch of 3 smooth, glossy whorls. Prominent, sharp, double keel at periphery of each spire whorl with lesser spiral ridges above and below. White with pale or dark brown spots and blotches. **Habitat:** offshore. **Distribution:** GO.

Lophiotoma indica (Röding, 1798): **754**
80mm. Thick, dull, elongate-fusiform, rounded whorls have sharp central keel, siphonal canal long and straight, spire almost half total length. Outer lip thin, posterior sinus broad and deep. Smooth spiral cords encircle all whorls. White blotched brown, markings sometimes flame-like, cords dotted brown; aperture and columella white. **Habitat:** offshore and beached. **Distribution:** GO.

Unedogemmula unedo (Kiener, 1840): **755**
80mm. Solid, dull, fusiform, upper part of whorls convexly sloping, lower part rounded, siphonal canal long and slightly curved. Posterior sinus deeply U-shaped. Sharp keel divides upper part of each whorl from lower part; surface encircled by sharp-edged ridges. White or pale yellow flecked and streaked with brown which is darkest on ridges; columella white. Syn: *Turris invicta* Melvill, 1910. **Habitat:** offshore. **Distribution:** GO.

Subfamily ZONULISPIRINAE
Ptychobela griffithii (Gray, 1834): **756**
35mm. Solid, dull, fusiform, last whorl about half total length, sutures appressed. Outer lip thickened behind, posterior sinus deep, callus on upper columellar wall joins outer lip to form short spout. Axial ribs form nodes at periphery of whorls, all whorls encircled by low ridges. Shades of brown, ribs paler. Sculpture varies considerably in strength. **Habitat:** intertidal and offshore. **Distribution:** NWG, SEG, GO, Mas.

Ptychobela opisthochetos Kilburn, 1989: **757**
21mm. Solid, biconic to elongate-fusiform, sutures appressed. Outer lip thickened behind, well curved in profile; deep posterior sinus, callus on upper columellar wall joins outer lip to form spout (extending to shoulder of last whorl). Prominent rounded nodes at periphery of each whorl, low ridges encircle all whorls. Colour forms include: a) pale to dark brown, with paler nodes and dark brown apex; b) yellowish, white or grey, dark brown apex and between nodes; c) white with brown apex. **Habitat:** intertidal and offshore. **Distribution:** Mas.

Lophiotoma acuta 753

Ptychobela griffithii 756

Turricula nelliae spuria 752

Lophiotoma indica 754

Unedogemmula unedo 755

Ptychobela opisthochetos 757

SEASHELLS OF EASTERN ARABIA

Superfamily: **CONOIDEA**

Family TEREBRIDAE
Auger shells, so called because of their long, acutely pointed spires, are predatory sand dwellers present in the warm waters of most seas and are familiar objects between tide marks. Like the cones some auger shells have a poison apparatus connected to harpoon-like radular teeth through which they inject venom into prey before swallowing it. The shell has a thin outer lip and the columella usually has a prominent fold. Many species are colourful and glossy, many plain and dull, but most have axial ribs and a prominent sub-sutural cord. There is no umbilicus. The thin, chitinous operculum is variously shaped and attached to the small, muscular foot. The only review of eastern Arabian members of the family, by Melvill, appeared in 1917 so the status and systematic position of several species described and illustrated here is tentative.

Duplicaria duplicata (Linnaeus, 1758): **758**
50mm. Solid, glossy, straight sided, whorls slightly stepped at sutures; globular protoconch of 2 smooth whorls. Strong axial ribs on all whorls, a spiral subsutural groove resembling a secondary suture on each whorl. Uppermost of 2 basal ridges continuous with outer lip. White, cream, brown or bluish with pale spiral band containing isolated brown spots; apex blue. **Habitat:** intertidal in sand. **Distribution:** NG, SEG, GO, Mas, SO.

Duplicaria spectabilis (Hinds, 1844): **759**
25mm. Solid, glossy, straight sided but earlier whorls expanding more rapidly than later ones, last whorl roundly keeled at base, sutures deep; protoconch of about 2 rounded whorls. Deep subsutural groove has crowded, straight, axial ridges above and below on all whorls, smooth between ridges. Siphonal canal strongly recurved. Amber with white band above sutures, columella white. Syn: *Terebra edgarii* Melvill, 1898, *T. remanalva* Melvill, 1910. **Habitat:** intertidal and offshore in mud. **Distribution:** NG, Mas.

Hastula nana (Deshayes, 1859): **760**
10mm. Thin, glossy, needle-like with straight-sided whorls, shallow sutures and broad siphonal canal; blunt-topped protoconch of 2.5 whorls. Widely spaced, scarcely raised, axial ridges crossed by fine scratches. Cream or white with two spiral rows of rectangular, brown spots between ridges, a third row at base of last whorl. **Habitat:** intertidal in sand. **Distribution:** GO, Mas.

Impages hectica (Linnaeus, 1758): **761**
55mm. Solid, glossy, straight sided, sutures finely incised, last whorl gently rounded below; pointed protoconch of 3 smooth whorls. Weak axial ribs on early spire whorls, later whorls have fine growth striae. Basal ridge does not reach columella. Greyish yellow with white band above sutures, purplish brown flames and streaks below them; protoconch amber. Syn: *T. caerulescens* Lamarck, 1822. **Habitat:** intertidal in sand. **Distribution:** GO, Mas, SO.

Terebra ambrosia Melvill, 1912: **762**
16mm. Solid, silky, narrowly elongate, about 11 slightly convex whorls, stepped at sutures, base gently rounded; protoconch of 2 smooth, rounded whorls. Prominent subsutural groove cuts through sharp, slightly curved, axial ribs; coarse spiral threads between ribs; thin basal ridge continuous with outer lip. Purplish brown to golden brown. **Habitat:** intertidal and offshore in sand. **Distribution:** GO, Mas.

Terebra anilis (Röding, 1798): **763**
50mm. Solid, moderately glossy, very narrow, up to 25 concave-sided whorls, sutures shallow, base slightly keeled. Each whorl has 5 low, spiral ridges and many more axial ridges giving cancellate effect; above these ridges is a broad, nodular, subsutural cord with lesser cord below. Strongly twisted siphonal canal bears a thin basal ridge continuous with outer lip. Dark brown fading to amber. **Habitat:** intertidal in sand. **Distribution:** NWG, GO, Mas, SO.

Terebra anilis 763

Duplicaria spectabilis 759

Duplicaria duplicata 758

Terebra ambrosia 762

Hastula nana 760

Impages hectica 761

170 SEASHELLS OF EASTERN ARABIA

Superfamily: **CONOIDEA**

Terebra babylonia Lamarck, 1822: **764**
65mm. Solid, dull, straight sided, sutures impressed, base rounded. Early whorls have 2 rows of broad nodules below sutures which fuse into one row on later whorls; deep, curved, axial and spiral grooves conspicuous on later whorls. Off-white with orange grooves and aperture. **Habitat:** offshore and intertidal. **Distribution:** NWG, GO.

Terebra bathyrhaphe Smith, 1875: **765**
17mm. Thin, silky, broadly needle-like, spire whorls slightly concave sided, base rounded, aperture narrow, sutures deep; pointed protoconch of about 3 smooth, glossy whorls. Widely spaced, slightly curved, axial ribs which are depressed centrally and crossed by fine, irregular threads, these being stronger around base. Greyish white or fawn. Syn: *T. persica* Smith, 1877. **Habitat:** intertidal and offshore. **Distribution:** NG, Mas.

Terebra cinctella Deshayes, 1859: **766**
30mm. Thick, dull, straight sided, sutures deep, base slightly angled, outer lip sinuous in profile; protoconch of 2 rounded whorls. Strong, sinuous, axial ribs interrupted by strong subsutural groove on all spire whorls; weak spiral ridges at base, basal ridge strong and sharp edged. White. **Habitat:** intertidal in sand. **Distribution:** Mas.

Terebra fuscobasis Smith, 1877: **767** NOT ILLUSTRATED
10mm. A variable species with arched axial ribs and white with a brown zone subsuturally and another below periphery of last whorl. Smaller than **T. nassoides** (see below), with more arched ribs and lacking the purple columella. Syn: *Strioterebrum (Partecosta) wilkinsi* Dance & Eames, 1966. **Habitat:** offshore. **Distribution:** NG.

Terebra helichrysum Melvill & Standen, 1903: **768**
25mm. Solid, glossy, straight sided, sutures incised; pointed, smooth protoconch. Coarsely beaded subsutural cords become prominent on later whorls; evenly spaced, smooth, crescentic, axial ribs with 3 evenly spaced, spiral riblets between each pair of ribs. Yellowish white with widely spaced, pale brown patches on subsutural cord. **Habitat:** offshore. **Distribution:** GO.

Terebra loisae Smith, 1903: **769**
30mm. Solid, glossy, straight sided, sutures impressed, base roundly angled; pointed protoconch of 2 smooth whorls. Weak, sinuously axial ribs interrupted by weak subsutural groove on all spire whorls; basal ridge continuous with outer lip. White tinged beige and blue with isolated brown spots on subsutural band, sometimes yellowish white; apex purplish. **Habitat:** intertidal in sand. **Distribution:** Mas.

Terebra macandrewi Smith, 1877: **770**
15mm. Solid, moderately glossy, whorls slightly convex, sutures deep, last whorl gradually rounded below; protoconch of 3 smooth whorls. Sharp, axial, slightly sinuous ribs, indented below sutures; fine spiral striae between ribs. Basal ridge reaches columella. Whitish below sutures, otherwise whorls are pale fawn with 2 reddish brown, spiral bands. **Habitat:** offshore and beached. **Distribution:** NG, GO, Mas.

Terebra maculata (Linnaeus, 1758): **771**
115mm. Thick, heavy, silky gloss, slightly convex sided, sutures almost stepped, sharply pointed protoconch. Very thick columella has 2 broad, spiral grooves. Smooth except for fine growth striae. Cream with spiral rows of dark brown streaks and beige bands; columella white. Mostly worn shells only; illustrated from Maldive Islands specimens. **Habitat:** beached. **Distribution:** Mas, SO.

Terebra cinctella **766**

Terebra loisae **769**

Terebra maculata **771**

Terebra helichrysum **768**

Terebra babylonia **764**

Terebra macandrewi **770**

Terebra bathyrhaphe **765**

SEASHELLS OF EASTERN ARABIA

Superfamily: **CONOIDEA, RISSOELLOIDEA, ARCHITECTONICOIDEA**

Terebra nassoides Hinds, 1844: **772**
20mm. Solid, glossy, almost straight sided, sutures well incised, last whorl gently rounded below; pointed protoconch of 2 smooth whorls. Axial, sometimes sinuous, ribs, thickest at their tops and depressed centrally. Uppermost of 2 basal ridges sharp and continuous with outer lip. Yellowish or bluish, sometimes with 2 spiral bands of faint brown spots between ribs; columella, basal ridges and apex purple. **Habitat:** intertidal in sand. **Distribution:** NG, SEG, GO, Mas.

Terebra nebulosa Sowerby, 1825: **773**
50mm. Solid, glossy, almost straight sided, sutures incised, last whorl strongly rounded below; protoconch not seen. Thick, sinuous, axial ribs encircled below sutures by a groove punctured at intervals between ribs. White with irregular, reddish-orange blotches, punctured grooves and aperture orange. **Habitat:** beached. **Distribution:** GO.

Terebra triseriata Gray, 1834: **774**
75mm. Solid, moderately glossy, very long and narrow with up to 40 straight-sided whorls, deep sutures, small aperture with strongly twisted columella; protoconch not seen. Subsutural, spiral row of large nodules, smaller row below; another still smaller, presutural row; each whorl has weak axial ridges crossed by 4-6 strong spiral cords. Yellow to brown. Record based on a fresh shell, lacking early whorls, from Khor Fakkan, collected by H. Kauch; photo is of 2 complete shells from Japan. **Habitat:** offshore. **Distribution:** GO.

Superfamily RISSOELLOIDEA
Discoidal or short-spired snails with translucent, nondescript shells seldom exceeding two millimetres. Two families accommodate the few known species which live in shallow water among algae.

Family OMALOGYRIDAE
Minute snails coiled in one plane so that the protoconch is visible from either side. All species have a sculpture of coarse axial ridges and growth lines. They occur in tide pools among algae.

Omalogyra japonica (Habe, 1972): **775**
0.42mm. Fragile, dull, discoidal, umbilicus very wide, sutures deep, aperture circular; protoconch of 1 whorl. Coarse axial ridges on last whorl. Brown. Probably the smallest gastropod in eastern Arabia. **Habitat:** tide pools among algae. **Distribution:** GO, Mas.

Family RISSOELLIDAE
Rissoellid snails are minute, their fragile, smooth shells seldom exceeding 2 millimetres in length. The semicircular operculum has a blunt, peg-like projection on the edge nearest the columella. Those species which have been observed alive have variously pigmented bodies, some being spotted or banded. They are found among seaweeds in shallow water where they feed on various organisms, including diatoms.

Rissoella cf *atrimacula* Ponder & Yoo, 1977: **776**
1mm. Fragile, translucent, smooth, globose, short spired and umbilicate. The colour of the animal is unknown. **Habitat:** shallow water. **Distribution:** GO, Mas, SO.

Superfamily ARCHITECTONICOIDEA
Contains two families: the Architectonicidae with sturdy, top-shaped or discoidal shells which may occur intertidally, and the Mathildidae with thin, elongate shells, restricted to offshore waters.

Family ARCHITECTONICIDAE
The shells of this family, popularly known as sundials, are mostly low spired, about twice as wide as high, usually have a flat or slightly concave base and a wide, deep umbilicus, the nodular cord on its inner edge winding within like a coil of rope (in *Heliacus* a 2nd and sometimes a 3rd cord may be seen within). The tilted protoconch is sunk below level of succeeding whorl. The chitinous operculum may be flat and few whorled or plug-like and many whorled. Tropical and sub-tropical in distribution, sundials feed on corals, sea-anemones and other coelenterates and have a long, free-swimming larval stage. In eastern Arabia they are seldom found alive intertidally but may occur plentifully as empty shells on beaches.

Architectonica gualtierii Bieler, 1993: **777** NOT ILLUSTRATED
A large, conical species, similar to *A. stellata* (see below) but has a more spotted base and more reticulate sculpture above. No shells were available for illustration. **Habitat:** offshore on sand and mud. **Distribution:** GO, SO.

Terebra triseriata 774

Omalogyra japonica 775

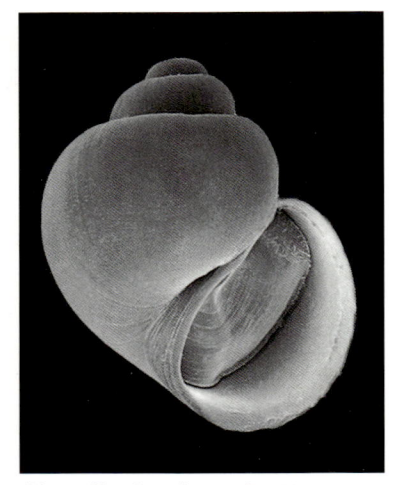

Rissoella cf *atrimacula* 776

Terebra nassoides 772

Terebra nebulosa 773

Superfamily: **ARCHITECTONICOIDEA**

Architectonica laevigata (Lamarck, 1816): **778**
30mm. Thick, silky, with slightly rounded spire whorls and convex base, sutures deep and distinct. 4 deep spiral grooves per spire whorl crossed by radiating grooves on earlier whorls; broad, deep umbilicus encircled by 2 grooves. Each whorl has lavender spiral band with white and brown-spotted bands above and below, base paler and spirally brown spotted. **Habitat:** intertidal and offshore. **Distribution:** GO, Mas.

Architectonica perspectiva (Linnaeus, 1758): **779**
40mm. Thick, glossy, with slightly rounded spire whorls and convex base; sutures fine. Each whorl has a broad spiral zone with a cord above and another below, crossed by radiating grooves; base has smooth, double rim; broad, deep umbilicus encircled by 2 grooves. Beige-pink with white post-sutural cord adjacent to brown band and brown-spotted pre-sutural band; base paler, brown spots around umbilicus and on basal rim. Flat operculum has eccentric nucleus. **Habitat:** on sand. **Distribution:** NG, SEG, GO, Mas.

Architectonica cf *reevei* (Hanley, 1862): **780**
28mm. Solid, glossy, spire elevated but flat topped, almost straight sided, base flattened, sutures fine. Each whorl has 1 broad zone crossed by radiating grooves with 2 cords below, earlier whorls appear beaded; base has smooth, double rim; moderately broad, deep umbilicus encircled by 1 cord, crenulate within. Pinkish cream with brown blotches on cords on spire; base paler with brown spots around umbilicus and on double rim. Single empty shell found by H. Kauch at Fujairah. **Habitat:** beached. **Distribution:** GO.

Architectonica stellata (Philippi, 1849): **781**
30mm. Thick, silky, spire dome-like, base flattened, sutures fine. Each whorl has 2 broad zones crossed by radiating grooves (absent from last whorl) with a cord above and another below; base has smooth, double rim; broad, deep umbilicus encircled by 2 cords (1 crenulate and 1 beaded). Pinkish cream with brown blotches on cords on spire; base paler, brown spots around umbilicus and at periphery. Syn: *Solarium abyssorum* Melvill & Standen, 1903 (juvenile form), *A. purpurata* (Hinds, 1844) sensu Bosch & Bosch, 1982:43, non Hinds, 1844. **Habitat:** offshore and beached. **Distribution:** GO.

Granosolarium asperum (Hinds, 1844): **782**
3mm. Thin, depressed-discoidal, with very wide open umbilicus, apertural lip fragile; protoconch deeply submerged in spire. Last whorl acutely keeled and crenulate, as is the keel surrounding the base. White. Syn: *Solarium (Torinia) admirandum* Melvill & Standen, 1903 (the accompanying figures, from Melvill, 1904e, illustrate this taxon). **Habitat:** deep water. **Distribution:** GO.

Architectonica perspectiva **779**

Architectonica laevigata **778**

Architectonica cf *reevei* **780**

Architectonica stellata **781**

Granosolarium asperum **782**

SEASHELLS OF EASTERN ARABIA

Superfamily: **ARCHITECTONICOIDEA**

Heliacus (Heliacus) areola
(Gmelin, 1791): **783** NOT ILLUSTRATED
12mm. Thick, dull, usually twice as wide as high, spire dome shaped, base convex. Each spire whorl has 4 deep spiral grooves; intervening ribs coarsely, radially grooved. 2 cords within the narrow umbilicus. Cream blotched and radially streaked blackish brown; white around umbilicus. Easily confused with ***H. variegatus*** (see below). **Habitat:** offshore and beached. **Distribution:** SEG, GO.

Heliacus (Heliacus) variegatus
(Gmelin, 1791): **784**
20mm. Solid, dull, high or low domed, usually twice as wide as high, almost flat topped, base concave. Each whorl has 4 deep, spiral grooves; intervening ribs cut by radiating grooves. 2 cords within the wide umbilicus. Dark or pale brown with white dashes on pre-sutural cords and sometimes irregularly streaked white; columella white. May be confused with ***H. areola***. **Habitat:** offshore and beached. **Distribution:** NG, GO, Mas, SO.

Heliacus (Teretropoma) infundibuliformis (Gmelin, 1791): **785**
10mm. Solid but brittle, dull, discoidal, about twice as broad as tall, aperture almost circular, sutures deep, umbilicus very wide; protoconch of 2 smooth, sunken whorls. Double keeled at periphery of last whorl, rest of shell spirally ribbed, keels strongly beaded, ribs weakly beaded. Reddish brown with paler streaks at periphery. **Habitat:** offshore and beached. **Distribution:** Mas.

Heliacus (Torinista)* cf *caelatus
(Hinds, 1844): **786**
10mm. Thin, discoidal, low-domed spire, base flat, sutures deep, aperture 5-sided, umbilicus wide open; protoconch of 1.5 smooth, sunken whorls. Spire whorls crossed by oblique ribs, beaded each end, spiral groove separating each row of beads from ribs. Double keeled in profile, a lesser keel and spiral grooves crossed by axial ribs on base; double keels visible in umbilicus. Whitish with occasional fawn spots on upper keel. The shells illustrated here are only tentatively assigned to this species. **Habitat:** sand offshore. **Distribution:** NG.

Heliacus (Torinista) cerdaleus
(Melvill & Standen, 1903): **787**
7mm. Thin, discoidal, low-domed spire, base rounded, sutures inconspicuous, aperture 5-sided, umbilicus wide open; protoconch of 1.5 smooth, sunken whorls. Spiral and axial grooves on spire whorls produce spiral rows of beads, 3 rows around flattened periphery (none around umbilicus). Whitish with occasional brownish beads on spire, pinkish around umbilicus. **Habitat:** offshore. **Distribution:** NG.

Heliacus (Torinista) implexus
(Mighels, 1845): **788**
12mm. Thick, dull, twice as wide as high. Each spire whorl has 4 shallow spiral grooves and a broad subsutural rib; regular, radiating grooves cover upper and lower surfaces. Aperture corrugated all around. 2 cords within umbilicus. Operculum conical and multispiral. Cream mottled and barred with shades of brown, umbilicus paler. Syn: *Solarium homalaxis* Melvill, 1893. **Habitat:** offshore, deep water and beached. **Distribution:** GO, Mas.

Heliacus (Torinista) rotula
Kilburn, 1975: **789** NOT ILLUSTRATED
4.4mm. A small species superficially resembling *H. (T.) caelatus* (see above). Known from SEG and GO.

Pseudomalaxis zanclaeus meridionalis (Hedley, 1903): **790**
NOT ILLUSTRATED
Listed, as *P.* cf *zancleus* (Philippi, 1844), by Glayzer et al. 1984:319, from Kuwait (NWG) and by Smythe, 1979a:65, from Dubai (SEG). A Pacific species, possibly a misidentification for ***P. obolos*** (Barnard, 1963) for which there is a record, from the Arabian Gulf, by Bieler, 1993:318.

Psilaxis radiatus (Röding, 1798): **791**
12mm. Solid, glossy, discoidal, base rounded, periphery keeled; sutures fine. Double cord above each suture; triple cord at periphery; surface almost smooth; wide, deep umbilicus encircled by 2 cords (1 crenulate, 1 beaded). Dark brown with interrupted white cords; crenulate umbilical cord white. Syn: possibly the species listed as *Philippia hybrida* (Linnaeus, 1758) by Melvill, 1928:100, from Muscat (GO).

Heliacus (Torinista) implexus **788**

Heliacus (Heliacus) variegatus **784**

Heliacus (Torinista) cerdaleus **787**

Heliacus (Torinista) cf *caelatus* **786**

Heliacus (Teretropoma) infundibuliformis **785**

Heliacus (Torinista) implexus **788**

Habitat: offshore and beached. **Distribution:** SEG, GO, SO, Mas.

Spirolaxis cornuammonis (Melvill & Standen, 1903): **792**
5mm. A fragile, semitransparent, loosely coiled tube, 6 keels seen in cross section, descending in a low spiral; smooth, glossy protoconch of about 2 whorls, demarcated from spire by a thin ring. Spire whorls encircled by weak axial threads which develop sharp points where they cross keels. Milky white. **Habitat:** deep water. **Distribution:** GO.

Spirolaxis rotulacatharinea (Melvill & Standen, 1903): **793**
3mm. Thin, fragile, translucent, discoidal, its whorls disconnected from the clearly defined end of the 2-whorled protoconch. Aperture roundly rectangular in outline. Tube has a strong, regularly ribbed, spiral keel above and below which helps shell to stand up edge-wise. Milky white. **Habitat:** deep water. **Distribution:** GO.

Family MATHILDIDAE
Elongate shells with a heterostrophe protoconch which immediately distinguishes them from shells of the otherwise similar *Turritella*. There is no umbilicus but an operculum is present. The few species live offshore and are nowhere common.

Mathilda carystia Melvill & Standen, 1903: **794**
11mm. Thin, dull, elongate, sutures impressed; heterostrophe protoconch of 1 smooth whorl. Each spire whorl has about 4 prominent spiral keels (about 6 on last whorl); keels continue as spiral ridges on base. Regular axial bars join keels. Amber to brown, protoconch colourless. **Habitat:** offshore in muddy sand. **Distribution:** NWG.

Mathilda gracillima Melvill & Standen, 1901: **795**
11mm. Thin, semitranslucent, elongate, sutures deep; complete protoconch not seen. Each spire whorl has 2 strong and 2 weaker spiral keels (2 weak extra keels on last whorl). Fine axial threads overall. Yellowish white. Incomplete protoconch on holotype suggests that a heterostrophe element was lacking and so species is probably not a *Mathilda*. **Habitat:** offshore in sand and mud. **Distribution:** GO.

Mathilda telamonia Melvill, 1912: **796**
13mm. Thin, moderately glossy, narrowly elongate, sutures deep; heterostrophe protoconch of 2 smooth whorls. The middle of each spire whorl has a bold spiral keel, another bold keel above suture and about 4 lesser keels. Between keels are regular, delicate, axial riblets. Whitish. **Habitat:** offshore. **Distribution:** GO.

Mathilda zmitampis Melvill & Standen, 1901: **797**
10mm. Thin, moderately glossy, broadly elongate, sutures impressed; protoconch of 2 smooth, glossy, heterostophe whorls. Each spire whorl has 4 strong, spiral keels linked by strong axial riblets giving cancellate effect, weak spiral threads on base. White. **Habitat:** deep water in sand. **Distribution:** GO.

Superfamily PYRAMIDELLOIDEA

Family PYRAMIDELLIDAE
A large, complicated family of parasitic gastropods bearing elongate and tapering, often minute and usually glossy shells. The animal, which lacks a radula, feeds by sucking blood and fluids from the bodies of its invertebrate hosts with which it is always closely associated when alive. Most of the species recorded from eastern Arabia are known only from empty shells, their hosts being unknown. The shell has a small aperture and there are usually folds on the columella, but the differences between species are often slight, so their identification is often a matter for specialists. The protoconch, which is usually missing, is unusual in that its whorls coil in a direction opposite to those of the spire, a condition known as heterostrophy. A few are described and illustrated here but most eastern Arabian species have been inadequately studied by specialists. Some are well illustrated by Melvill, 1910a.

Mathilda gracillima **795**

Mathilda telamonia **796**

Mathilda carystia **794**

Mathilda zmitampis **797**

Psilaxis radiatus **791**

Spirolaxis cornuammonis **792**

■ *When Captain F. W. Townsend examined the mud brought up on the anchor of his ship from a depth of 156 fathoms off Muscat he knew it contained many tiny shells. He may have expected some of them to be unusual. He could not have been prepared, however, for the discovery of anything so unusual as the felicitously named* Spirolaxis cornuammonis. *At once exquisite and curious, a ram's horn in miniature, it reminds us that many molluscan species have developed external coverings far removed in appearance from the popular conception of a seashell.*

Spirolaxis rotulacatharinea **793**

Superfamily: **PYRAMIDELLOIDEA**

Subfamily ODOSTOMIINAE

Chrysallida* cf *fischeri (Hornung & Mermod, 1925): **798**
4mm. Elongate-conic, spire whorls gently rounded, sutures impressed; protoconch of about 2 smooth whorls; tiny umbilicus. Straight axial ribs the same width as their intervals which are delicately, spirally striate. White. **Habitat:** offshore and beached. **Distribution:** GO.

Gurmatia pulchrior (Melvill, 1904): **799**
3.5mm. Thin, translucent, globose-ovate, whorls slightly ledged at sutures; protoconch of 1.5 obliquely immersed whorls. Columella slightly sinuous, umbilicus tiny. Close-set, smooth, spiral cords separated by narrower grooves; fine axial threads override spiral sculpture. White. This species was placed in ***Gurmatia***, when that genus was introduced, by Dance & Eames, 1966: 41. **Habitat:** offshore and beached, among coral sand. **Distribution:** SEG.

Hinemoa* cf *indica (Melvill, 1896): **800**
6mm. Solid, moderately glossy, needle-like, each of the seemingly straight-sided spire whorls strengthened by 3 strong, rounded, spiral ribs, the last whorl having an extra rib around the gently rounded base; protoconch submerged in previous whorl. Fine, axial threads between ribs; columella straight and smooth. Yellowish or white. **Habitat:** offshore and beached. **Distribution:** NWG, Mas.

Miralda scopulorum (Watson, 1886): **801**
1.5mm. Solid, narrowly ovate; protoconch of 1 smooth whorl. Each spire whorl has 2 spiral rows of coarse beads, below them a strong spiral cord (3 on last whorl). Columella has a small median fold. White. **Habitat:** intertidal. **Distribution:** GO, Mas.

Mumiola spirata (A. Adams, 1853): **802**
3mm. Solid with stepped whorls and lacking a fold on the columella, it was described (as *M. carbasea* Melvill, 1904) from 156 fathoms off Muscat (GO), was listed by Melvill, 1910a: 193, from Bushehr (NG) and Musandam (GO) and occurs intertidally at Masirah.

Odostomia eutropia Melvill, 1899: **803**
6mm. Thin, glossy, smooth, ovate or oblong-fusiform, spire whorls straight sided, last whorl sharply or weakly keeled, sutures deeply channelled; heterostrophe protoconch deeply sunk in succeeding whorl. Strong fold on columella; aperture lirate deep within. **Habitat:** intertidal and offshore in mud. **Distribution:** all.

Oscilla* cf *jocosa Melvill, 1904: **804**
2.5mm. Solid, strongly, spirally ribbed and with a columella oblique to the shell's axis, it was described from 156 fathoms off Muscat (GO) and also occurs intertidally at Masirah.

Pyrgulina callista Melvill, 1893: **805**
4mm. A semitransparent shell listed by Melvill, 1910a:196, from Qeshm Island (NG) and Musandam (GO); also known from Masirah.

Syrnola* cf *aclis (A. Adams, 1854): **806**
11mm. Solid, glossy, broadly needle-like, whorls almost straight sided, sutures well defined; protoconch usually missing. Columella has a strong fold at its upper end. Widely spaced lirae deep within aperture. Surface smooth. Golden brown. Identification of illustrated shells is tentative only. **Habitat:** offshore and beached. **Distribution:** NWG, SO.

Superfamily: **PYRAMIDELLOIDEA**

Syrnola cf *brunnea* (A. Adams, 1854): **807**
10mm. Thin, glossy, needle-like, whorls straight sided, sutures deep; protoconch of about 1.5 obliquely immersed whorls. Columella has a strong fold at its upper end. Yellowish brown to golden brown. Smooth except for very fine growth marks. Aperture smooth deep within. **Habitat:** intertidal under rocks. **Distribution:** NWG, GO.

Subfamily PYRAMIDELLINAE
Eulimella venusta Melvill, 1904: **808**
7mm. Described from 40 fathoms in the Arabian Sea, it occurred also at 156 fathoms off Muscat and has been dredged off As Sib (GO).

Otopleura mitralis (A. Adams in Sowerby, 1854): **809**
17mm. Solid, dull, elongate-ovate to elongate-fusiform, whorls convex and usually slightly stepped at sutures, seemingly telescoped. Outer lip thin, columella with 3 folds; no umbilicus. Irregular axial ribs with coarse malleations. Cream blotched and mottled brown and greyish brown. Varies greatly in size, shape and strength of sculpture. **Habitat:** intertidal and offshore in coral sand. **Distribution:** NG, GO, Mas.

Pyramidella acus (Gmelin, 1791): **810**
45mm. Solid, glossy, elongate, spire almost straight sided; apical whorls usually missing. Aperture usually lirate within, columella with 3 folds, the uppermost the strongest; no umbilicus. Tops of whorls narrowly ledged and finely corrugated. White or bluish white with purplish or brown blotches and brown dots (5 rows of dots on last whorl); aperture and columella white. **Habitat:** offshore and beached. **Distribution:** NG, GO, Mas.

Pyramidella maculosa Lamarck, 1822: **811**
45mm. As previous species but sometimes with shallow spiral grooves, sutures within V-shaped intervals between whorls. White or beige with pale, orange-brown blotches which often form axial streaks; aperture and columella white. Syn: *Obeliscus sulcatus* A. Adams, 1854. **Habitat:** offshore among sand, mud and stones. **Distribution:** NG, GO, Mas.

Pyramidella dolabrata (Linnaeus, 1758) var. *terebelloides* (A. Adams, 1854): **812** NOT ILLUSTRATED
14mm. Thin, semitranslucent, glossy, elongate-conic, spire almost straight sided; protoconch of 1.5 partly submerged, heterostrophe whorls. Small, deep umbilicus; columella has 1 to 4 folds (usually 3), outer lip thin. Growth lines and fine spiral scratches. Bluish white to yellowish with reddish brown spiral lines (about 5 on last whorl). Much smaller and more fragile than *P. maculosa*. The only shells available for illustration were mislaid. **Habitat:** offshore. **Distribution:** GO.

Subfamily TURBONILLINAE
Tropaeas brunneomaculata (Melvill, 1897): **813**
15mm. Solid, dull, narrowly elongate, whorls straight sided, sutures well defined; protoconch about same height as first spire whorl. Columella has a low fold at its upper end. Flattened axial ribs as wide as their intervals; close-set, fine, spiral striae encircle all whorls. White with scattered mahogany-brown spots and dashes. Probably not a *Tropaeas*. **Habitat:** offshore in muddy sand and beached. **Distribution:** NG, GO.

Syrnola cf *brunnea* 807

Eulimella venusta 808

Pyramidella acus 810

Otopleura mitralis 809

Pyramidella maculosa 811

Tropaeas brunneomaculata 813

SEASHELLS OF EASTERN ARABIA

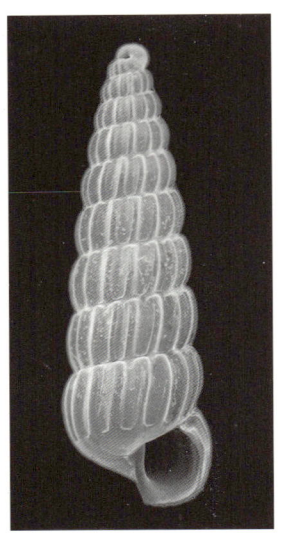

Superfamily: **PYRAMIDELLOIDEA, PHILINOIDEA**

Turbonilla icela Melvill, 1910: **814**
5mm. Delicate, narrowly elongate with well-rounded spire whorls and deep sutures; protoconch of about 2 smooth, heterostrophe whorls. Obliquely axial ribs narrower than their smooth intervals. Milky white. **Habitat:** offshore to deep water in sand. **Distribution:** NG, GO.

Turbonilla linjaica Melvill & Standen, 1901: **815**
4mm. Delicate, narrowly elongate with gently rounded spire whorls and impressed sutures; protoconch of two smooth, heterostrophe whorls. Almost straight, crowded, axial ribs with fine, crowded, spiral striae between them. Reddish to almost white. **Habitat:** offshore and intertidal in sand. **Distribution:** NG, NWG.

Turbonilla stylifera Thiele, 1925: **816**
8mm. Thin, glossy, initially needle-like, then gradually expanding towards flat base, sutures deep; protoconch of about 2 smooth, heterostrophe whorls. Columella has 1 strong fold. First 3 spire whorls smooth and straight sided, fourth whorl weakly axially ribbed, succeeding whorls with strong, obliquely axial ribs; between each pair of ribs is 1 puncture mid-whorl and 1 above suture. White. Most distinctive pyramidellid of eastern Arabia. **Habitat:** offshore in muddy sandy. **Distribution:** GO.

Family AMATHINIDAE
The limpet-like shell shows that the animal clings to smooth surfaces such as boulders and other shells. In this respect it is unlike that of most families in the Pyramidelloidea, but the heterostrophe protoconch is typical for the superfamily.

Amathina tricarinata (Linnaeus, 1758): **817**
27mm. Solid, dull, roughly oval in outline, narrower posteriorly; protoconch heterostrophic. From the backwardly recurved apex 3 strong, rounded and occasionally scaly ribs radiate to the anterior edge where they project slightly; low, wavy, axial ribs occupy posterior slope. White outside tinted yellow inside. Has a brown, fibrous periostracum when fresh. Syn: *Patella tricostata* (Gmelin, 1791). **Habitat:** intertidal attached to other objects. **Distribution:** NWG.

Leucotina gratiosa Melvill, 1898: **818**
12mm. Thin, translucent, glossy, attenuately fusiform, spire whorls ledged at sutures then gently rounded, aperture elongate, columella straight; heterostrophe protoconch submerged in succeeding smooth whorl. Spire whorls encircled by evenly spaced, punctate grooves (15 on last whorl). Milky white. **Habitat:** offshore and beached. **Distribution:** NG, NWG, GO.

Subclass OPISTHOBRANCHIA
With few exceptions all opisthobranchs (so called in allusion to the posterior position of the mantle cavity) occur in marine biotopes. Many species, generally known as nudibranchs or sea slugs, have no shell and so have no place in this book. Those with a shell make up a small proportion of the eastern Arabian molluscan fauna. Most of them have a parasitic or carnivorous habit. Broadly speaking the species have thin, globose shells and are known as bubble shells. A thin, chitinous operculum is present in some groups and often there are calcareous gizzard plates to help the animals crush the organisms on which they feed. Systematists argue endlessly about the relationships of groups within this subclass and the arrangement adopted below is not meant to be a serious contribution to the argument.

Superfamily PHILINOIDEA
Most of the groups in this superfamily have thin, few-whorled, globose shells but some, such as the Ringiculidae and many of the Acteonidae, have sturdy shells with several whorls. In many species the shell is completely or partly concealed by the animal's mantle which may be voluminous compared with the size of the shell and may outshine it in beauty. The superfamily is well represented in eastern Arabia and many so-called new species have been described from the region. Many are described or listed by Melvill and Standen, 1901:453-455, but only the names of the larger, more familiar ones may be accepted with confidence. The taxa they list under the generic names *Tornatina* and *Cylichna*, for instance, belong to several different families. A thorough systematic revision of the eastern Arabian members of this superfamily is overdue.

Family ACTEONIDAE
Under strong magnification the thin, sturdy shells of many members of this family display spiral rows of grooves which are punctate or contain thin axial bars. A thin, chitinous operculum may be present. The entire animal is contained within the shell.

Turbonilla icela 814

Turbonilla linjaica 815

Turbonilla stylifera 816

Leucotina gratiosa 818

Punctacteon eloiseae 820

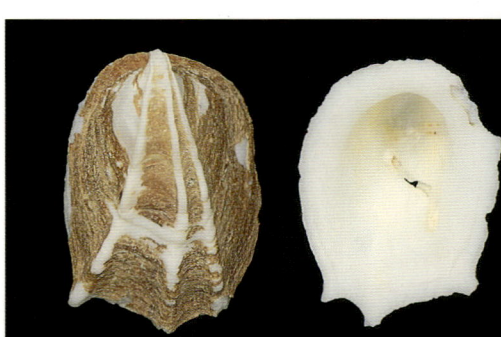

Amathina tricarinata 817

Superfamily: **PHILINOIDEA**

Acteon sieboldii (Reeve, 1842): **819**
15mm. Thin, glossy, elongate-ovate, last whorl well rounded or straight sided, sutures deep. Columella twisted, base of aperture produced downwards. Lower half of last whorl spirally grooved, smooth above except for slightly curved growth lines. Spire whorls milky, last whorl and aperture pinkish; columellar fold white. **Habitat:** beached. **Distribution:** GO, Mas.

Punctacteon eloiseae (Abbott, 1973): **820**
30mm. Solid, moderately glossy, broadly ovate, last whorl straight sided, sutures deep; tiny, deep umbilicus. Strongly twisted columella has 1 fold; parietal wall callused. Regular, close-set, spiral grooves. White with bold pattern of widely separated, black-bordered, orange-red, often crescent-shaped blotches, arranged (on last whorl) in 3 spiral bands; aperture white. Operculum elongate, amber. **Habitat:** mud flats. **Distribution:** Mas.

■ *Although now a familiar object in collections of exotic shells the existence of* Punctacteon eloiseae *was unknown to collectors before 1970 when Donald Bosch obtained some specimens from local fishermen at Masirah. Until then that small island, now famous for its remarkable shell fauna, had been almost inaccessible to those with a scientific interest in shells. Appropriately, the scientific name of this delightful shell commemorates that of the discoverer's wife, Eloise Bosch. Locally it is known as "the Eloise" and its principal haunt is often called "Eloise beach".*

Punctacteon fabreanus (Crosse, 1874): **821**
13mm. Thin, dull, elongate-ovate, spire short (but longer with increasing size), sutures deep. Columella twisted; umbilical chink. Regular, punctate, spiral grooves and fine growth lines. Pinkish with irregular white and lavender mottlings or streaks arranged in spiral rows (3 rows on last whorl); aperture and columella white. Syn: *P. flammeus* (Gmelin, 1791) may be an earlier name. **Habitat:** offshore and intertidal in coral sand. **Distribution:** NG, GO, Mas.

Pupa affinis (A. Adams, 1855): **822**
20mm. Thin, glossy, elongate-ovate, last whorl straight sided, sutures deep, base obliquely sloping below the double columellar fold. All whorls encircled by regular, shallow grooves. White with 2 or 3 broad, spiral bands of greyish-brown dashes which avoid the grooves; aperture and columella white. Reddish brown periostracum. **Habitat:** intertidal in sand. **Distribution:** NWG, SEG, GO, Mas, SO.

Family BULLIDAE
Members of this family have smooth, brittle, globose or ovate shells with a sunken spire and an inverted, or involute, apex. The animal, which lacks an operculum, can withdraw completely into the shell. The few species have gizzard plates to help them pulverise their food.

Bulla ampulla Linnaeus, 1758: **823**
40mm. Thin, brittle, glossy, globose, spire intorted and resembling a narrow umbilicus, outer lip uniformly curved. Thickened lower part of outer lip continuous with columella; callused parietal wall. Fine growth lines. Creamy or white blotched, spotted and mottled with purple, violet, and shades of brown. Intorted apex white edged. **Habitat:** intertidal in sand and beached. **Distribution:** all.

Family BULLINIDAE
As in the Acteonidae the thin, globose and usually red-lined shells of the bullinids display rows of spiral grooves which may be punctate or may contain axial bars. An operculum is present.

Bullina lineata (Wood, 1828): **824**
15mm. Thin and flexible, globose-ovate, spire short and slightly sunken, sutures incised. Columella straight; umbilicus minute. Regular spiral grooves containing fine, axial threads encircle all whorls. White with 2 red, spiral bands and sinuously axial, red lines on last whorl. **Habitat:** beached. **Distribution:** GO, Mas, SO.

Family HAMINOEIDAE
The fragile, smooth, globular or cylindrical shells have a sunken or covered spire and are almost incapable of containing the retracted animal. The animals lack an operculum but have gizzard plates to help them grind up minute organisms before digesting them.

Atys cylindrica (Helbling, 1779): **825**
13mm. Thin, brittle, translucent, glossy, cylindrical-ovate, outer lip projecting above intorted spire and forming an upward-pointing gutter. Top and bottom encircled by unequally spaced, shallow grooves. Thin parietal glaze, columella thickened at base. Milky white. **Habitat:** beached. **Distribution:** all.

Acteon sieboldii 819

Atys cylindrica 825

Pupa affinis 822

Punctacteon fabreanus 821

Bullina lineata 824

Bulla ampulla 823

SEASHELLS OF EASTERN ARABIA

Superfamily: **PHILINOIDEA**

Atys pellyi **826**

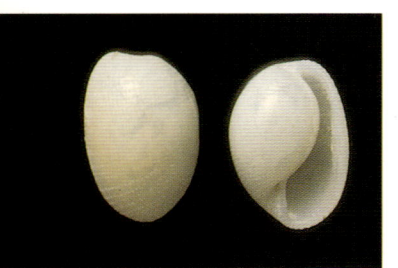

Haminoea crenilabris **828**

Haminoea cf *vitrea* **829**

Philine cf *aperta* **833**

Hydatina physis **830**

Hydatina zonata **831**

Micromelo undatus **832**

Atys pellyi (Smith, 1872): **826**
5mm. Thin, silky, translucent, pear shaped, spire deeply sunken, aperture broadly expanded below and narrowing sharply above. Columellar lip reflected over small and deep umbilicus. Fine axial and spiral scratch marks. White. **Habitat:** deep and shallow water, often beached. **Distribution:** NG, NWG, GO.

Diniatys dentifera (A. Adams in Sowerby, 1850): **827** NOT ILLUSTRATED
Listed by Smythe, 1979a:68, from U. A. E. (SEG).

Haminoea crenilabris (Melvill & Standen, 1901): **828**
3.5mm. Solid, glossy, elongate-globular, spire deeply sunken, umbilicus small and deep. Columella slightly swollen at base; outer lip crenulate along its length. Entire surface covered with close-set, spiral rows of evenly spaced punctae. **Habitat:** deep water in mud. **Distribution:** GO.

Haminoea cf *vitrea* (A. Adams in Sowerby, 1850): **829**
14mm. Thin, brittle, silky, globose-ovate, no spire, aperture longer than rest of last whorl. Fine spiral scratches and coarse growth lines, thinly glazed parietal wall, columella thickened at base. Orange-yellow, columella and glaze white. **Habitat:** offshore and beached. **Distribution:** all.

Family HYDATINIDAE
The few species in this family have globose, thin, flexible and mostly colourful shells which do not fully contain the animal. There is no operculum. They feed on polychaete worms.

Hydatina physis (Linnaeus, 1758): **830**
23mm. Thin and flexible, smooth and silky, globose-ovate, spire intorted, sutures channelled, aperture almost reaching top of last whorl. Thin, broadly expanded parietal callus; thin, reflected inner lip. White to amber encircled by fine, brown, ripply lines; callus and aperture milky with lines showing through. **Habitat:** intertidal and beached. **Distribution:** all.

Hydatina zonata (Lightfoot, 1786): **831**
25mm. Thin and flexible, smooth and glossy, globose-ovate, spire intorted, sutures channelled, aperture longer than rest of last whorl. Thin parietal glaze; weak, punctate, spiral grooves sometimes present. Narrow, central, white, spiral band; a thin brown line each side, another around top of last whorl and another around base, very fine, axial, brown lines occupying intervening spaces. Syn: *H. velum*. (Gmelin, 1791). **Habitat:** intertidal and beached. **Distribution:** all.

Micromelo undatus (Bruguière, 1792): **832**
14mm. Thin and flexible, glossy, ovate, spire sunk below top of last whorl; aperture wider towards base. Thin columellar callus. Surface encircled by microscopic, close-set, punctate lines; occasional growth breaks. White flushed pink, with net-like pattern, last whorl having 4 spiral, dark brown lines connected by sinuous or zigzag lines; aperture milky with lines showing through. Record based on 2 shells found by M. Tuhkanen at Qurm beach. **Habitat:** beached. **Distribution:** GO.

Family PHILINIDAE
The shell is calcareous but always fragile and usually colourless. The voluminous animal has 3 broadly triangular, calcareous plates in its gizzard with which it crushes molluscs, worms and other organisms before digesting them. There is no operculum.

Philine cf *aperta* (Linnaeus, 1758): **833**
20mm. Fragile, flexible, dull, translucent, oval and wide open, consisting of one loosely coiled whorl, so no spire. Columella slightly thickened. Coarse growth lines the only ornament. Milky white. Shell completely covered by living animal's mantle. **Habitat:** offshore in sand and beached. **Distribution:** all.

Family RETUSIDAE
The mostly tiny species of this family have cylindrical, colourless shells with a short spire and a rounded, flattened or lop-sided protoconch which is partially submerged in the succeeding whorl. They have gizzard plates for crushing minute molluscs and other organisms but most species lack an operculum. Some Arabian Gulf species have been reviewed by Smythe, 1979b.

Superfamily: **PHILINOIDEA**

Retusa bysma Melvill, 1904: **834**
NOT ILLUSTRATED
4mm. Described by Melvill, 1904c:168, from 156 fathoms off Muscat (GO).

Retusa tarutana Smythe, 1979: **835**
2.2mm. Solid for its size, glossy, cylindrical, short spired, sutures deep, protoconch rounded and partly sunk into succeeding whorl. Shoulders of whorls rounded, columella weakly twisted. Fine axial striations. White. Syn: *R. turrigera* Melvill, 1910, may be an earlier name. **Habitat:** intertidal among algae. **Distribution:** NG, NWG, SEG.

Retusa sp: **836**
1.5mm. Solid for its size, dull, cylindrical and slightly concave sided, spire sunk below level of top of last whorl, aperture constricted and equivalent of maximum shell length. Fine axial and spiral threads produce cancellate sculpture overall. White. **Habitat:** offshore. **Distribution:** GO.

Family RINGICULIDAE
Little is known about members of this neglected small family, which would reward the attention of a dedicated systematist. They are said to suck in tiny creatures through their siphon while gliding through muddy sand. The shell is small, colourless, has rounded whorls, a thickened outer lip and a twisted, notched columella which becomes continuous with the outer lip. There is no operculum.

Scaphander bushirensis **839**

Ringicula propinquans Hinds, 1844: **837**
5mm. Solid to very thick, glossy, translucent, ovate, last whorl globose, spire whorls rounded, sutures deep, protoconch submerged in previous whorl. Outer lip greatly thickened and bearing a large central tooth on inner edge; 2 thick folds on columella; thick, ridge-like callus on parietal wall. Widely spaced spiral grooves on all whorls. Milky white. Syn: *R. acuta* Philippi, 1849, a preoccupied name, sensu Melvill & Standen, 1901:456. **Habitat:** offshore to deep water and beached. **Distribution:** NWG, SEG, GO.

Family SCAPHANDRIDAE
Most eastern Arabian species in this family are small to minute, colourless and smooth. Most are known from only a few specimens collected in deeper water and do not differ markedly from each other. Gizzard plates are known to be present in some species but none has an operculum. The generic names employed here are tentative.

Cylichna collyra Melvill, 1906: **838**
3mm. Described from 156 fathoms off Muscat (GO).

Scaphander bushirensis (Melvill & Standen, 1901): **839**
8mm. Thin, fragile, dull gloss, ovate-cylindrical, aperture expanded below, narrow above, spire deeply sunken. Columella gently curved, no umbilicus. Surface encircled by evenly spaced, obscurely punctate striae. White. **Habitat:** offshore and deep water. **Distribution:** NG, GO.

Tornatina inconspicua H. Adams, 1872: **840**
3.3mm. Differs from *T. involuta* (see below) by its smaller size, more rounded and comparatively broader last whorl, and shorter, flat-topped spire. Milky white when fresh. Syn: *Retusa omanensis* Melvill & Standen, 1903. **Habitat:** deep to shallow water. **Distribution:** NG, SEG, GO.

Tornatina involuta (Nevill & Nevill, 1871): **841**
6mm. Thin, translucent, smooth, glossy, cylindrical-ovate, short spired, sutures deeply channelled, protoconch comparatively large, rounded, sunk into succeeding whorl. Thin parietal callus, 1 columellar fold. Milky white. **Habitat:** shell sand. **Distribution:** NG, NWG, SEG.

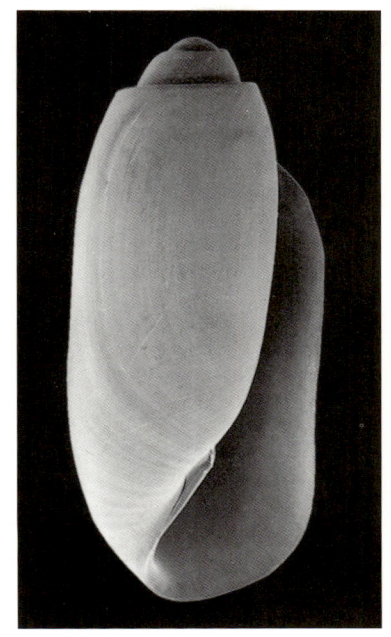

Tornatina involuta **841**

Retusa tarutana **835**

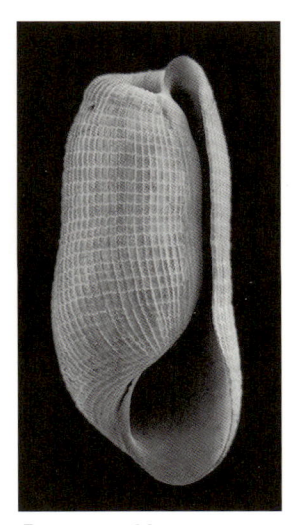

Retusa sp. **836**

Cylichna collyra **838**

Tornatina inconspicua **840**

Ringicula propinquans **837**

SEASHELLS OF EASTERN ARABIA

Superfamily: **UMBRACULOIDEA**, *Order:* **THECOSOMATA**

Diacavolinia flexipes **844**

Umbraculum umbraculum **843**

Cavolinia uncinata **845**

Tornatina persiana **842**

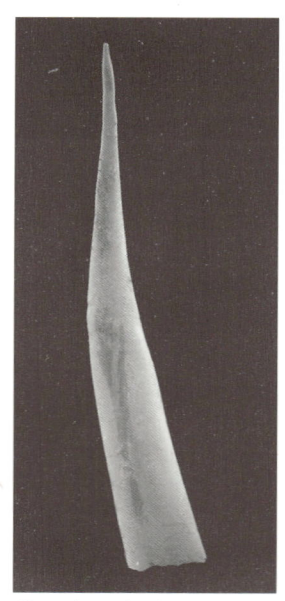

Creseis acicula **846**

Tornatina persiana Smith, 1872: **842**
2mm. Similar to *T. involuta* but shorter, twice as long as wide, with angled shoulders. **Habitat:** shell sand. **Distribution:** NG, NWG, SEG, GO.

Superfamily UMBRACULOIDEA
For description see under next family.

Family UMBRACULIDAE
Small family of opisthobranchs associated with pools at low tide where they may feed on sponges. The flattened, limpet-like shell sits incongruously on top of the much larger body. The large, thick foot has coarse tubercles and a pair of slender tentacles. Usually the shells occur empty.

Umbraculum umbraculum (Lightfoot, 1786): **843**
65mm. Solid, dull externally, glossy internally, flattened, oval, saucer-like with eccentric, pointed nucleus. Externally there are concentric ripples and coarse, broad, radial ridges. Internally smooth. White or yellow internally with central, brown blotch and sometimes concentric brown rings around it. Externally white with brown nucleus. Syn: *U. sinicum* (Gmelin, 1791). **Habitat:** live in intertidal rock pools, shells only beached. **Distribution:** Mas.

Order THECOSOMATA
Pteropods, also called sea butterflies because the animals have wing-like extensions and flit about in the upper layers of the sea, have fragile, transparent shells and lack an operculum. Needle shaped, cigar shaped or globular, they feed on minute organisms. Alive, they occur in countless millions in the surface waters. Dead, they may form a layer, called "pteropod ooze", on the sea floor in tropical regions, including the Gulf of Oman. Several species drift up onto sandy beaches in eastern Arabia.

Family CAVOLINIIDAE
Shells in this family may be triangular, conical, broadly needle-like, swollen or flattened, but they are always bilaterally symmetrical. The animal lacks an operculum.

Subfamily CAVOLINIINAE
Diacavolinia flexipes (Van der Spoel, 1993): **844**
6mm. Fragile, glassy, transparent, the upper side is bubble-like and very finely ridged. The longer, less inflated underside has a central ridge with a pointed, spout-like extension and 2 lesser extensions at the other (rear) end. Colourless or amber, sometimes with a brown blotch. **Habitat:** floating, settled on the sea floor, sometimes beached. **Distribution:** GO.

Cavolinia uncinata (Rang, 1829): **845**
7mm. Fragile, glassy, transparent, the upper side is bubble-like and very finely wrinkled or ridged. The underside is flatter, curves over the upper at the front end, has about 5 rounded ridges and 3 pointed extensions at the rear end. Colourless or amber. **Habitat:** floating or settled on the sea floor. **Distribution:** GO.

Subfamily CLIONAE
Creseis acicula (Rang, 1828): **846**
6mm. Fragile translucent, glossy, needle-like, sharply pointed. **Habitat:** floating or beached. **Distribution:** Mas.

182 SEASHELLS OF EASTERN ARABIA

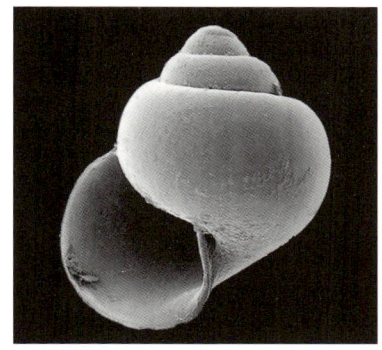

Superfamily: **ELLOBIOIDEA**

Limacina (Limacina) retroversa **847**

Limacina (Munthea) bulimoides **848**

Allochroa bronnii **849**

Cassidula cf labrella **850**

Cassidula nucleus **851**

Ellobium aurisjudae **852**

Family LIMACINIDAE
Shells in this family are sinistral, thin and transparent. The animal has a thin, chitinous operculum.

Limacina (Limacina) retroversa (Fleming, 1823): **847**
1mm. Fragile, translucent, glossy, sinistral, globose-ovate, sutures deep, aperture rounded. Fine, spiral lines overall. Brown columella. **Habitat:** floating or beached. **Distribution:** SO.

Limacina (Munthea) bulimoides (Orbigny, 1836): **848**
1.3mm. Fragile, glossy, transparent, sinistral, ovate, sutures deep, aperture produced below but often broken. Brown at sutures. **Habitat:** floating or beached. **Distribution:** SO.

Subclass PULMONATA
The pulmonates differ from the prosobranchs principally by the alteration of the mantle cavity to form a pulmonary sac without gills. They are air breathers, enabling them very successfully to colonise lakes, rivers and the land itself. Most pulmonates live well away from salt water but the few described here justify their inclusion by being always, if only marginally, associated with a saltwater or a brackish-water environment. Like prosobranchs their shells are variously formed, some coiled, some limpet shaped. None has an operculum. Most marine pulmonates are herbivorous.

Superfamily ELLOBIOIDEA
Apart from two small families, not represented in eastern Arabia, this superfamily contains only the family Ellobiidae, described below.

Family ELLOBIIDAE
An extensive family of air-breathing gastropods characteristic of muddy, brackish-water biotopes, such as mangrove swamps, where they may be abundant. Their shells are mostly covered with a thick, brown periostracum; this is usually missing from beach-worn shells. The aperture often has conspicuous teeth. Only a few species in eastern Arabia, which reflects the relative scarcity of suitable biotopes.

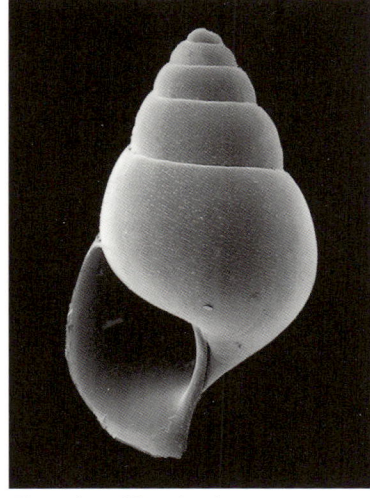

Subfamily ELLOBIINAE
Allochroa bronnii (Philippi, 1846): **849**
7mm. Solid, dull, ovate, spire straight sided, sutures incised. Outer lip thickened behind and bearing a spiral rib centrally; columella has 1 fold; 2 plate-like teeth on parietal wall. Regular, close-set, spiral striae on all whorls, often eroded. Creamy yellow with broad, brown, spiral zones, often faded. **Habitat:** pools at high-tide mark and beached. **Distribution:** Mas.

Cassidula cf labrella (Deshayes, 1830): **850**
12mm. Solid, dull, ovate; short spired, aperture half total length, thick ridge around base. Thickened outer lip turned inwards at top, 2 unequal teeth within; columella with 2 folds. Irregular spiral striae overall. Pale brown. **Habitat:** tidal flats and beached. **Distribution:** GO, Mas.

Cassidula nucleus (Gmelin, 1791): **851**
17mm. Thick, dull, ovate, slightly angled at shoulder; aperture 3 times longer than spire, ridge around base. Thickened outer lip has deep notch a short distance below upper end; columella has 2 folds. Irregular spiral striae overall. Shades of brown, lip paler. **Habitat:** khors and beached. **Distribution:** SEG, Mas.

Ellobium aurisjudae (Linnaeus, 1758): **852**
50mm. Solid, dull, elongate-ovate and slender, spire short, sutures incised, apex usually eroded. Lower two thirds of outer lip thickened and reflected, columella with 3 folds, no umbilicus. Spiral rows of granules most prominent below sutures; these override low, sinuous, axial ridges. Cream-beige, apertural lips pale brown; covered with brown periostracum when fresh. **Habitat:** mud flats, mangroves and beached. **Distribution:** GO.

■ *The shell of* Ellobium aurisjudae *illustrated at the left of our picture belonged, like its partner, to a creature which lived out its life in a muddy place exposed regularly at low tide, for it required easy and constant access to the air it once breathed. Unlike its partner this shell, bleached but otherwise in good condition, was dredged up from a depth of one hundred feet off the coast of Oman, at As Sib, an improbable — indeed impossible — habitat for this air-breathing mollusc. This shows to what depths water currents and wave action may transport molluscan shells from their original shallow-water sites.*

Superfamily: ELLOBIOIDEA, SIPHONARIOIDEA

Subfamily MELAMPODINAE
Melampus castaneus (Mühlfeld, 1816): **853**
13mm. Thin, glossy, smooth, ovate, short spired, last whorl slightly angled, base of aperture produced. Outer lip with up to a 13 elongate teeth, columella with 3 folds. Shades of brown with a paler band encircling shoulder of last whorl, but colour and pattern variable. **Habitat:** mud flats. **Distribution:** NG, SEG, GO, Mas, SO.

Subfamily PEDIPEDINAE
Laemodonta monilifera (H. & A. Adams, 1854): **854**
4mm. Thick, dull, roundly ovate, short spired; protoconch of 2.5 smooth whorls. Thickened outer lip has 2 stout teeth, columella has 3 folds, parietal callus well developed, umbilicus a mere chink. Strong, close-set ribs encircle whole shell. Uniformly amber, aperture paler. Periostracum forms small spiral frills. **Habitat:** intertidal rocks and beached. **Distribution:** SEG, Mas.

Laemodonta rapax (Dohrn, 1860): **855**
5.5mm. Description as for ***L. monilifera*** (see above) but spire longer, umbilicus closed and ribs thicker and fewer. **Habitat:** intertidal rocks and beached. **Distribution:** Mas.

Laemodonta cf ***sykesii*** (Melvill, 1897): **856**
4mm. Solid, dull, ovate, short spired; protoconch not seen. Thickened outer lip has 2 stout teeth, columella has 1 fold, parietal wall 2 folds. Fine spiral striae, but central area of last whorl almost smooth. Yellowish with white teeth and folds. **Habitat:** intertidal rocks and beached. **Distribution:** Mas, SO.

Pedipes sp: **857**
13mm. Solid, dull, ovate, short spired, sutures appressed, very large last whorl with inner lip narrowly separated from adjacent wall, umbilicus very narrow; protoconch not seen. Sharp fold on parietal wall, 2 rounded, columellar folds. Unevenly spaced, spiral threads crossed by fine, obliquely axial lines. Orange-brown fading to white. May be an undescribed species. **Habitat:** beached, probably from mud flats. **Distribution:** GO, Mas, SO.

Superfamily SIPHONARIOIDEA
For details see description of next family.

Family SIPHONARIIDAE
Superficially the shells of the false limpets resemble those of the true limpets (Patelloidea) but they are not even distantly related. An internal furrow on the right side of the false limpet shell covers a short siphon leading to the pulmonary chamber, a feature absent from true limpets which, of course, do not breathe air. In other respects, however, the two groups behave similarly, living on rocks intertidally and browsing only when submerged by the tide. Of the few eastern Arabian species one, *Siphonaria compressa*, is unusual because it occurs in a salt-water lagoon (khor) at Juzor al Halaaniyaat (Kuria Muria Islands) where it lives among fronds of algae. False limpet shells are dull and often eroded externally but are highly polished internally.

Siphonaria ashgar Biggs, 1958: **858**
30mm. Thick, depressed-conical, oval in outline with scarcely discernible siphonal groove at margin. About 46 thin ribs which are stronger posteriorly but not thicker either side of siphonal groove. Ribs off-white, interstices brown; internally dark brown with white radial streaks at margin. **Habitat:** intertidal on rocks. **Distribution:** NG, SEG, GO.

Pedipes sp. 857

Laemodonta rapax 855

Melampus castaneus 853

Laemodonta cf *sykesii* 856

Laemodonta monilifera 854

Superfamily: **SIPHONARIOIDEA**

Siphonaria belcheri Hanley, 1858: **859**
20mm. Thick, depressed-conical to almost flat, oval in outline but margin often distorted, siphonal groove projects prominently. About 14 coarse, equidistant, thick ribs with lesser ribs between them, 2 nearly fused ribs at site of siphonal groove. Ribs off-white, interstices brown; internally whitish rays have dark or paler brown between, muscle scar orange. **Habitat:** intertidal on rocks. **Distribution:** GO, Mas, SO.

Siphonaria carbo Hanley, 1858: **860** NOT ILLUSTRATED
5mm. Resembles a small, elevated **S. belcheri** (see above). Thick, low- to high-conical with central apex, oval in outline, siphonal groove scarcely discernible at margin. About 20 thick, equidistant ribs. All brown or brown with yellowish white ribs; internally golden brown with whitish radial streaks at margin and white muscle scar. **Habitat:** intertidal on rocks. **Distribution:** Mas.

Siphonaria* cf *compressa Allanson, 1958: **861**
7mm. Thin, high-conical, irregularly oval in outline, with pointed, recurved apex near posterior margin, siphonal groove scarcely discernible at margin. 2 or 3 weak ribs from apex to margin, fine growth lines. Beige with isolated brown spots and blotches; internally white. Probably an undescribed species. **Habitat:** among weed and on rocks in a khor (lagoon) at Juzor al Halaaniyaat (Kuria Muria Islands). **Distribution:** SO.

Siphonaria kurracheensis Reeve, 1856: **862**
25mm. Solid, flattened, apex to left of centre and turned to left, the irregularly oval outline modified by unequally spaced, projecting radial ribs, siphonal groove prominent. About 12 strong, radial ribs with lesser ribs between them. Shades of brown, ribs darker; internally blotched chocolate, beige and cream, margin paler. **Habitat:** intertidal on rocks. **Distribution:** SEG, GO, Mas, SO.

Siphonaria belcheri **859**

Siphonaria kurracheensis **862**

Siphonaria ashgar **858**

Siphonaria cf *compressa* **861**

SEASHELLS OF EASTERN ARABIA

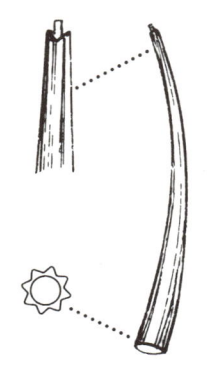

Superfamily: **SIPHONARIOIDEA, AMPHIBOLOIDEA**, *Class:* **SCAPHOPODA**

Dentalium octangulatum **865**

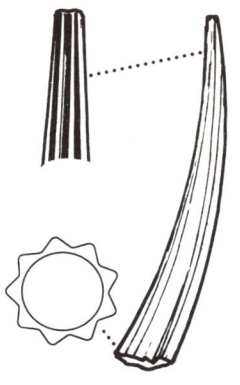

Dentalium reevei forma *lineolatum* **866**

Siphonaria savignyi Krauss, 1848: **863**
18mm. Solid, high-conical, apex central and turned to left, oval in outline, siphonal groove prominent at margin. Unequally spaced, strong and weak ribs varying in number. Very variable in colour and pattern, beige to white usually with darker ribs; internally white, cream or orange with brown blotches or rays which are sometimes fused; muscle impression cream or rose. Syn: *S. rosea* Hubendick, 1943. **Habitat:** intertidal on rocks. **Distribution:** NG, NWG, GO, Mas, SO.

Superfamily AMPHIBOLOIDEA
For details see description of next family.

Family AMPHIBOLIDAE
Of the two genera in this family one has thick shells and is absent from eastern Arabia. The other genus has thin, globose shells and is represented by one species. Unusually for a pulmonate gastropod it has a chitinous operculum. Although associated with muddy, brackish water (or wherever there is a significant influx of fresh water) its light, buoyant shell is usually found drifted up on beaches.

Salinator fragilis (Lamarck, 1822): **864**
10mm. Thin, fragile, silky; globose-ovate, spire whorls sometimes slightly keeled above periphery, sutures deep, umbilicus wide and deep; protoconch inclined and partially submerged in first spire whorl. Entire surface covered by fine, sinuous striae. Uniformly beige or brown, sometimes with 2 darker brown, spiral bands above periphery. Thin, brown operculum. **Habitat:** black mud often near mangroves and beached. **Distribution:** SEG, Mas.

Class SCAPHOPODA
A small, widely distributed class, the tusk shells are immediately recognisable by their tubular, generally plain and colourless shells. The popular name aptly describes the appearance of most species. Each end of the shell is open, the hole at the smaller end allowing the passage of water, the larger containing the animal's foot. Tusk shells live buried in sand where they feed on minute organisms captured by fine filaments called captaculae. No tusk shell has an operculum. The names of some of the species listed here are tentative as identification, particularly of beach-worn shells, is difficult, even for a specialist.

Family DENTALIIDAE
Most of the few tusk shells picked up casually on the beach or in shallow water are likely to belong to this family. The shell tapers gradually towards the narrower end where there is a simple hole which may be slit for a short distance or notched, or may encircle a short pipe. The tube may be smooth, ribbed longitudinally or encircled by fine ridges. The shells are usually obtained from shell sand.

Dentalium octangulatum Donovan, 1803: **865**
30mm. Solid, silky, rapidly enlarging in diameter, the curvature becoming almost constant. Apex has a small, V-shaped notch and a short pipe. Mostly with 8 strong ribs but may have 7 or 9. Spaces between ribs may be concave, flat or even slightly convex. The 8-ribbed form may have intermediate riblets. Milky white. **Habitat:** intertidal and offshore in sand. **Distribution:** all.

Dentalium reevei Fischer, 1871 forma ***lineolatum*** Cooke, 1885: **866**
30mm. Solid, dull, rapidly enlarging in diameter, the curvature very gradual. Apex not seen. There are 9 strong ribs with 3 riblets between each pair. Interspaces are concave. Pale yellow. Known from a worn shell found by H. Henseler. **Habitat:** beached. **Distribution:** Mas.

Dentalium tomlini Melvill, 1918: **867**
28mm. Solid, glossy, gradually enlarging in diameter, the curvature gradual. Apex has a small V-shaped notch. About 12 low, flat-topped ribs with a riblet between each pair. Interspaces are concave. White, yellow, pink or orange. **Habitat:** offshore and beached. **Distribution:** NWG, SEG, Mas.

Fissidentalium cf ***perinvolutum*** Ludbrook, 1954: **868**
40mm. Solid, dull, slowly enlarging in diameter, only slightly curved. Apex not seen. About 30 close-spaced, low,

Salinator fragilis **864**

Siphonaria savignyi **863**

Dentalium octangulatum **865**

186 SEASHELLS OF EASTERN ARABIA

Class: **SCAPHOPODA**

rounded riblets which become progressively weaker with growth, crossed by irregularly spaced growth rings. Off-white. Known from a single shell found by H. Henseler. **Habitat:** beached. **Distribution:** Mas.

Graptacme acicula (Gould, 1859): **869**
20mm. Thin, fragile, translucent, polished, very gradually expanding so that its mature diameter is only about twice that at apex, slightly and evenly curved. Smooth except for very fine, longitudinal scratch marks at apex and faint growth rings. Colourless or milky white. **Habitat:** beached. **Distribution:** GO.

Graptacme acutissima (Watson, 1879): **870**
55mm. Thin, fragile, translucent, polished, gradually then rapidly expanding, almost straight initially then evenly curved. Apex circular in cross section with a short, V-shaped notch. Smooth except for fine longitudinal riblets (which become mere scratch marks about half way along tube) and occasional growth ridges. Colourless, cloudy white or flesh coloured. **Habitat:** beached. **Distribution:** GO, Mas.

Tesseracme quadrapicalis (Sowerby, 1869): **871**
30mm. Solid, semitranslucent, glossy, rapidly expanding, evenly curved. Apex diamond shaped in cross section, the points corresponding to 4 sharp-edged, longitudinal ribs which become obsolete a short distance along tube. Fine intermediate riblets between ribs, whole tube encircled by crowded, fine growth lines. Milky white. **Habitat:** beached. **Distribution:** NWG, GO.

Family LAEVIDENTALIIDAE
Large tusk shells with apical notch and lacking distinct longitudinal sculpture.

"Laevidentalium" cf *curvotracheatum* (Plate, 1908): **872**
75mm. Thin, semitranslucent, glossy, expanding slowly at first then more rapidly so that mature diameter is much larger than at apex, evenly curved. Apex has broadly V-shaped notch and a short pipe. Fine longitudinal scratch marks, these becoming obsolete half way along or earlier. Irregular growth marks along tube. Horn coloured often with opaque white bloom. Systematic position of Masirah specimens uncertain. **Habitat:** beached. **Distribution:** Mas.

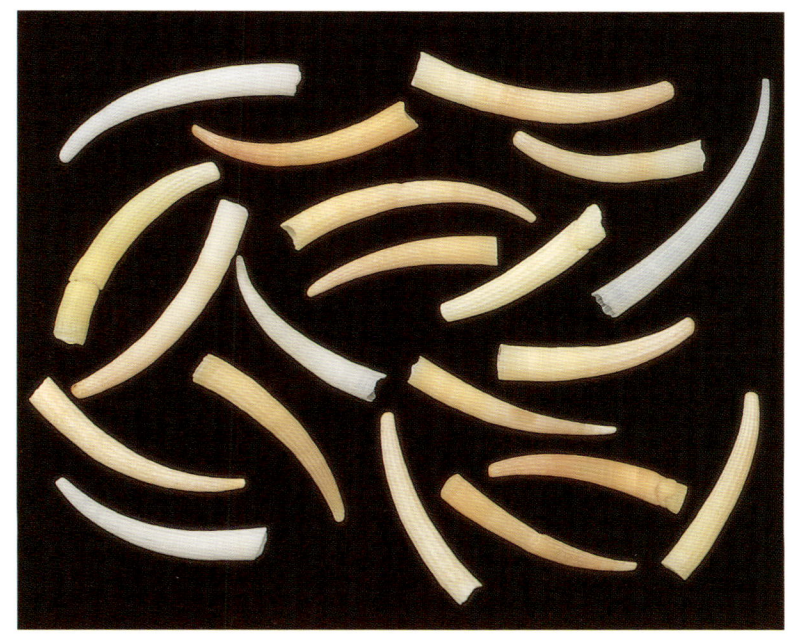

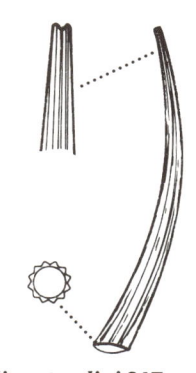

Dentalium tomlini **867**

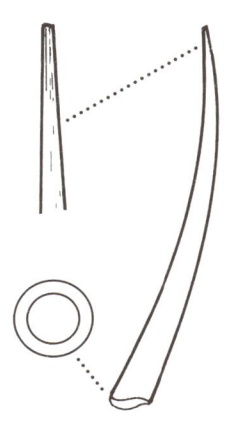

Fissidentalium cf *perinvolutum* **868**

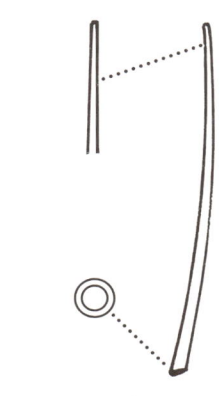

Graptacme acicula **869**

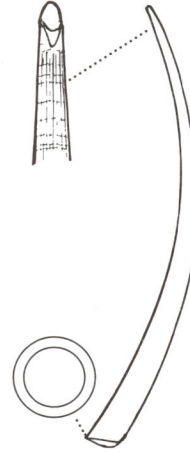

Dentalium tomlini **867**

Graptacme acutissima **870**

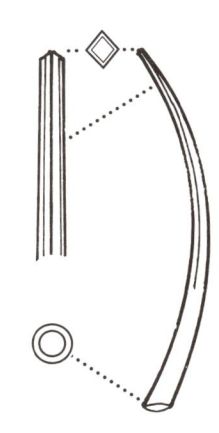

Tesseracme quadrapicalis **871**

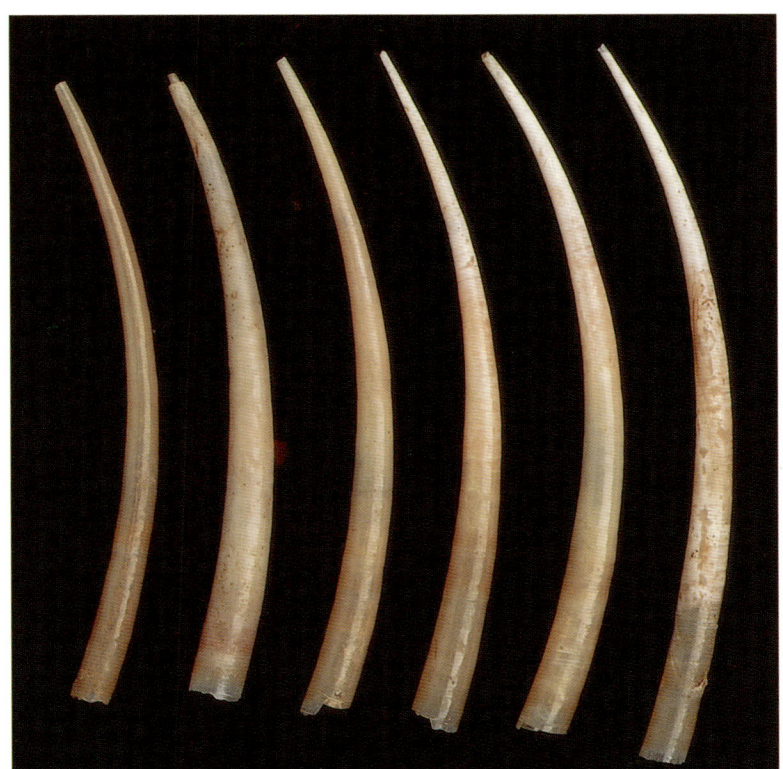

"Laevidentalium" cf *curvotracheatum* **872**

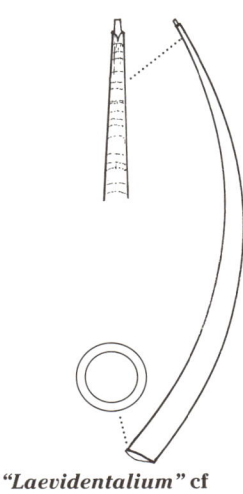

"Laevidentalium" cf *curvotracheatum* **872**

SEASHELLS OF EASTERN ARABIA 187

Class: **SCAPHOPODA**

Laevidentalium longitrorsum
873

Cadulus euloides **876**

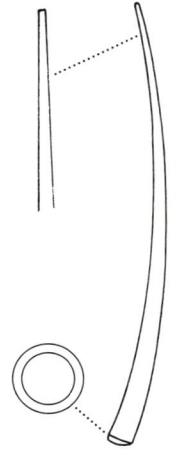

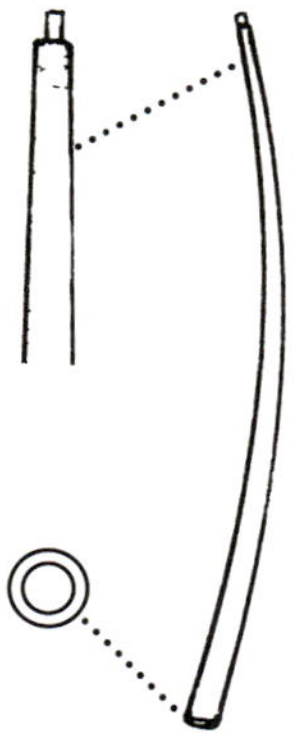

Episiphon sewelli **874**

Dischides prionotus **875**

Laevidentalium longitrorsum
(Reeve, 1843): **873**
75mm. Description as for *"L."* cf
curvotracheatum (described
above) but tube expands less rapidly
and lacks riblets. Records based on
a few worn, beached shells only.
Habitat: beached. **Distribution:**
NWG, GO.

Laevidentalium longitrorsum **873**

Family GADILINIDAE
Long and narrow shells with a short apical pipe.

Episiphon sewelli Ludbrook,
1954: **874**
25mm. Thin, glossy, expanding
slowly so that mature diameter is
only about 3 times that at apex,
slightly but evenly curved. A short
pipe emerges from apex. Smooth
except for very fine growth ridges.
Colourless, milky white or beige.
Apertural pipe and lack of
longitudinal ornament distinguish it
from species of *Graptacme*. **Habitat:**
beached. **Distribution:** NWG, GO,
Mas.

Family SIPHONODENTALIIDAE
Tiny, fragile shells which are difficult to identify as they are almost featureless. Generic distinctions depend on the modification of the apex by horizontal and axial slots which produce lobes of different shapes. The apex is usually worn so slots and lobes are seldom present.

Dischides prionotus (Watson,
1879): **875**
4mm. Thin, fragile, translucent,
expanding slowly so that mature
diameter (which is slightly
constricted) is scarcely twice that at
apex, only slightly curved. Apex has
two horizontal slots (but rim is
usually damaged). Smooth
throughout length. Colourless
sometimes with white bloom which
picks out growth rings. **Habitat:**
offshore. **Distribution:** GO.

Family GADILIDAE
See comments under previous family.

Cadulus euloides Melvill &
Standen, 1901: **876**
10mm. Thin, fragile, glossy, orifice at
apex less than half that at the
slightly constricted aperture, slightly
curved. Tube smooth throughout.
Colourless or obscurely ringed with
greyish white. **Habitat:** deep water
in mud. **Distribution:** GO.

Class POLYPLACOPHORA
Known popularly as chitons or
coat-of-mail shells, members of this
class bear little resemblance to any
other molluscan group. They are
bilaterally symmetrical, flattened
dorso-ventrally, and have an oval
or elongate-oval outline. A large,
creeping foot dominates the ventral
area, the head lacking both
tentacles and eyes. The foot
exercises a powerful suction force
enabling a chiton to cling firmly to
rocks and stones, the largest
eastern Arabian species,
Acanthopleura vaillantii, being a
familiar sight on rocks exposed to a
constant battering from the
elements. Chitons are exclusively
marine, occur in deep water as well
as intertidally and browse upon
algae and other encrusting
organisms.
Typically covering the dorsal
surface are eight articulated shell
valves which are more or less
arched along the mid-line (median
ridge) and are variously sculptured.
The surface layer of the valves
upon which ornamental features
are impressed is the tegmentum
and the porcellaneous under layer
is the articulamentum. Eye spots
(ocelli) may be visible on the
tegmentum. The valves are held
together by a muscular girdle
which often bears different kinds of
spicules and/or scales (spiculation).
Between the head valve and the tail
valve are six intermediate valves
each displaying a triangular area
either side (lateral areas); the
remaining triangular portion is the
median area the central part of
which is the jugal area. The tail
valve comprises a jugal area with
the equivalent of the median area
posterior and lateral to this; the

meeting point of these areas is the mucro. Marginal extensions (insertion plates) are inserted into the girdle and these may have lateral notches or slits. Associated slit rays may be visible on the under surface. The number of slits on each valve is characteristic of a species and is expressed as a slit formula, e.g. 8-10/1/6 means there are typically 8-10 slits on the head valve, one on either side of each intermediate valve and six on the tail valve. Except for the head valve all valves also have articulating anterior extensions (sutural laminae). Chitons are difficult to identify but detailed descriptions and line illustrations of most of those described below are available in a report by Kaas and Van Belle (1988).

Order NEOLORICATA
Chitons with the articulamentum projecting beneath the tegmentum, either as sutural laminae only or as sutural laminae and insertion plates on one or more of the eight valves.

Family ISCHNOCHITONIDAE
The dorsal sculpture of the valves varies from almost smooth to coarsely granulose. Girdle spiculation ranges from minute, smooth spicules to large, ribbed scales. Typically the insertion plates are smooth. Those of the head and tail valves have several slits; those of the intermediate valves have one or more pairs of slits.

Subfamily CALLOCHITONINAE
Callochiton vanninii Ferreira, 1983: **877** NOT ILLUSTRATED
12mm. Oval to elongate oval. Tegmentum glossy with very fine, granular appearance. Head valve semicircular, distinctly V-shaped posterior margin. Intermediate valves have convex anterior margin, rounded sides and a prominent apex. Tail valve semicircular, mucro slightly anterior of centre. Well-developed insertion plates. Slit formula 12-16/2/12-16. Girdle wide, encroaching on sutures, densely covered in long, sharply pointed spicules, marginal fringe of long, curved spines. Dark red or brown. Specimens not available for illustration. **Habitat:** under rocks. **Distribution:** Mas.

Subfamily ISCHNOCHITONINAE
Ischnochiton winckworthii Leloup, 1936: **878** NOT ILLUSTRATED
10mm. Oval. Tegmentum has irregular rows of variously raised granules. Head valve semicircular, widely V-shaped posterior margin. Intermediate valves rectangular, anterior and posterior margins almost straight with rounded sides and almost no apex. Tail valve almost semicircular, mucro anterior of centre. Insertion plates short. Slit formula 8-11/1/9-10. Girdle moderately wide, covered by ribbed, overlapping scales. Colour and pattern variable, beige, olive green often blotched white, girdle often banded yellow and olive green. Specimens not available for illustration. **Habitat:** on and under stones. **Distribution:** NWG, SEG.

Ischnochiton yerburyi (Smith, 1891): **879**
15mm. Elongate oval. Tegmentum has coarse, reticulate sculpture. Head valve semicircular, widely V-shaped posterior margin. Intermediate valves rectangular, anterior margin slightly convex, sides rounded, posterior margin straight, no apex. Tail valve almost semicircular, mucro posterior of centre. Insertion plates short. Slit formula 10-11/1/11-13. Girdle narrow, covered by finely ribbed scales; margin has short, strongly ribbed spines. Colour and pattern variable, buff, brown, olive green, etc. **Habitat:** under stones. **Distribution:** all.

Lepidozona luzonica (Sowerby, 1842): **880** NOT ILLUSTRATED
9mm. Oval. Tegmentum distinctly sculptured. Head valve, posterior area of tail valve and lateral areas of intermediate valves have radiating, granulose riblets. Central area of intermediate valves and anterior area of tail valve have parallel riblets. Head valve semicircular, widely V-shaped posterior margin. Intermediate valves broadly rectangular, rounded sides and no apex. Tail valve almost semicircular, inconspicuous mucro anterior of centre. Insertion plates short. Slit formula 11-14/1/10-13. Girdle narrow, covered by weakly ribbed, rectangular scales. Yellowish or greenish with occasional darker streaks or bluish-green spots. Specimens not available for illustration. **Habitat:** on and under rocks. **Distribution:** NWG, SEG.

Subfamily CALLISTOPLACINAE
Callistochiton omanensis Kaas & Van Belle, 1994: **881**
25mm. Oval. Tegmentum heavily sculptured. Head valve semicircular, widely V-shaped posterior margin with many pronounced radial ribs. Intermediate valves roughly rectangular, median area with many longitudinal ribs, lateral areas with 2 pronounced ribs, no apex. Tail valve with distinct median mucro, antero-median area ribbed, postero-median area with pronounced diagonal ridges. Well-developed insertion plates. Slit formula 8-10/1/9. Girdle narrow, covered by oval, short-ribbed scales, margin has short, blunt spines. Beige flecked dark brown. **Habitat:** under rocks. **Distribution:** Mas, SO.

Family CHITONIDAE
Dorsal sculpture of the valves varies from smooth to coarse, typically with distinct lateral areas. Girdle spiculation varies from overlapping scales to short spicules or long spines. Comb-like insertion plates are typical, those of the head and tail valves having several slits, those of the intermediate valves having one or more pairs of slits.

Ischnochiton yerburyi **879**

Callistochiton omanensis **881**

Order: **NEOLORICATA**

Subfamily CHITONINAE

Chiton fosteri Bullock, 1972: **882**
25mm. Elongate oval. Tegmentum almost smooth, lateral areas of intermediate valves very slightly raised, but surface features often eroded. Head valve semicircular, widely V-shaped posterior margin. Intermediate valves roughly rectangular with straight sides, slight apex. Tail valve large with mucro anterior of centre. Insertion plates strong and crenulated. Slit formula 9/1/16. Girdle broad, covered by moderately sized, almost smooth scales. Purplish pink with distinctive narrow, concentric, brown bands; girdle has wide bands of alternating colours. **Habitat:** under rocks. **Distribution:** Mas, SO.

Chiton peregrinus Thiele, 1910: **883**
30mm. Oval. Tegmentum has fine, wavy lines, lateral areas with two radial ribs, but surface features often eroded. Head valve semicircular. Intermediate valves roughly rectangular with pronounced apex. Tail valve mucro is posterior of centre. Insertion plates strong and crenulated. Slit formula 8-10/1/12-14. Girdle broad, often banded, covered by large, bluntly pointed, overlapping scales. Colour and pattern variable, brown, black, greyish green, beige, etc. Syn: *C. iatricus* Winckworth, 1930, *C. lamyi* Dupuis, 1917. **Habitat:** on and under rocks. **Distribution:** NWG, SEG, GO, Mas, SO.

Chiton (Rhyssoplax) affinis Issel, 1869: **884**
12mm. Elongate oval. Tegmentum coarsely sculptured, head and tail valves with pronounced radial ribs, lateral areas of intermediate valves with 2 or 3 ribs, sculpture of central areas finer. Head valve semicircular. Intermediate valves roughly rectangular with straight sides, no apex. Tail valve very large, mucro central. Robust insertion plates. Slit formula 10/1/10. Girdle narrow, covered by overlapping, ribbed scales. Colour variable but patterned with grey, green, cream or pink, or any of these colours predominating; girdle often banded alternately light and dark. **Habitat:** on and under rocks. **Distribution:** NWG, SEG, GO, Mas.

Subfamily ACANTHOPLEURINAE

Acanthopleura vaillantii Rochebrune, 1882: **885**
75mm. Oval. Tegmentum has raised, irregularly distributed, oval papillae, but surface is often severely eroded. Small, obscure, extra-pigmented ocelli widely distributed on head and tail valves and most of lateral area of intermediate valves. Head valve semicircular. Intermediate valves rectangular to V-shaped, their edges rounded, apex blunt. Tail valve less than semicircular, mucro posterior of centre. Short, comb-like insertion plates. Slit formula 8-10/1/9-11. Girdle wide, covered by large, blunt spines interspersed among fine spicules, marginal spines pointed and variously ribbed. Chocolate with some lighter grey markings; girdle banded alternately dark brown and beige or uniformly dark brown. Syn: *A. haddoni* Winckworth, 1927. **Habitat:** on and under rocks. **Distribution:** NWG, SEG, GO, Mas, SO.

Subfamily TONICIINAE

Onithochiton erythraeus Thiele, 1910: **886** NOT ILLUSTRATED
30mm. Elongate oval. Tegmentum on head valve, posterior portion of tail valve and lateral areas of intermediate valves have concentric, undulating sculpture and rows of ocelli. Central areas of intermediate valves and anterior area of tail valve more or less smooth but with longitudinal riblets towards anterior edge. Head valve semicircular. Intermediate valves roughly rectangular, sides slightly rounded, apex reduced. Tail valve triangular, mucro terminal. Insertion plates on

Chiton fosteri **882**

Chiton peregrinus **883**

Chiton (Rhyssoplax) affinis **884**

Chiton (Rhyssoplax) affinis **884**

Acanthopleura vaillantii **885**

*Order **NEOLORICATA**, Class: **CEPHALOPODA***

A friendly Omani fisherman picks up Robert Moolenbeek and a Dutch colleague, Peter van Pel (in green cap) after a successful collecting trip at Juzor al Halaaniyaat (Kuria Muria Islands) in March 1995.

head and intermediate valves robust and crenulated. Slit formula 8/1/0. Girdle covered by cylindrical, grooved, bluntly pointed spicules. Colour variable, cream to reddish, mottled orange, brown and grey. Specimens not available for illustration. **Habitat:** under rocks. **Distribution:** GO, Mas, SO.

Tonicia (Lucilina) sueziensis (Reeve, 1847): **887** NOT ILLUSTRATED
15mm. Elongate oval. Tegmentum has concentric, slightly raised ridges on head valve, lateral areas of intermediate valves and post-mucral part of tail valve; remaining areas have longitudinal, parallel riblets. Radiating lines of black ocelli on head valve and along line of slit rays on intermediate valves. Head valve semicircular. Intermediate valves roughly rectangular, pronounced apex. Tail valve small, mucro posterior of centre. Insertion plates comb-like. Slit formula 8-9/1/10-12. Girdle broad, covered by minute, ribbed spicules. Mottled with shades of pink and brown; girdle banded alternately light and dark. Specimens not available for illustration. **Habitat:** under rocks. **Distribution:** NWG, SEG, Mas, SO.

Family ACANTHOCHITONIDAE
Dorsal surface of valves usually coarsely sculptured. The girdle, which envelops the valves of species in this family, is covered by spicules and spines of varying lengths and shapes; often there are tufts of spines or bristles around the head valve and at each of the sutures between valves. Sutural laminae and insertion plates usually large and robust, the head valve typically having five slits, the other valves two each.

Acanthochitona mastalleri Strack, 1989: **888** NOT ILLUSTRATED
15mm. Oval to elongate oval. Tegmentum has irregularly distributed, oval, concave granules; jugal area narrow, raised and distinctly separated. Head valve semicircular. Intermediate valves much reduced, jugal area prominent, anterior margin of lateral areas deeply concave, distinct apex. Tail valve same as intermediate valves with reduced anterior area, mucro posterior of centre. Massive sutural laminae and insertion plates. Slit formula 5/1/2. Girdle very broad, covered by densely packed, minute, pointed, smooth spicules interspersed among slightly longer, curved spines; sutural tufts of spines weakly developed. Whitish mottled brown and grey. Specimens not available for illustration. **Habitat:** under rocks. **Distribution:** GO, Mas.

Acanthochitona woodwardi Kaas & Van Belle, 1988: **889** NOT ILLUSTRATED
8mm. Elongate oval. Tegmentum has flat, roundish granules, jugal areas weakly, longitudinally ribbed. Head valve semicircular. Intermediate valves twice as wide as long, front margin slightly convex, pronounced apex. Tail valve oval, prominent mucro posterior of centre. Insertion plates strong. Slit formula 4-5/1/2. Girdle covered by small pointed spicules and there are 18 sutural tufts of slender, sharp spines. Colour variable, may be strikingly patterned. Specimens not available for illustration. **Habitat:** under rocks. **Distribution:** NWG, SEG, Mas.

Notoplax arabica Kaas & Van Belle, 1988: **890** NOT ILLUSTRATED
10mm. Elongate oval. Tegmentum coarsely sculptured with large, elevated, elongate pustules. Jugal area of intermediate valves and tail valve distinctly raised, rounded and lacking pustules. Head valve semicircular, widely V-shaped posterior margin. Intermediate valves roughly triangular, anterior margins concave. Small, oval tail valve, prominent mucro posterior of centre. Sutural laminae and insertion plates very prominent and strong. Slit formula 5/1/6. Girdle covered by small, sharply pointed spicules. Brownish yellow, greyish or orange. Specimens not available for illustration. **Habitat:** under stones. **Distribution:** NWG, SEG.

Class CEPHALOPODA

(Subclass Coleoidea)
Squids, cuttlefish, octopuses, argonauts and nautiluses are known collectively as cephalopods (head-footed ones). Only the nautiluses, members of the subclass Nautiloidea, have true external shells but they are not recorded from eastern Arabia. The other cephalopod species have eight or ten sucker-bearing arms. Some of these species are included here because they have either an internal, shelly plate (cuttle bone) or secrete a calcareous, superficially shell-like and detachable receptacle to house their egg masses. The living animals may occur inshore but are difficult to see. Cuttle bones and egg cases often drift onto beaches but are seldom found undamaged. Squids have a fragile internal pen (or gladius) but we have no reliable records for eastern Arabia. Octopuses are recorded but they lack shells.

Class: **CEPHALOPODA**

Family SEPIIDAE
The predatory cuttlefish have ten sucker-bearing arms, broad, flattened bodies and side fins. They are capable of remarkable and rapid colour changes and can conceal themselves in a cloud of blackish ink expelled through a funnel, this same funnel expelling water to propel them backwards.

Their internal shells, or cuttle bones, are narrowly or broadly leaf shaped and are often found drifting in the sea or washed up on beaches. The identification of the various species is partly based on these cuttle bones, but they have been very little studied and so are worth collecting, especially if complete. The cuttle bones of two species recorded from eastern Arabia are described below. Four others occur in the region: *S. brevimana* Steenstrup, 1875, *S. omani* Adam & Rees, 1966, *S. savignyi* Blainville, 1827, *Sepiella inermis* Férussac & Orbigny, 1835.

Sepia pharaonis Ehrenberg, 1831: **891**
190mm (shell only). Thick but lightweight, moderately glossy, elongate-oval in outline, surrounded by a broad chitinous margin. Dorsal surface granular with 3 low, longitudinal ribs. Ventral surface has a broad, shallow, longitudinal groove dividing a striated zone; posterior third not striated. Anteriorly there is a short, smooth, rounded spine which is split longitudinally. Chalky white.
Habitat: animal free swimming; shells drift onto beaches.
Distribution: GO, Mas, SO.

Sepia cf *trygonina* (Rochebrune, 1884): **892**
100mm (shell only). Thick but lightweight, dull, narrowly elongate in outline and conspicuously curved

Sepia pharaonis **891**

Sepia cf *trygonina* **892**

192 SEASHELLS OF EASTERN ARABIA

Argonauta hians **894**

Argonauta argo **893**

downwards anteriorly, surrounded by a chitinous margin. Dorsal surface granular with a prominent longitudinal rib and a lesser rib each side. Ventral surface striated along most of its length, with a shallow, longitudinal groove. Anteriorly there is a short, smooth, upwardly curved spine. **Habitat:** animal free swimming; shells drift onto beaches. **Distribution:** GO, Mas.

Superfamily ARGONAUTOIDEA
There are four families in this superfamily but three of them lack shells and so do not feature here. The fourth family includes species which secrete an egg case resembling a delicate shell. For details see next family description.

Family ARGONAUTIDAE
Female members of this small family secrete a calcareous case carried aperture-side up and retaining the egg mass. This case is not attached to the mollusc which may discard it at will. Fragile and lightweight, such cases drift up onto beaches where they disintegrate speedily. Large, perfect examples occur but rarely. There is a considerable size difference between sexes, the female argonaut being about six times larger than the male.

Argonauta argo Linnaeus, 1758: **893**
120mm. Thin, brittle, semitranslucent, this egg case is laterally compressed, the aperture being widest adjacent to the incurved apex; the apertural lip has a spur which projects backwards from each side towards and past the apex. Double keel of pointed tubercles extends from apex to forward edge of lip; rounded ridges radiate from direction of apex, each ridge ending with a tubercle. Milky white but early portions of keels stained rusty brown. **Habitat:** egg cases drift onto beaches. **Distribution:** GO, Mas, SO.

Argonauta hians Lightfoot, 1786: **894**
90mm. Thin, brittle, semitranslucent, laterally compressed, the aperture being broadly open, the lip often having a well developed spur at right angles on each side. Double keel of large, rounded tubercles extends from apex to forward edge of lip; rounded ridges radiate from direction of apex, each ridge ending with a tubercle. Shades of brown, rusty brown on keels and around spurs. Syn: *A. gondola* Dillwyn, 1817. **Habitat:** egg cases drift onto beaches. **Distribution:** SEG, GO, Mas, SO.

BIVALVES
(BIVALVIA)
By P. Graham Oliver

***Servatrina pectinata* 983**
(see page 224)

The Bivalvia are readily recognised by the two-part shell joined by a flexible ligament. They are an ancient group and appear in the fossil record in the Middle Cambrian some 550 million years ago. It is believed that a cap shaped shell resembling a limpet evolved into a folded form which became hinged along the midline. The hinge is operated by the elastic ligament acting against the muscles holding the valves together. To prevent shearing of the valves a variety of hinge teeth have been evolved. The nature of the ligament and the hinge teeth are among the major characters used in classifying bivalves.

Unlike the majority of their snail-like relatives the bivalves have forgone a creeping existence and taken up a burrowing or sessile way of life. The animal becomes completely enclosed by the shell and the foot instead of having a large flat sole is blade-shaped. Nearly all bivalves feed by filtering particulate matter, either as sediment particles or micro-organisms suspended in the water. Primitive bivalves belonging to the Protobranchiata use palps and palp tentacles to collect food but in more advanced groups the gill becomes the most important feeding structure.

The functional radiation of the bivalves has two main components. The first is the diversification of burrowing forms related to the rate of burrowing, depth of burrowing and the type of substrate. The constant problem for the burrowing bivalve is how to maintain the water flow into and out of the mantle cavity. The evolution of retractible siphons allowed bivalves to burrow deeper and deeper. Consequently animals with no siphons tend to live close to the surface whereas those with long siphons burrow deeply. Compressed, glossy, smooth shells tend to be rapid burrowers whereas tumid, ribbed shells are poor burrowers. All types of substrate have been colonised by bivalves although most burrowers live in mud and sand. A few bivalves have solved the problem of boring into hard substrates such as peat, wood and rock. This is done by using the shell as a rasp (piddocks) or by the use of acid secretions to soften calcareous rocks (date mussels).

The second is the radiation away from the burrowing habit towards an epifaunal way of life. The most common epifaunal habit is the use of the byssus to secure the animal to rocks, to each other or other hard surfaces (mussels, pearl oysters, some scallops). In many an intermediate stage is reached where the animal is byssally attached but remains half buried in soft sediments (pen shells, horse mussels). The other method of attachment is by cementing one valve directly to the substrate (oysters, thorny oysters). In the scallops some species have lost all attachment and live freely on the surface. In order to escape predators they are able to swim by clapping the valves.

These radiation patterns are not rigidly linked to any one taxonomic group so it is not possible to use the general outline of the shell as a guide to identification. Mussel shaped shells, for example, are of course typical of the Mytiloidea but can also be found in the Arcoidea, Carditoidea and Arcticoidea. These parallel radiations cause considerable confusion to those who are unfamiliar with the bivalves. Environmental factors such as temperature and salinity can cause subtle changes in shell shape giving rise to confusions in identifications. Consequently bivalves can be difficult to determine and in an attempt to overcome this, a full guide to identification structures is included.

General External Features

The bivalve shell typically consists of two calcareous, convex valves which are hinged dorsally and free ventrally. The hinge margin is typically united by a non-calcified ligament and a set of articulating hinge teeth. The valves lie on the left and right sides of the animal. The shell grows co-marginally around the post-larval shell which remains as a beak somewhere along the dorsal margin. The area surrounding the beak is known as the umbo although some authors regard these terms as interchangeable. Each valve can consequently be regarded as having dorsal, ventral, anterior and posterior margins. The beaks effectively divide the dorsal margin into anterodorsal and posterodorsal parts. If the beaks are situated centrally the valves are equilateral but if the beaks lie in front of or behind the midline then inequilateral. The beaks may face each other across the dorsal margin, i.e. orthogyrate but more commonly they point in the anterior, prosogyrate or posterior opisthogyrate directions. In a few bivalves they may actually be coiled. The external features of the bivalve shell other than the sculpture are relatively few, but some explanation of the areas of the shell is necessary (Figure 1)

Many bivalves have areas

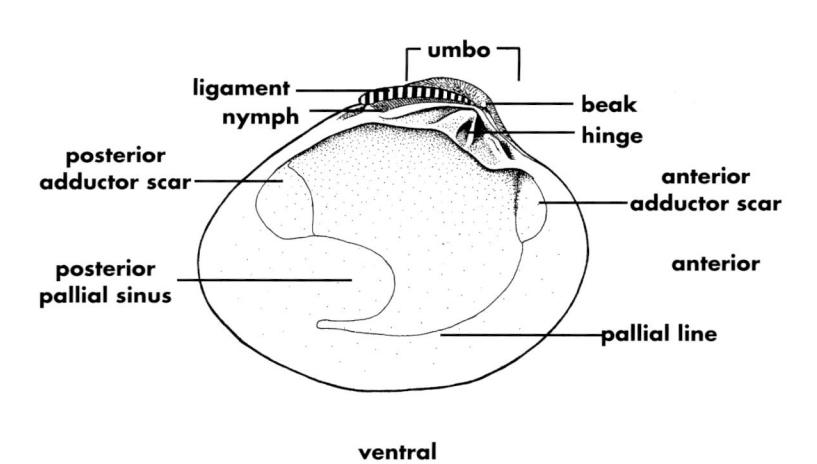

Figure 1 Internal features of a typical bivalve

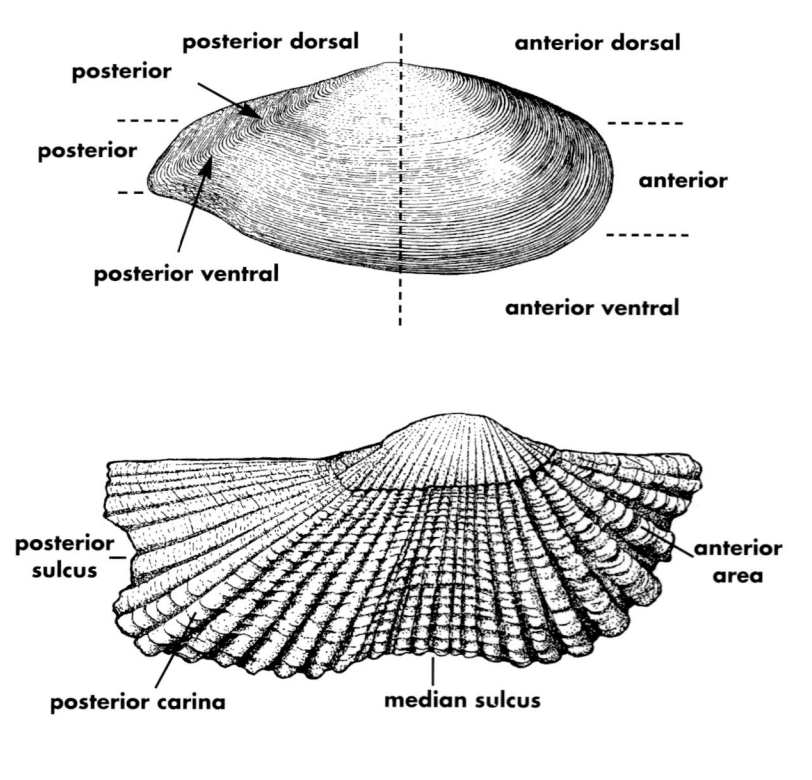

Figure 2 Some external features of the bivalve shell

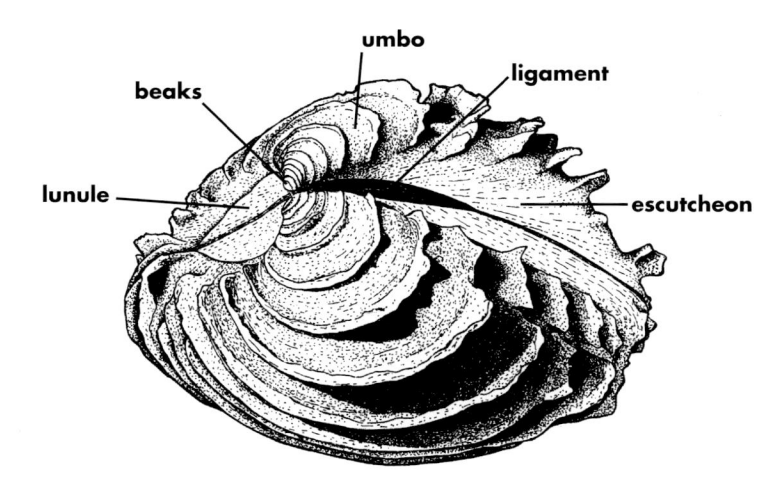

Figure 3 Some features of the dorsal area of a bivalve

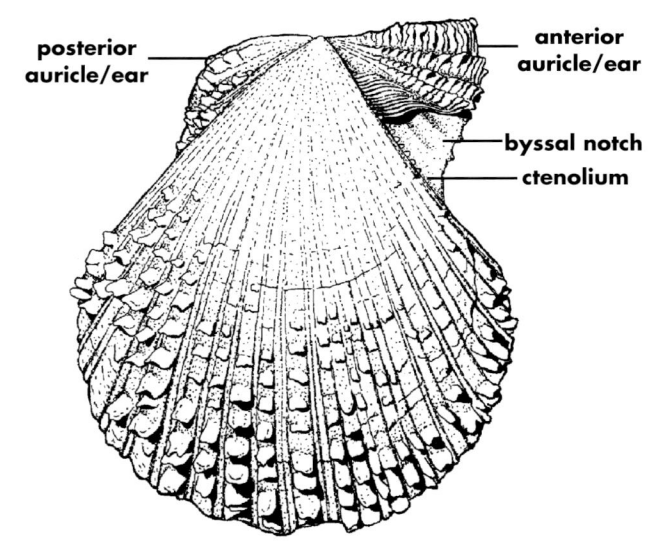

Figure 4 External features of a scallop

demarcated by angles or ridges which radiate from the umbos. The most common of these is a posterior angle or posterior ridge running over the umbos to the posterior ventral margin. If the ridge is elevated it is known as the posterior carina or keel. The area behind this angle is known as the posterior area or posterior slope. In some groups a radial depression or groove may accompany the posterior angle, this is termed the posterior sulcus. Less frequently and to a lesser degree there may be a similar anterior angle demarcating the anterior area. In between these angles lies the median area. In a few forms especially where a large byssus is employed the median area may be depressed to form a median sulcus. (Figure 2)

Two other important features are found on the dorsal margin. Anterior to the beaks there may be a defined, roughly heart shaped area which is known as the lunule. Posterior to the beaks there may be a similar structure which is usually elongate in outline, often depressed and known as the escutcheon. (Figure 3)

In many bivalves the valves do not meet at all points around the margins. These gapes vary in position and most commonly occur posteriorly where it reflects the presence of large siphons. An anterior or anteroventral gape is usually for the extension of the foot and is referred to as the pedal gape. Many bivalves employ byssus threads extruded from the foot to attach themselves to their substrate. If this byssus is large there is often a ventral or byssal gape. As many byssally attached species are reduced anteriorly the byssal gape is often in an anteroventral position. The majority of features described so far are typical of bivalves such as *Venus*, *Tellina* and *Lucina*. Some groups differ radically in outline and have a number of features peculiar to them.

Scallops and scallop-like forms such as *Spondylus* and *Lima* have shells which possess ear like extensions of the dorsal margin, called auricles. In *Pteria* auricles may be greatly extended into wings. In scallops the valves remain the right and left sides although the animal lives in the opposite plane lying on one valve. The byssus is accompanied by a gape but in the right valve only, where it is often accompanied by a series of small teeth on its inner margin, the ctenolium. (Figure 4)

The description of margins in species which are strongly inequilateral needs clarification as the posterodorsal edge itself can be further divided. In mytilids the ligament is greatly extended along the dorsal margin and this portion termed the ligament margin can be differentiated from the remaining dorsal margin by a distinct change in angle or curvature.

In the giant clams (*Tridacna*) the orientation is completely altered as the byssal gape is on the same margin as the hinge i.e. the dorsal margin. Other groups in which orientation is problematic are cemented forms such as *Ostrea* and *Chama*. Oysters are invariably cemented by the left valve but in *Chama* by either left or right.

In cemented forms the valves are typically unequal, inequivalve a condition which is present to some degree in many species. It may be extreme as in *Corbula* where one valve fits into the other or it may be subtle as in tellins where it is reflected in small differences in the convexity of the valves. Valves may be variously twisted, not only posteriorly as in tellins but around the umbos as in *Trisidos*.

On the inner surface of the valves a number of important characters relevant to identification are present.

Muscle Scars

Most bivalves exhibit scars on the interior of the valves which result from the attachment of muscles. These reflect the gross anatomy of the animal and are important in classification.

The valves are closed by adductor muscles, typically two, situated medially close to the anterior and posterior margins. The adductor scars are circular to oval rarely elongate or crescentric in outline. Where both anterior adductor and posterior adductor scars are present the condition is termed dimyarian and when of equal size homomyarian. In groups which become sessile and attached by a byssus the anterior portion of the shell is often reduced, resulting in the diminution of the anterior adductor. When the adductor scars are of unequal size the term is heteromyarian and if this process continues until a single adductor is present then monomyarian.

Most bivalves possess a foot which can be used for burrowing or crawling and from which the byssus is extruded. It is attached to the shell by a number of pedal protractor and retractor muscles. In burrowing homomyarian forms the pedal muscle scars are small but in byssally attached heteromyarian forms the posterior pedal retractor scar is large and significant. This scar is situated close to the posterior adductor scar. In many groups such as *Arca*, *Pteria* and *Pinna* it is also employed as a byssus retractor but in mytilids and *Anomia* a separate byssus retractor is present. The relative size and positioning of pedal and byssal muscle scars can be very important in separating closely related species of mytilids, giant clams and anomiids.

One of the most important features of the bivalve shell is the pallial musculature. The mantle tissue which secretes the shell is attached by a series of muscles close to the margins of the shell. These typically leave a linear scar known as the pallial line, occasionally this may take the form of an interrupted series or may be added to by accessory scars. When the pallial line follows the margins it is said to be entire but it is frequently indented posteriorly. This indentation or pallial sinus reflects the development of the siphons and may be shallow or extend right up to the anterior adductor scar.

The description of the pallial sinus is important. It can be said to have dorsal, anterior and ventral sections, if the ventral section is fused with the pallial line it is confluent if not then free. The dorsal section may be ascending (directed towards umbo) or descending (towards ventral margin). The anterior section may be rounded (sinus will be oval), straight (sinus will be rectangular) or the dorsal section may meet the ventral section directly (sinus will be acute). For the ventral section other than the point of confluence it is also important to note the angle at which it meets the pallial line.

Cruciform muscles are present only in tellins and are situated close and ventral to the end of the pallial line. Unfortunately, the scars are not always well impressed but if so form two small subcircular depressions. Their presence is a simple character for the recognition of the Tellinacea which is otherwise quite a variable group.

Hinge

The articulating dorsal margins of each valve are usually thickened hinge plate and bear an assortment of teeth or ridges collectively termed hinge teeth. If none are developed the hinge is said to be edentulous. In those forms with teeth, their form, number and disposition are frequently of systematic significance. There are four basic forms of hinge.

(a) Taxodont: The teeth are numerous, all taking the form of simple projections and situated in a row along the dorsal margins. This form is typical of *Nucula*, *Arca* and *Limopsis*.

(b) Isodont: This hinge has only a few teeth placed symmetrically either side of the ligament. The teeth are ridge like or rectangular and are called crura. They may interlock so well that the valves cannot be separated without breaking the teeth. Isodont hinges are found only in a few groups such as *Spondylus* and *Plicatula*. In the Anomiacea the crura become elongated into long narrow diverging ridges.

(c) Dysodont: This describes the condition where there are no true teeth only a few small ill-defined denticles situated either side of the ligament. They resemble the marginal denticles of many bivalves and are found most frequently in the Mytiloidea.

(d) Heterodont: This is the hinge form found in the majority of bivalves. The teeth are few in number and differentiated in two forms. Cardinal teeth which number from 1-3 in each valve radiate from the beaks. Lateral teeth which number from 1-2 either side of the cardinal set are situated at some distance from the beaks and are subparallel to the shell margin. The combinations of cardinal and lateral teeth are great and attempts have been made to classify and formulate these. These are, however, not straightforward and are not included here. Confusion can also arise through fusion of some teeth, or through erosion. The latter is important in some lucines and tellins where some teeth may be represented only by denticles or weak ridges. In fragile shells, hinge teeth are easily fractured and care should be taken to note if there are any fracture lines present. (Figure 5)

Ligament

The ligament joins the valves along the hinge margin, is elastic and serves to open the shell when the adductor muscles relax. It is an important diagnostic character but is a complex structure and often difficult to interpret.

The ligament always lies between or posterior to the umbos. It may be visible when the valves are closed, therefore external or if not then internal. Some ligaments have both external and internal components. Some external ligaments may be sunken between the dorsal margins but still visible when viewed from above. Other external ligaments which are set in a deeply cleft dorsal area are often worn dorsally and, therefore, difficult to observe until the valves are separated.

Ligaments are composed of fibrous and lamellar layers and the disposition of these determine the type formed. A terminology has developed for describing ligaments but intermediates occur and the terms are not exact. A flattened ligament lying on an area between the beaks is termed alivincular. In its simplest form it is a triangular structure composed of a median fibrous section bounded by equal sized lamellar sections. When symmetrical it is called amphidetic and when projected posteriorly opisthodetic. External alivincular ligaments grow by ventral extension of the dorsal area which may become deeply cleft or wide and flat. In oysters the cemented valve may exhibit extraordinary extension.

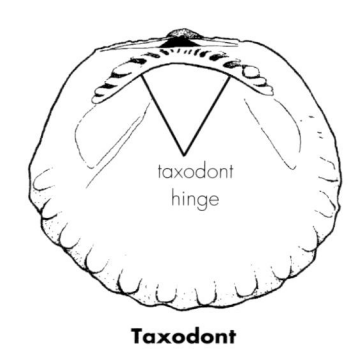

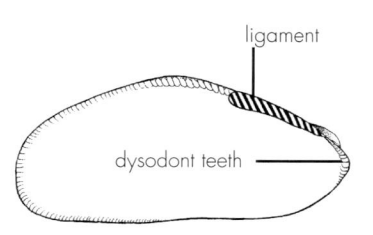

Taxodont

Dysodont

Isodont

Heterodont

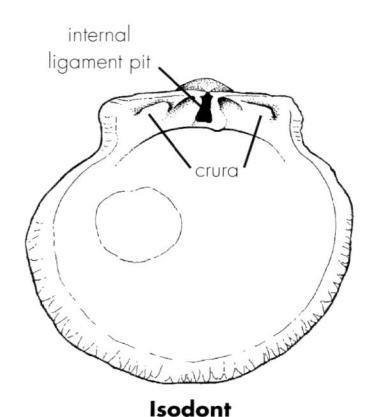

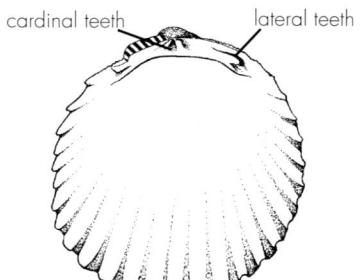

Figure 5 Major forms of hinge dentition

When internal the fibrous layer predominates and the ligament takes the form of a single triangular sheet between the beaks.

Ventral growth of the cardinal area weakens the hinge and alternatively the ligament grows in a posterior direction but remains between the valve margins. This is typical of *Mytilus* and *Pinna* and is termed transverse.

Derived from the transverse form is a much shorter but still posteriorly directed ligament which differs in being strongly arched and protruding above the valve margins. This parivincular ligament is often supported by a calcareous plate extending from the dorsal margin termed a nymph. This ligament is typical of *Cardium*, *Venus* and *Tellina*. The multivincular ligament takes the form of a series of triangular fibrous sections and is essentially a serial repetition of the alivincular type. It is of restricted occurrence and found only in the Isognomonidae. The duplivincular ligament takes the form of a series of lamellar bands on the cardinal area, typically placed obliquely they give a chevroned appearance but occasionally they are vertically aligned. This ligament is typical of *Arca*, *Glycymeris* and the vertical form of *Noetia*.

When internal the ligament is often associated with a number of shell structures. The recess into which the ligament is attached is called the resilifer and if it is hollowed out or projecting it is known as a chondrophore. The distinction between these is not always clear and they are often considered interchangeable The chondrophore may be supported by a ridge extending from the umbonal cavity and this is termed a resilial buttress. Occasionally associated with internal chondrophores the growth of the ligament causes stress in the umbonal region of the shell such that it may become cracked. This is seen in *Laternula* and *Periploma* and should not be confused with an accidental fracture. In order to strengthen an internal ligament a small calcareous plate, the lithodesma, situated ventrally may be present. (Figure 6)

Orientation of the valves

The identification of any bivalve requires us to know if we are observing the right or left valve and which is the anterior and posterior of the valves. It has already been stated that the hinge line is dorsal and that most ligaments and certainly all transverse and parivincular ligaments lie posterior to the umbo. Consequently if such a bivalve is held dorsal side up and the ligament lies between the beaks and the observer then the left valve will be in the left hand and the right valve in the right hand. Another simple indication is the pallial sinus if present it is always posterior. However in forms with amphidetic alivincular ligaments orientation is not so clear.

In species with a byssal notch this is always in an anterior or anteroventral position, so in scallops, wing and hammer oysters this will give the correct orientation.

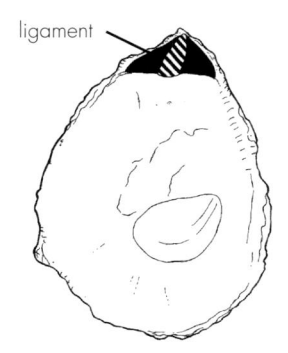

Alivincular ligament

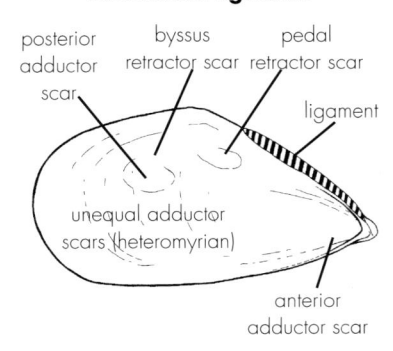

Transverse ligament

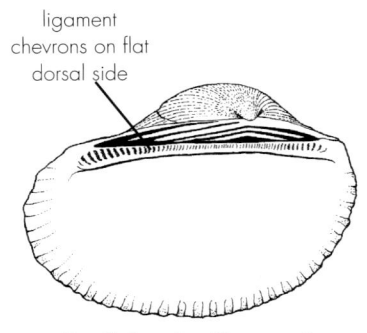

Duplivincular ligament

Multivincular ligament

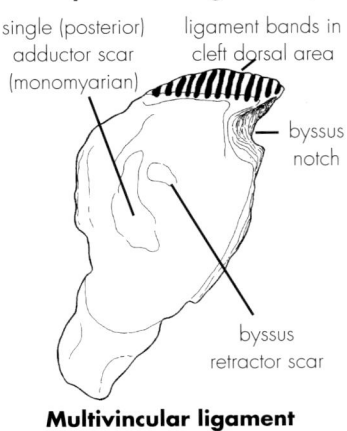

Figure 6 Ligament types and musculature

In some arcoid bivalves there is neither byssal notch nor pallial sinus and the orientation of these is difficult to deduce from the shell alone. With living material the anatomy is straight forward, the toe of the foot is anterior, so are the labial palps and mouth, the anus is posterior.

Shell shape

Shells are three dimensional objects and their verbal description is complex. Terminology is vast and variable and reference to good illustrations is always advised.

Dimensions

Three main dimensions are employed: Length, height and breadth.

Length: This is taken to be the maximum dimension along the longitudinal axis, i.e. from anterior to posterior margins.

Height: This is taken to be the maximum dimension along the vertical axis, i.e. from dorsal to ventral margins.

Breadth: This is a measurement of the maximum inflation of the valves when joined, other terms such as tumidity or convexity can be used but not thickness as this strictly applies to the thickness of the shell layers in a single valve.

Both valves can also be interpreted by reference to the umbo and an imaginary axis through the line of the adductor scars. Given the variety of shell forms these axes are not always definable. However if accurate comparison of shell form is to be made through measurement then a constant orientation has to be found.

Outline

The interpretation of outline is essential to species discrimination especially where other shell features are not critical. Outlines have often been described using terms most successfully developed for describing leaf shape in botany. In botany however there is a fixed axis marked by the stem insertion. The position of the beak in a bivalve does act as a fixed point but for example an oval shell may have the umbos on the long or the short axis. This is clarified by stating the relative dimensions of the shell e.g. "height much greater than length". From here the general terms can be applied but it will immediately be noticed that very few bivalves are of these classical shapes. Consequently they are qualified or used in combination. To detail the outline further each margin may be described and attempts at defining

BIVALVIA INTRODUCTION

Divergent

Pustulose

Scissulate

Cancellate

Divaricate

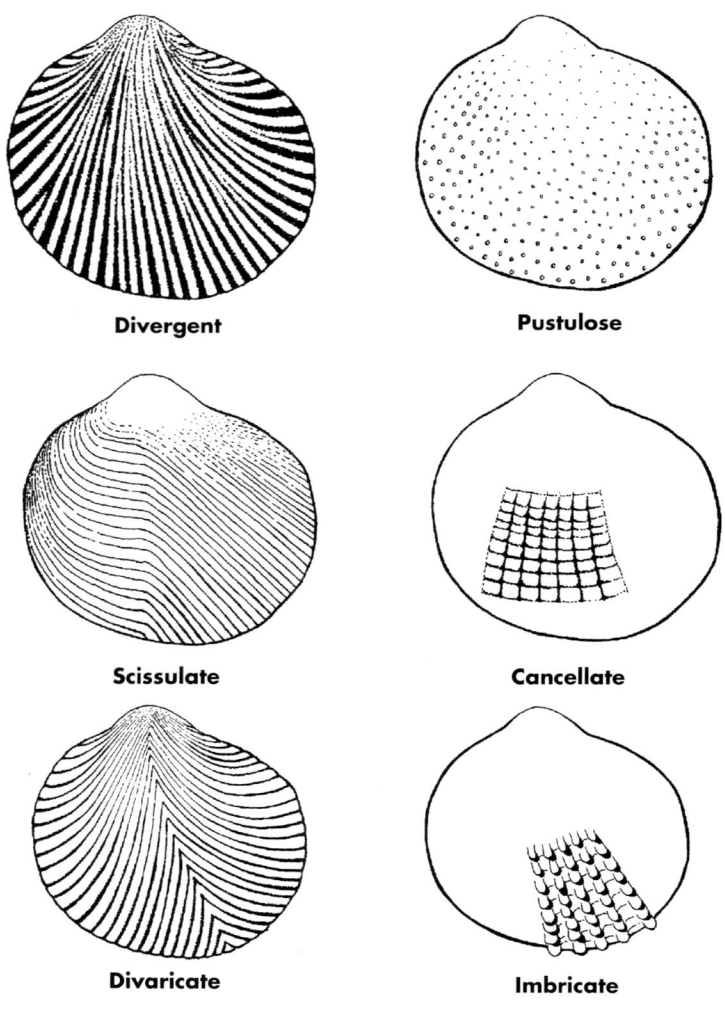

Imbricate

Figure 7 Some atypical and compound sculptures.

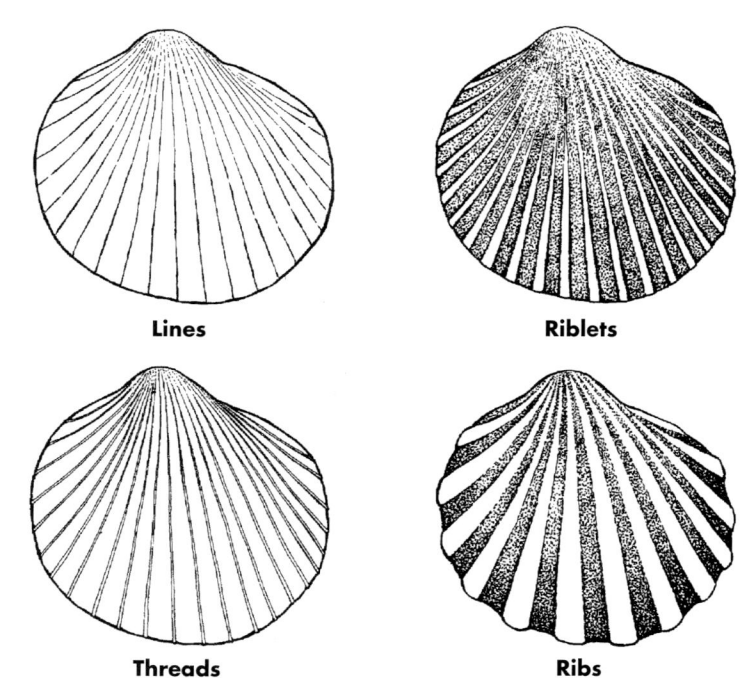

Lines

Riblets

Threads

Ribs

Figure 8 Basic patterns of radial sculpture

curvature and angles are often made. Angular measurements can be made but curvature requires complex mathematics and bivalve descriptions will remain with numerous terms such as "anterior margin broadly rounded, posterior margin narrowly rounded". This is an unfortunate necessity as such terms are inexact and primarily comparative. Where they are used comparative figures should always be presented or consulted.

Shell sculpture

Shell sculpture is a valuable taxonomic character especially at the species level. It is however only so if it is fully and accurately described. Most descriptions give only a general outline and often refer to number of ribs or concentric ridges. Using numbers alone can be unsatisfactory as often closely related species have overlapping ranges. It is also often difficult to decide what to count, for example close to the dorsal margins where the radial elements become obscure discerning the sculpture can be problematic. However the shape and details of the sculpture elements can provide decisive indications and the following introduction is aimed to guide the observer towards those features.

Orientation

Shell sculpture is broadly divided into two elements, radial and concentric = comarginal. Radial sculpture radiates from the umbos towards the margins, concentric sculpture follows the margins. Occasionally other patterns may be observed the most common of these being what were initially concentric elements becoming oblique to the comarginal line. This is termed oblique or acentric sculpture and if particularly dense as in some tellins then scissulate. Another deviation can be found with radial sculpture where the primary elements divide and the divisions are not truly radial in orientation. This is termed divergent. In some groups oblique sculpture is not derived from concentric elements but is angled oppositely on the anterior and posterior halves of the valve. The change in angle may occur on any part of the valve but is typically along a median or posteromedian axis. This divaricate pattern is found in many taxonomic groups and like most broad sculptural patterns they are not indicative of broad taxonomic divisions. (Figure 7)

Radial sculpture

Radial sculpture varies from surface lines through a variety of increasingly raised forms to ribs and occasionally to a much lesser degree of incised forms to grooves. Lines are not visibly raised, threads are raised lines and these are followed by riblets and ribs. Incised sculpture is described as striations or grooves. The distinction between consecutive terms is not definite and when of taxonomic value they should be

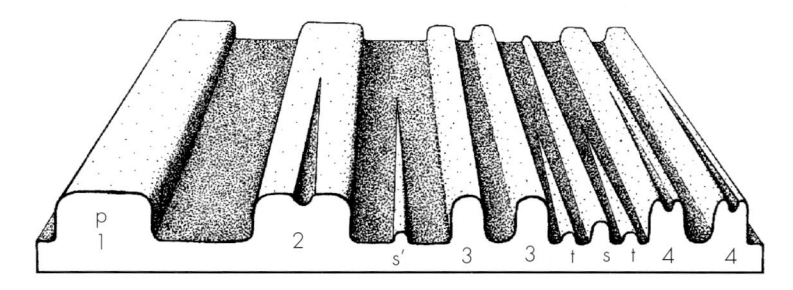

p - primary rib
s' - emerging secondary riblet
s - secondary riblet
t - emerging tertiary riblet

1 - entire primary rib
2 - bifurcating primary
3 - fully split primary
4 - second bifurcation

Figure 9 Patterns of development and addition of secondary riblets

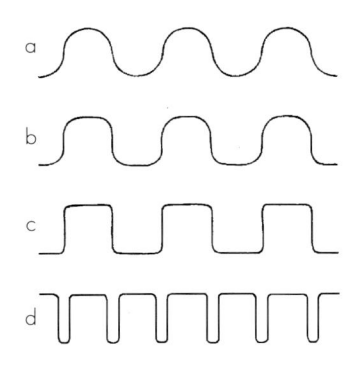

a - rounded ribs and interspaces
b - round topped, vertical sided ribs: interspaces flat
c - rectangular ribs: interspaces = ribs
d - rectangular ribs: interspaces narrow

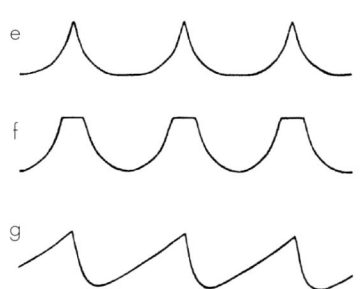

e - acute ribs
f - tabulate ribs
g - skewed ribs

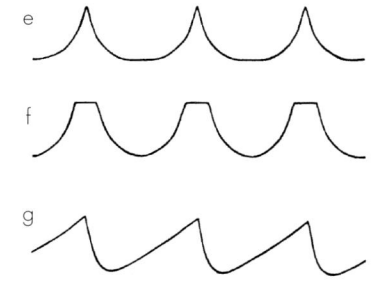

Figure 10 Cross sectional patterns in radial sculpture

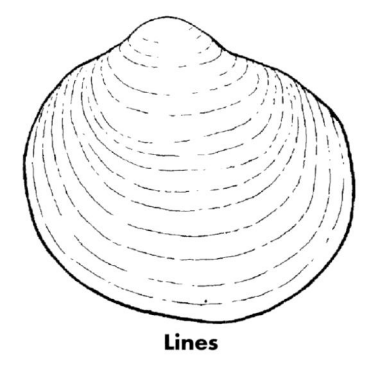

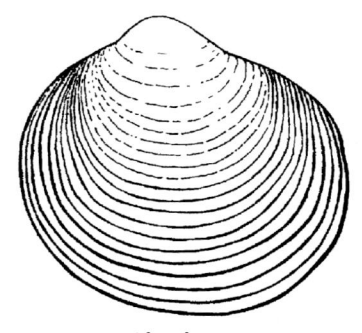

Lines

Lirations

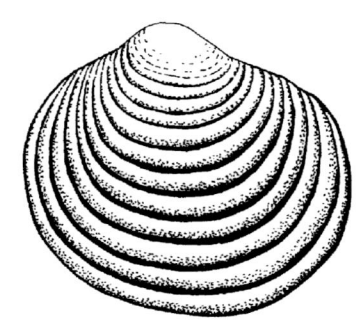

Ridges

Foliaceous

Figure 11 Patterns of concentric sculpture

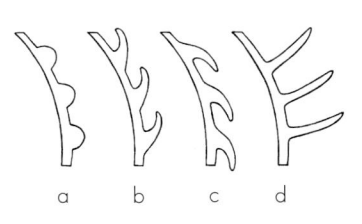

a - rounded edges
b - recurved lirae
c - downcurved lirae
d - vertical lamellae

Figure 12 Cross sectional patterns in concentric sculpture

given measured or given as numbers per unit length of shell at a given point from the umbo. The terms costellae and costae are interchangeable with riblets and ribs. (Figure 8)

In the simplest form radial ribs consist of equally sized elements, but this is seldom found. It is common to have a sequence of primary ribs with riblets and raised threads in each primary interspace. In such cases it is necessary to clarify the pattern and the terms primary, secondary and tertiary elements are used. (Figure 9)

Other than the size and number of radial elements it is also of value to note their cross-sectional shape and the relative size of the interspaces. Most descriptions quote only rib number but this is not always reliable as many species undergo primary rib bifurcation or the insertion of new ribs in between the primaries at later stages of growth. It is always useful to quote the size of each shell when the rib numbers are given as diagnostic characters.

The basic rib patterns and shapes are illustrated. (Figure 10)

Concentric sculpture
Most shells exhibit some trace of concentric sculpture at least in the form of growth lines. These indicate periodic changes in the rate of shell growth and are often if sometimes mistakenly taken for annual rings. They are not evenly spaced, being closer together towards the margins of adult shells. Usually they take the form of incised lines but occasionally they can be stronger.

Regular concentric sculpture like the radial forms is continuously variable. Elements that increase in width are termed lines, lirations, ridges and undulations. These elements can also increase vertically, typically they remain thin and when moderately developed are termed lamellae but if greatly developed into leaf like projections the pattern is foliaceous. Concentric elements do not always project vertically from the shell and more often either recurved towards the umbos or extended towards the margins. (Figure 11)

Spacing is also important, as is the density of the concentric elements. It is necessary to measure these at a fixed point from the beaks as both can vary with the growth of the shell. (Figure 12)

Combined sculpture:
Although simple concentric and radial patterns are frequently encountered, in many bivalves both elements combine to produce a wide

Superfamily: **SOLEMYOIDEA**

Fluted   **Smooth**   **Serrate**

Denticulate   **Crenulate**

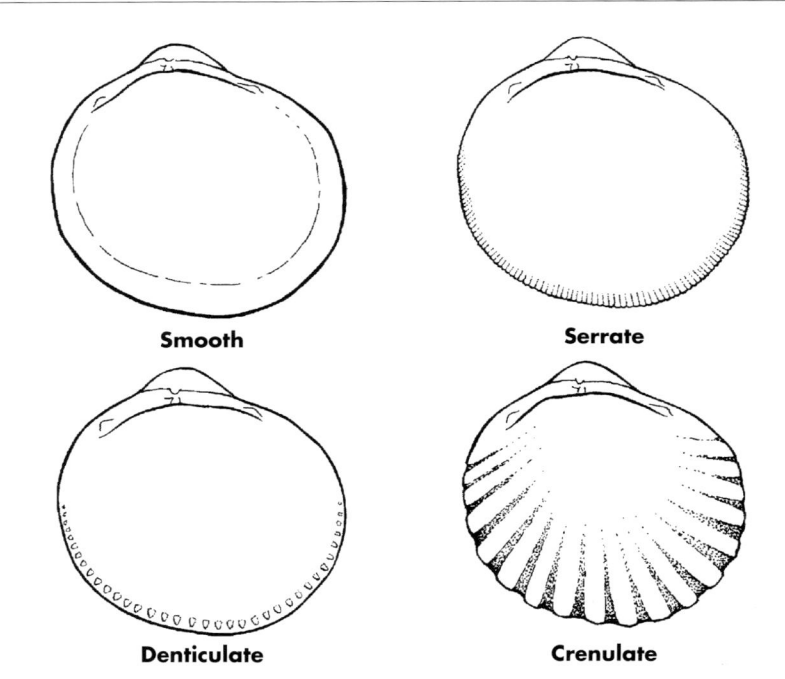

Figure 14 Types of inner margin sculpture

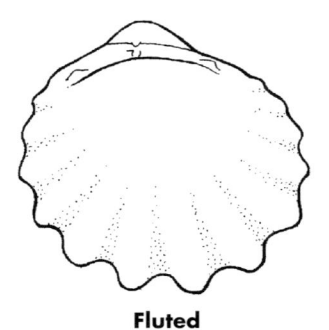

variety of sculptures. Frequently one of the major elements dominates so that for example in broadly ribbed forms the concentric lines are apparent only in the interspaces. If the concentric element is stronger a variety of tubercles, pustules or bars may develop where the concentric and radial elements cross. Similar patterns can be produced where a dominant concentric sculpture is crossed by a weaker radial element. In general where the sculpture takes the form of a network of intersecting concentric and radial elements it is termed reticulate. In combined patterns the concentric element may form flattened interrupted lamellae which are projected ventrally, this is termed imbricate, but often the projections are erect and form scales, if these are pointed then spines. If both elements are equally developed and interrupt each other to form small blocks of various shapes the patterns are termed decussate (diamond-shaped intervals) or cancellate (rectangular intervals). (Figure 13)

Sculptural changes with growth: During growth some species undergo remarkable changes in sculpture which can lead to great taxonomic confusion. In *Circe* for example the umbonal region has a series of divaricate ridges which are absent from the rest of the shell which is concentrically sculptured.

Periostracum
The periostracum is the outer non-calcified layer of the shell and is often projected into hairs, bristles or is flattened into lamellae. These can be useful characters especially in some of the Arcoidea and Mytiloidea and to a lesser extent across the whole of the Bivalvia.

Marginal structures
The inner margins of the valves although often smooth also can be a series of interlocking projections which usually reflect the surface radial sculpture. If the margin is simply undulating then it is termed fluted but if there are a series of wide projections and corresponding sockets then it is termed crenulated. Marginal crenulations are not uniform and variations in their forms can be of systematic valve. In other bivalves the inner margin may be serrated or denticulate that is with either a series of fine sharp or fine blunt projections. These may be truly marginal and visible when the valves are closed or they may lie a little distance in where they are seen only in the open valves.
In inequivalve shells co-marginal ridges may be present especially in the larger valve. (Figure 14)

Colour
Many bivalves are strikingly coloured or patterned and while this may be of systematic value in some groups it is notoriously variable in others. Colour is considered not to be functional in most bivalves especially those which are burrowing forms. The pigments are products from metabolism and are deposited in the shell for disposal. These pigments are probably dependant on diet and the environment and consequently it is possible to find colours dominating in one population but not in another. One only has to look at the colour range of *Circenita callipyga* to realise that colour should not be used as a primary character for identification.

Class BIVALVIA
Subclass PROTOBRANCHIA
The structure of the gills, palps and foot of the protobranchs suggests that they are among the most primitive of living bivalves. They are represented by four superfamilies, the Solemyoidea, Nucinelloidea, Nuculoidea and Nuculanoidea. The protobranch shell is not uniform in appearance, for example the Solemyoidea lack hinge teeth whereas the Nuculoidea have many. Protobranchs all live by taking in sediment particles and extracting food from them, so they tend to be found in muddy environments and are more common in deep water rather than in intertidal habitats.

Superfamily SOLEMYOIDEA
The oblong shells of Solemyoidea are distinctive with the broad fringe of shiny periostracum overlapping the margins. The oblong, almost cylindrical outline

cross bars   tubercles   scales   spines

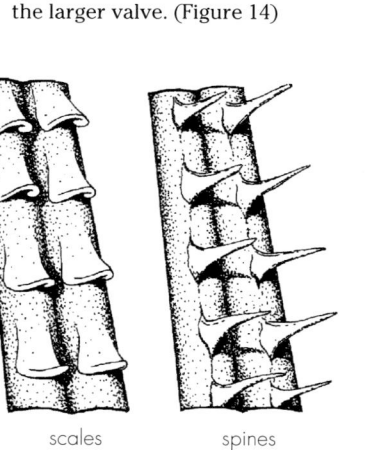

Figure 13 Examples of rib sculpture

Superfamily: **SOLEMYOIDEA, NUCINELLOIDEA, NUCULOIDEA, NUCULANOIDEA**

Nuculoma layardii **899**

resembles that of a small razor clam but the beaks are close to the posterior margin not the anterior. The hinge lacks teeth but has a strong and prominent ligament. Only one species is known from the eastern Arabian region.

Solemya africana Martens, 1879: **895**
To 30mm. Fragile, compressed. Cylindrical; anterior a little broader, lateral margins rounded. Ligament partly external, partly internal, oblique chondrophore supported by two weak buttresses. Periostracum shiny olive brown to black with faint radial rays. **Habitat:** offshore, occasionally beached. **Distribution:** GO, Mas.

Superfamily NUCINELLOIDEA
The Nucinelloidea are recognised by the presence of both a short series of taxodont teeth and a single lateral tooth. The small seed-shaped shell and the hinge teeth often lead to confusion with small *Nucula* species.
They inhabit offshore mud and sand sediments to considerable depths and are represented in eastern Arabia by one species.

Huxleyia diabolica (Jousseaume, 1897): **896** NOT ILLUSTRATED
To 3mm. Obliquely ovate; posterior margin angulate; anterior margin long almost straight. Hinge with a short taxodont series of teeth dorsally and a single anterior lateral tooth. Ligament internal impinging on cardinal area. Sculpture smooth. White. **Habitat:** offshore, occasionally beached. **Distribution:** GO, Mas.

Superfamilies NUCULOIDEA and NUCULANOIDEA
The Nuculoidea and Nuculanoidea are instantly recognisable by the form of the hinge. The teeth are all alike (taxodont), many, typically V-shaped in section, long and sharply pointed. The ligament is usually internal, set on a shallow resilifer. The shells are generally rather small, subovate-trigonal (Nuculoidea) or elongate, elliptical to rostrate (Nuculanoidea) in outline. They should not be confused with the other taxodont bivalves (Arcoidea and Limopsoidea) which have larger, more blunt teeth and an external ligament set on a flat dorsal area between the beaks.
The Nuculoidea are represented by the Nuculidae (nut shells) which are ovate-triangular in outline and have a persistent greenish periostracum. The inner margin may be smooth or serrated.
The Nuculanoidea are represented by the Nuculanidae and Yoldiidae. The Nuculanidae are strongly or partly rostrate with a strong sculpture. The Yoldiidae are generally elliptical, smooth and shining.

Family NUCULIDAE
Nucula consentanea Melvill & Standen, 1907: **897**
To 9mm. Solid. Beaks behind mid-line. Ovate-trigonal, anterior slope much longer than posterior. Escutcheon slightly domed. Surface matt;

Nucula consentanea **897**

sculpture of low dense concentric ridges, irregularly rugose over anterior area; faint radial striae overall. Inner margin denticulate. Greenish, often with rusty deposits. **Habitat:** deep water. **Distribution:** GO.

Nucula cf *rugulosa* Sowerby, 1833: **898** NOT ILLUSTRATED
To 2mm. Thin, translucent. Beaks just behind mid-line. Narrowly trigonal, height and length equal; anterior slope only slightly longer than posterior. Neither escutcheon nor lunule projecting. Sculpture of evenly spaced concentric ridges with radial lines of shell structure showing through. Inner margin denticulate. White. **Habitat:** in muddy shell gravel, offshore. **Distribution:** NWG, GO.

Nuculoma layardii (A. Adams, 1856): **899**
To 12mm. Beaks well behind mid-line. Obliquely oval; posterior narrowly rounded, anterior broad. Escutcheon short, scarcely domed. Lunule long, projecting as a low ridge. Sculpture smooth except for growth lines. Inner margin smooth.

Solemya africana **895**

■ *Felix Jousseaume originally called this little shell* Diabolica diabolica *and used the same frivolous nomenclatural convention with other new species, e.g.* Extra extra. *J. C. Melvill, who had received a classical education, was so incensed that he wrote, in 1898, "The names.... should be disallowed as offending nearly all the canons of the laws of nomenclature". The names, however, were published correctly and cannot be discounted.*

Shiny, greenish periostracum. **Habitat:** in mud, offshore. **Distribution:** NG, NWG, SEG, GO.

Family NUCULANIDAE
Nuculana bellula (A. Adams, 1856): **900**
To 9mm. Beaks slightly behind midline. Elongate oval with a pointed rostrum, demarcated by a sharp ridge close to dorsal margin. Sculpture of numerous low concentric ridges. Hinge plate not massive, teeth in two nearly equal series. White. **Habitat:** offshore. **Distribution:** GO.

Nuculana brookei (Hanley, 1860): **901**
To 8mm. Beaks almost central. Transversely oval; anterior rounded, posterior acute with a low ridge close to the dorsal margin. Sculpture of evenly spaced, elevated, concentric ridges. Hinge massive. White. **Habitat:** offshore. **Distribution:** GO.

Nuculana sculpta (Issel, 1869): **902**
To 7mm. Beaks towards the anterior. Subelliptical, posterior margin narrow, truncate, slightly auriculate. Sculpture in 3 parts: anterior area demarcated by a low radial rib and with 2 - 3 faint riblets all slightly nodulose; median area with widely spaced oblique lines; posterior area weakly sulcate, 2 posterior dorsal ribs, 1 riblet in sulcus and 2 ribs at junction with median area, all with erect widely spaced scales. White. **Habitat:** offshore. **Distribution:** GO, Mas.

Family YOLDIIDAE
Yoldia tropica Melvill, 1897: **903**
To 18mm. Thin, translucent. Beaks slightly in front of mid-line. Subelliptical; dorsal margins long sloping very gently; anterior rounded; posterior dorsal junction subacute, posterior margin short almost straight and slanting inwards. Surface shiny, obscure sculpture of weak oblique lines stopping along posterior slope. White with a thin yellowish periostracum. **Habitat:** in mud, offshore. **Distribution:** GO.

Subclass PTERIOMORPHA
Superfamily ARCOIDEA
The arcoid hinge is long with two series of small teeth (taxodont) which increase in size towards the margins. The dorsal area is correspondingly long and carries a ligament made up of alternating elastic and fibrous bands (duplivincular), a type shared only with the Glycymerididae. The arcoid outline is approximately trapezoidal, varying from subovate, subrectangular to quadrate, whereas the Glycymerididae have circular to oval shells. The surface sculpture is variously radially ribbed. The periostracum is often well developed and extended into concentric lamellae, hairs or scales. The adductor scars are generally subequal, the pallial line is entire and the inner margin may be smooth or deeply crenulate. As

Yoldia tropica **903**

Nuculana brookei **901**

Nuculana bellula **900**

Nuculana sculpta **902**

Superfamily: **ARCOIDEA**

the Arcoidea are suspension feeders they have large simple gills and small palps. The mantle edge is free and siphons are never formed; in epifaunal species it is often coloured and bears eye spots.

Three families are represented in the eastern Arabian fauna: The Arcidae is by far the best represented family and is characterised by obliquely aligned ligament bands which form a distinctive chevron pattern. They fall into two ecological groups and so have been divided into two subfamilies: Arcinae and Anadarinae. The differences between them may be obscure and it is doubtful whether these subfamilies are valid. The Arcinae attach themselves to hard surfaces (epifaunal) by a sheet of byssus (epibyssate) and are found in a variety of littoral and sublittoral habitats. The shell has a ventral byssal gape which may be very large (e.g. *Arca*) or narrow (e.g. *Barbatia*) and the weak sculpture is in the form of numerous riblets. In exposed littoral habitats some species are almost mussel-shaped with a large byssus and these contrast with those from sheltered crevice sites where the outline is rectangular and the byssus greatly reduced. The Anadarinae are either free living in soft sediments (infaunal) or possess a weak byssus of threads attached to sand grains or gravel within the sediment (endobyssate). They are slow burrowers and may live only partly buried (semi-infaunal) in the sediment and because of this they are most frequently found in sheltered conditions. The shell differs from that of the Arcinae in being more inflated, less elongate and with prominent radial ribs forming deep marginal crenulations.

The Noetiidae are much less diverse with only five small species in eastern Arabia. The ligament has vertically aligned bands and the range of shell form is more conservative than in the Arcidae. Ecologically they may also be divided into epifaunal and infaunal groups but the shell sculpture is always of radial riblets regardless of habitat. The byssal gape is absent in all but *Sheldonella* despite *Striarca* and *Didimacar* also being epifaunal. The Cucullaeidae have the chevron ligament but the hinge teeth are subparallel and the adductor muscles are attached to elevated buttresses. Only one species is found in our region and is a large, inflated shell with fine radial sculpture. All recent Cucullaeidae are infaunal.

Family ARCIDAE
Subfamily ARCINAE

Arca symphenacis Oliver & Chesney, 1994: **904**
To 50mm. Subrectangular; median area depressed; posterior flared, margin truncate, keel rounded. Sculpture of radial ribs and median riblets. Zigzag patterned, off-white and brown. **Habitat:** strandline only, probably attached to sublittoral rocks. **Distribution:** Mas, SO.

Arca avellana Lamarck, 1819: **905**
To 40mm. Irregularly boat shaped. 4-6 riblets on posterior area, elsewhere finely decussate. Ligament covering whole of dorsal area. Serrated, spathulate periostracal bristles on prominent sharp keel. Buff anteriorly, darker shades of brown posteriorly. **Habitat:** crevices of rocks and corals, lower shore and below. **Distribution:** GO, Mas, SO.

Arca avellana **905**

Arca symphenacis **904**

Superfamily: **ARCOIDEA**

Arca ventricosa **906**

Arca ventricosa Lamarck, 1819: **906**
To 120mm. As *Arca avellana* but: 2-4 ribs on posterior area. Ligament confined to anterior inter-umbonal area. Posterior dorsal area with brown transverse bars; posterior half of shell with V-shaped brown/black stripes. **Habitat:** crevices of sublittoral rocks and corals. **Distribution:** SEG.

Arca acuminata dayi Oliver & Chesney, 1994: **907**
To 30mm. As *Arca avellana* but narrower and may be very irregular, finely ribbed all over, dark spiky bristles on sharp keel and with fine hairs elsewhere. Adductor scars with raised flanges. Off-white tinged rusty brown. **Habitat:** strandline only, probably attached to sublittoral rocks. **Distribution:** Mas, SO.

Acar plicata (Dillwyn, 1817): **908**
To 30mm. Subrectangular often irregular, anteriorly reduced. Sculpture coarse, of lamellae crossed by radial riblets. Adductor scars slightly raised. White to buff, tinged orange or pink. **Habitat:** attached to corals and rocks, lower shore and below. **Distribution:** all.

Arca acuminata dayi **907**

Acar plicata **908**

206   SEASHELLS OF EASTERN ARABIA

Superfamily: **ARCOIDEA**

Barbatia foliata 911

Acar abdita Oliver & Chesney, 1994: **909**
To 15mm. As **A. plicata** but beaks more central, sculpture more delicate and twice as dense. Larval shell large. Dirty white. **Habitat:** under rocks embedded in sand and gravel, upper shore. **Distribution:** GO.

Barbatia obliquata (Wood, 1828): **910**
To 60mm. Mytiliform. Sculpture of many riblets, more prominent posteriorly. White tinged with brown posteriorly. Periostracum of brown/black appressed lamellar bristles. **Habitat:** attached to rocks and in crevices, littoral and sublittoral. **Distribution:** SEG, GO, Mas, SO.

Barbatia foliata (Forsskål, 1775): **911**
To 120mm. Subrectangular, some more square and trigonal. Ligament between beaks. Sculpture of slightly nodulose, narrow, elevated riblets. White. Periostracum dense, lamellar bristles dark brown. **Habitat:** attached to rocks and corals, lower shore - sublittoral. **Distribution:** all.

Barbatia obliquata **910**

Acar abdita **909**, with juvenile *Acar plicata* (bottom right) for comparison

SEASHELLS OF EASTERN ARABIA

Superfamily: **ARCOIDEA**

Barbatia parva 914

Barbatia perinesa 915

Barbatia setigera 913

Barbatia decussata 912

Barbatia decussata (Sowerby, 1833): **912**
To 60mm. Subrectangular, anterior and posterior margins rounded. Sculpture of low riblets with narrower interspaces, scarcely nodulose. White. Periostracum confined to rib interspaces, spiky, dark brown. **Habitat:** under rocks, upper shore. **Distribution:** GO.

Barbatia setigera (Reeve, 1844): **913**
To 50mm. Almost modioliform, anterior subtruncate. Riblets raised on anterior, low on posterior, interrupted by widely spaced radial grooves. Pale purple brown anteriorly, dark purple-black posteriorly. Long narrow dark periostracal bristles emerging from grooves. **Habitat:** attached to rocks, lower shore - sublittoral. **Distribution:** all.

Barbatia parva (Sowerby, 1833): **914**
To 25mm. Narrowly subrectangular. Sculpture of 5-7 low broad riblets on the posterior area; 70-80 narrow subcancellate riblets on remainder. Chestnut brown or dark beige with a paler buff median zone; umbones rose or pink. **Habitat:** attached to rocks, lower shore - sublittoral. **Distribution:** NWG, SEG, GO, Mas.

Barbatia perinesa Oliver & Chesney, 1994: **915**
To 30mm. Narrowly subrectangular. Numerous radial riblets with a few widely spaced radial grooves. Pale with beige zones and internal margins flushed with purple. **Habitat:** strandline only, probably attached to sublittoral rocks. **Distribution:** GO, Mas.

Barbatia cibotina (Melvill & Standen, 1907: **916** NOT ILLUSTRATED
To 6mm. Roundly quadrate, anterior shortened. Ligament posterior. Sculpture of about 35 primary riblets multiplying to 60 mostly on median area; riblets over keel strongest, posterior dorsal riblets divergent. White. **Habitat:** deep water. **Distribution:** GO.

208 SEASHELLS OF EASTERN ARABIA

Superfamily: **ARCOIDEA**

Trisidos tortuosa 919

Barbatia avellanaria (Melvill & Standen, 1907): **917** NOT ILLUSTRATED
To 5mm. Roundly subrectangular, anterior shortened. Ligament posterior. Sculpture of numerous radial threads, strongest on keel. White. **Habitat:** deep water. **Distribution:** GO.

Bentharca requiescens (Melvill & Standen, 1907): **918** NOT ILLUSTRATED
To 7mm. Trapezoidal with a prominent median sulcus. Sculpture of 24-27 nodulose riblets. Teeth almost vertical, in two sets, anterior teeth few on a separate swollen plate. Ligament posterior. White. **Habitat:** deep water. **Distribution:** GO.

Trisidos tortuosa (Linnaeus, 1758): **919**
To 120mm. Very strongly twisted. Median sinus deep. Posterior keel on left valve acute, on right valve rounded. **Habitat:** Half-buried on sandy bottoms, offshore. **Distribution:** NG, NWG, Mas.

Superfamily: ***ARCOIDEA***

Anadara antiquata **921**

Anadara birleyana **920**

Anadara ehrenbergi **924**

Anadara pesmatacis **925**

Anadara uropigimelana **922**

Anadara erythraeonensis **923**

Subfamily ANADARINAE

Anadara birleyana (Melvill & Standen, 1907): **920**
To 70mm. Wedge shaped with long straight posterior margin. Sculpture of about 45, closely spaced, flattened ribs; anterior set with a median groove, developing up to four incised lines posteriorly. White with brown tinges. Periostracum velvety, rust brown. **Habitat:** in sand, offshore. **Distribution:** NWG, SEG.

Anadara antiquata (Linnaeus, 1758): **921**
To 100mm. Beaks in anterior third. Subrectangular, anterior subtruncate, posterior margin long. Ligament with outer chevron only. Sculpture of about 35 ribs, interspaces narrower, anterior ribs bisected. White. Periostracum dark brown.
This species is rarely found alive and most shells are apparently rather old, it is found widely in prehistoric sites. Either these shells were imported or the species has been greatly reduced.
Habitat: in muddy sand, intertidal and offshore. **Distribution:** SEG, GO.

Anadara uropigimelana (Bory de St. Vincent, 1824): **922**
To 80mm. Slightly inequilateral. Subquadrate almost as high as long. Ligament with outer chevron only. About 30 broad, low ribs, punctate around umbones, striate ventrally, interspaces narrow. White. Brown periostracum of dense short bristles. **Habitat:** in sand, offshore.
Distribution: NG, NWG, SEG, GO, Mas.

Anadara erythraeonensis (Philippi, 1851): **923**
To 100mm. Beaks just in front of mid-line. Subrectangular, anterior rounded, ventral margin curved. Ligament chevrons distinct. Sculpture of about 36 wide ribs, all but posterior few deeply bisected and some bisected again. White. Dark brown periostracum usually worn. **Habitat:** in sand offshore. **Distribution:** GO, Mas.

Anadara ehrenbergi (Dunker, 1868): **924**
To 70mm. Slightly inequivalve. Beaks just in front of mid-line. Elongate, subelliptical; anterior narrower than posterior. Ligament with outer chevron only. About 28 deeply incised, flat topped ribs, slightly scabrous, LV more so. White. **Habitat:** in sand and shell gravel, offshore. **Distribution:** NG, NWG, SEG, GO, SO.

Anadara pesmatacis Oliver & Chesney, 1994: **925**
To 50mm. As ***A. ehrenbergi*** but less elongate, with only 20-22 ribs which are sculptured with incised lines. **Habitat:** offshore. **Distribution:** Mas, SO.

Superfamily: **ARCOIDEA**

Scapharca natalensis (Krauss, 1848): **926**
To 50mm. Strongly inequivalve, beaks almost at mid-line. Quadrate, height equal to, or a little greater than length. Ligament with few distinct chevrons. Sculpture of 27-30 wide smooth ribs, interspaces just narrower than ribs. White. **Habitat:** in sandy mud and mud in shallow waters. **Distribution:** SEG, GO, Mas, SO.

Scapharca inflata (Reeve, 1844): **927**
To 150mm. As **S. natalensis** but with about 37 ribs, rounded in profile and equal to interspaces. Periostracal insertion marks seen as distinct concentric lines in interspaces. Ligament with many chevrons. White. **Habitat:** offshore. **Distribution:** SEG, GO, Mas, SO.

Scapharca indica (Spengler, 1789): **928**
To 50mm. Inequivalve. Beaks in anterior quarter. Subrectangular much longer than high. Sculpture of about 30 smooth ribs. Ligament mostly posterior. White. Olive-brown periostracum. **Habitat:** offshore. **Distribution:** GO.

Bathyarca anaclima (Melvill & Standen, 1907): **929** NOT ILLUSTRATED
To 4mm. Very thin, translucent, inequivalve. Semi-circular; dorsal margins long and straight, posterior broadly rounded, anterior narrower. Sculpture of few concentric lines and feeble radial threads bearing periostracal hairs. White. **Habitat:** deep water. **Distribution:** GO.

Scapharca indica **928**

Scapharca natalensis **926**

Scapharca inflata **927**

Superfamily: **ARCOIDEA**

Family NOETIIDAE

Striarca symmetrica (Reeve, 1844) **930**
To 12mm. Very slightly inequilateral. Subrectangular. Ligament in a small triangle not covering the whole area. Sculpture of weakly beaded riblets, those on posterior area most prominent. White. **Habitat:** under rocks, middle shore and below. **Distribution:** GO, Mas, SO.

Sheldonella lateralis (Reeve, 1844): **931**
To 30mm. Beaks in anterior quarter. Subtrapezoidal, anterior area small, posterior expanded. Ligament of few wide bars between beaks. Sculpture of low radial riblets, those on rounded posterior keel more developed and bifurcating. White. Periostracal bristles black. **Habitat:** attached to rocks in crevices, lower shore and below. **Distribution:** GO, Mas.

Arcopsis margarethae (Melvill & Standen, 1907): **932** NOT ILLUSTRATED
To 5mm. Almost equilateral. Subrectangular. Sculpture of lamellae crossed by radial riblets. Minute ligament in a triangular resilium between beaks. Hinge plate notched beneath ligament. White. **Habitat:** Deep water. **Distribution:** GO.

Didimacar tenebrica (Reeve, 1844): **933**
To 15mm. Beaks in anterior quarter. Subrectangular, rounded edges. Sculpture of numerous radial threads and riblets. Ligament behind the beaks. White. Periostracum dark brown. **Habitat:** Under stones, lower shore and below. **Distribution:** NWG.

Striarca symmetrica **930**

Didimacar tenebrica **933**   *Noetiella chesneyi* **934**

Sheldonella lateralis **931**

212   SEASHELLS OF EASTERN ARABIA

Noetiella chesneyi Oliver & Chesney, 1994: **934**
To 15mm. Compressed. Beaks slightly towards posterior. Anterior area more expanded than posterior; anterior margin rounded, posterior subacute. Ligament in front of beaks. Sculpture of many narrow riblets. White. **Habitat:** in soft mud, shallow water. **Distribution:** NWG.

Family CUCULLAEIDAE
Cucullaea labiata (Lightfoot, 1786): **935**
To 70mm. Slightly inequivalve, beaks almost central. Subquadrate, anterior margin rounded; posterior margin obliquely truncate. Sculpture of low radial ribs. Hinge teeth becoming subhorizontal. Adductor scars on raised buttresses. Reddish brown. **Habitat:** in sand, shallow water. **Distribution:** NWG, GO, SO.

Superfamily LIMOPSOIDEA
The Limopsoidea are represented by two quite distinct families, the Limopsidae and Glycymerididae. Both families have a taxodont hinge and their anatomy is similar to that of the Arcoidea. The Glycymerididae and the circular Limopsidae are poor burrowers usually living in mobile, fairly coarse sand and gravel, often in shallow water. The smaller ovoid-quadrate Limopsidae employ a weak byssus and attach themselves to the substrate (gravel, pebbles, rocks) usually in deep water.

The Limopsidae are generally small, (5-30mm) and subcircular to ovoid-quadrate in outline. The ligament is a simple, small, triangular structure lying in a shallow depression between the beaks and the sculpture varies from cancellate to weakly ribbed. The Glycymerididae by contrast are often much larger (to 80mm), circular to ovoid in outline with a duplivincular (chevrons) ligament. The sculpture varies from smooth, radially striate to strongly ribbed.

Family LIMOPSIDAE
Limopsis elachista Sturany, 1899: **936** NOT ILLUSTRATED
To 6mm. Ovoid-quadrate, posterior margin subtruncate. Sculpture weakly cancellate. Inner margin with weak crenulations. White with rust tinges. Periostracum of fine fringing bristles. **Habitat:** deep water. **Distribution:** GO.

Family GLYCYMERIDIDAE
Glycymeris livida (Reeve, 1843): **937**
To 65mm. Subcircular not becoming oblique. Sculpture of about 40 weak radial ribs all cut by numerous radial grooves. Hinge weakly arched; teeth large, up to 12 in each set. Inner margin crenulate. Uniformly reddish brown. Brown, velvety periostracum. **Habitat:** in clean sand and gravel, shallow water. **Distribution:** NWG, SEG, GO, Mas.

Glycymeris cf *arabica* (H. Adams, 1871): **938**
To 30mm. Subcircular, larger specimens slightly oblique. Sculpture of about 30 low, closely spaced, radial ribs all incised by radial grooves and pitted by insertion marks of periostracal bristles. Hinge strongly arched; teeth small with a maximum of 12 anterior and 14 posterior. Inner margin crenulate. White with reddish brown to dark brown tessellate markings or with larger coloured areas. Syn: *G. striatularis* (Lamarck, 1819) of authors; *G. spurca* (Reeve, 1843) of authors. **Habitat:** in clean sand, shallow water. **Distribution:** NG, NWG, SEG, GO, Mas.

Glycymeris pectunculus (Linnaeus, 1758): **939**
To 50mm. Ovoid, posterior margin subtruncate to subangular in adult, juveniles subcircular. Sculpture of about 20 rather broad, high rounded ribs crossed by closely spaced concentric lines. Inner margin strongly crenulate. Generally a pale off-white crossed by irregular bands of buff and brown, internally white with a deep brown or black area on the posterior margin. **Habitat:** in mixed sediments, offshore. **Distribution:** all.

Glycymeris maskatensis (Melvill, 1897): **940**
To 40mm. Outline as **G. pectunculus** (see above). Sculpture of about 20 rather broad, low ribs with more or less vertical sides, anterior ribs becoming bifurcate, all ribs slightly nodulose. Externally pale buff with large areas of brick red in blotches or bands; internally white except for buff on the hinge plate. **Habitat:** offshore. **Distribution:** GO.

Glycymeris livida **937**

Cucullaea labiata **935**

Glycymeris pectunculus **939**

Glycymeris cf *arabica* **938**

Glycymeris maskatensis **940**

Superfamily: **MYTILOIDEA**

Superfamily MYTILOIDEA

The Mytiloidea include the familiar mussels with their elongate, anteriorly beaked shells. The long and narrow ligament rests on a resilial ridge along the dorsal margin. Hinge teeth are absent but simple, small, feeble teeth (dysodont teeth) may be present in front of and/or behind the ligament. The anterior adductor muscle is small or absent. The shell is thin and most often smooth but some species have weak ribs. A strong byssus of numerous fine threads is often present.

Four subfamilies of the Mytilidae are represented in eastern Arabia. The Mytilinae are the typical mussels with beaks at the extreme anterior. They live attached to rocks, pilings etc. and may occur in dense patches. The shell may be smooth or with fine riblets all over. The Modiolinae are the horse mussels of northern latitudes and have the beaks set a short distance from the anterior end. They live partly buried in soft or mixed sediments with the byssus threads attached to sediment particles. Some may build a byssus nest which forms a cocoon within soft mud. The shells are generally smooth but the periostracum may be bristly.

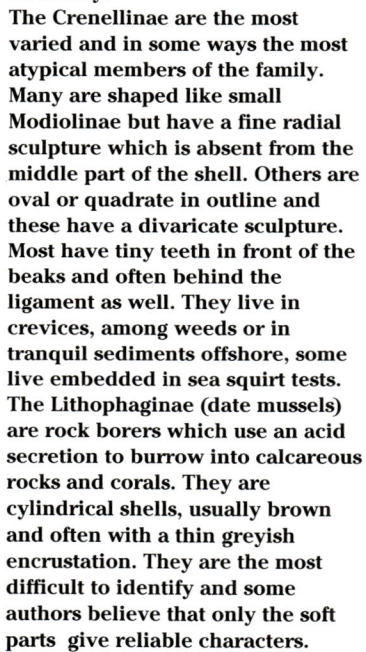

The Crenellinae are the most varied and in some ways the most atypical members of the family. Many are shaped like small Modiolinae but have a fine radial sculpture which is absent from the middle part of the shell. Others are oval or quadrate in outline and these have a divaricate sculpture. Most have tiny teeth in front of the beaks and often behind the ligament as well. They live in crevices, among weeds or in tranquil sediments offshore, some live embedded in sea squirt tests. The Lithophaginae (date mussels) are rock borers which use an acid secretion to burrow into calcareous rocks and corals. They are cylindrical shells, usually brown and often with a thin greyish encrustation. They are the most difficult to identify and some authors believe that only the soft parts give reliable characters.

Subfamily MYTILINAE

Perna picta (Born, 1778): **941**
To 80mm. Beaks terminal. Mytiliform. Lunule bent inwards forming 1-2 tooth like ridges. Sculpture smooth with growth lines only. Inner margin smooth. Anterior adductor scar absent. Ochre brown usually tinged with green. **Habitat:** attached to rocks, middle and lower shores. **Distribution:** GO, Mas, SO.

Septifer bilocularis var. *forskali* **942**

Perna picta **941**

Brachidontes variabilis **943**

■ Small species of Brachidontes *are difficult to separate as they are highly variable and our knowledge of their ecology and distribution is scant. The sculpture may vary from almost smooth to distinctly ribbed. Two forms are recognised here based on the arrangement of the small teeth adjacent to the beaks. They may be ecological or geographical variants.*

Septifer bilocularis var. ***forskali*** (Dunker, 1855): **942**
To 15mm. Mytiliform. Sculpture of numerous occasionally bifurcating riblets. Hinge with 2-3 large teeth below the beaks, 4-6, large dysodont teeth behind the ligament; shelf across umbonal cavity; margin elsewhere minutely crenulate. Bright green with reddish-brown spots and blotches. Periostracum of simple long fine bristles. **Habitat:** attached to rocks and shells, shallow water. **Distribution:** NWG, SEG.

Brachidontes variabilis (Krauss, 1848) : **943**
Form 1
To 15mm. Beaks terminal or nearly so. Narrowly wedged shaped to flared and arched. Anterior margin below beaks with 1 or 2 large teeth and a few tiny denticles. Sculpture of radial riblets, some bifurcating. Straw-coloured to purple black. As both *variabilis* Krauss, 1848, and the varietal name *semistriatus* Krauss, 1848 are preoccupied alternative names will be needed for this species complex. **Habitat:** attached to stones and rocks or among rock oysters and attached to roots in mangroves, upper shore. **Distribution:** NWG, SEG, GO.

Superfamily: **MYTILOIDEA**

Form 2
As above but sculpture finer and may be so weak as to appear smooth, marginal denticles remain prominent. Outline rarely flared. Anterior margin below beaks with denticles only. Shiny, brown-black. **Habitat:** attached to rock surfaces, upper and middle shores. **Distribution:** GO, Mas.

Subfamily MODIOLINAE
Modiolus auriculatus (Krauss, 1848): **944**
To 70mm. Beaks very close to anterior end. Modioliform; ligament and dorsal margins distinctly angled; dorsal and ventral margins parallel; dorsal margin concave. Purplish brown dorsally, paler often white ventrally; internally shades of purple, often very dark. Periostracum hairy, hairs smooth but often eroded, underlying periostracum shades of orange-brown to olive brown. **Habitat:** in shelly sand in crevices on rocky shores. **Distribution:** all.

Modiolus cf barbatus (Linnaeus, 1758): **945**
To 15mm. Beaks very close to anterior end. Modioliform; ligament and dorsal margins distinctly angled; dorsal and ventral margins diverging; dorsal margin straight or gently convex. Brownish red to vivid red. Periostracum of barbed hairs, underlying periostracum shades of orange-brown to olive-brown with red markings. **Habitat:** ? **Distribution:** GO, Mas.

Modiolus philippinarum (Hanley, 1843): **946**
To 60mm. Modioliform but distinctly wedge shaped; anterior area prominent, pointed; dorsal and ventral margins divergent; ligament and dorsal margins in a straight line; posterior margin subtruncate. Periostracum hairy, hairs simple usually eroded leaving a shiny smooth shell. Internal white with a dark, reddish-brown, posterior-dorsal area. Periostracum pale yellow-orange-brown, umbonal ridge commonly paler. **Habitat:** offshore. **Distribution:** NG.

Modiolus ligneus (Reeve, 1858): **947**
To 15mm. Umbones prominent. Modioliform, obliquely subcylindrical; ligament and dorsal margins slightly disjunct; dorsal margin curved, sloping and in a continuous curve with posterior margin. White. Periostracum smooth, shining, deep orange-brown to chestnut.
The eastern Arabian shells are rare and much smaller than in other parts of the Indian Ocean. **Habitat:** in a byssus nest in soft mud, offshore. **Distribution:** NWG, SEG, GO, Mas, SO.

Modiolus auriculatus **944**

Modiolus philippinarum **946**

Modiolus ligneus **947**    *Modiolus cf barbatus* **945**

SEASHELLS OF EASTERN ARABIA

Superfamily: **MYTILOIDEA**

Modiolus sirahensis (Jousseaume, 1891): **948**
To 20mm. Modioliform but very elongate, narrowly subrectangular; posterior area slightly expanded; ligament and dorsal margins in a straight line; posterior margin obliquely truncate. White. Periostracum smooth, brown to olive. The Arabian shells are rare and much smaller than in other parts of the Indian Ocean. **Habitat:** in soft sediments, offshore. **Distribution:** GO.

Amygdalum peasei (Newcomb, 1870): **949**
To 20mm. Fragile, very thin. Beaks subterminal. Modioliform but ventral and dorsal margins diverge equally to the posterior end. Periostracum smooth, glossy, and distinctively coloured white with irregular dark brown dendritic markings on the dorsal-posterior area. **Habitat:** in soft sediments, offshore. **Distribution:** NG.

Amygdalum watsoni (Smith, 1885): **950**
To 27mm. As *A. peasei* (see above) but expanding more rapidly so that the posterior forms a symmetrical broad curve. The periostracum is smooth, glossy, and white ventrally; translucent dorsally with small white markings. **Habitat:** in soft mud, deep water. **Distribution:** GO.

Subfamily CRENELLINAE

Solamen adamsianum (Melvill & Standen, 1907): **951**
To 10mm. Ovoid-rhomboidal, ligament margin very short, ventral margin longer than dorsal, both almost straight. Decussate sculpture of dense radiating threads and finer concentric lines. Anterior adductor scar ventral, well back from umbones. Inner margin denticulate. Straw coloured. **Habitat:** on sand and gravel, offshore. **Distribution:** NWG, SEG, GO, SO.

Solamen vaillanti (Issel, 1869): **952**
To 10mm. Modioliform almost transversely triangular, umbones prominent, ventral margin becoming arcuate. Sculpture of radiating riblets cut by widely spaced concentric lirae. Anterior adductor scar close to umbo. Inner margin denticulate. Straw coloured. **Habitat:** on shell sand, offshore. **Distribution:** GO.

Musculista perfragilis (Dunker, 1856): **953**
To 35mm. Fragile. Narrowly modioliform, umbones low, posterior not greatly expanded, dorsal margin very long, ventral margin long, shallowly curved. Surface shiny, sculpture smooth, radial lines are colour marks only. Dysodont teeth very few and faint. Pale green with reddish brown radial lines or bands down the posterior area, some also with transverse, zigzag, brown bands. **Habitat:** offshore. **Distribution:** SEG, GO.

Musculista senhousia (Benson, 1842): **954**
To 20mm. Fragile, rather tumid, umbones prominent. Modioliform, dorsal margin quickly merging with long posterior curve, ventral margin straight or slightly concave. Surface shiny-matt, sculpture appears smooth but dense, fine, concentric, striae present; radial lines are colour marks only. Dysodont teeth few. Pale green with transverse, zigzag, brown bands and reddish-brown radial lines or bands down the posterior area. **Habitat:** on sand/mud flats, subtidal. **Distribution:** SEG.

Musculus calceatus (Melvill & Standen, 1907): **955** NOT ILLUSTRATED
To 4mm. Modioliform but posterior not much deeper than anterior; posterior ridge high and broad. Sculpture of about 30 posterior riblets with both median and anterior areas smooth; internally anterior area reveals sculpture by presence of about 12 marginal crenulations. Hinge with 2-3 anterior and up to 10 posterior, dysodont teeth. **Habitat:** deep water. **Distribution:** GO.

Amygdalum watsoni 950

Amygdalum peasei 949

Solamen vaillanti 952

Musculista senhousia 954

Modiolus sirahensis 948

Solamen adamsianum 951

Musculista perfragilis 953

Superfamily: **MYTILOIDEA,**

Musculus cumingianus (Reeve, 1857): **956**
To 40mm. Inflated. Beaks subterminal. Modioliform but ovoid. Sculpture of 4-8 anterior riblets, median area smooth, 20-30 posterior riblets. Inner margins crenulate except on the median edge. Reddish brown, internally with an almost black ventral area. Periostracum shiny, dark brown. **Habitat:** in sea squirt tests, offshore. **Distribution:** GO, Mas, SO.

Musculus coenobitus (Vaillant, 1856): **957**
To 15mm. As *M. cumingianus* but less inflated; sculpture of 9-17 anterior riblets, 20-30 posterior riblets. Periostracum greenish brown tinged pink. **Habitat:** presumably with sea squirts. **Distribution:** SEG.

Musculus* cf *costulatus (Risso, 1826): **958**
To 12mm. Compressed. Modioliform, transversely ovoid. Sculpture of 10-14 anterior ribs, median area smooth, 16-20 posterior ribs distinctly scaled and with distinct cross marks in the interspaces. Anterior dysodont teeth (3-4) prominent, posterior dysodont teeth present. Inner margin crenulate except along edge of median area. Reddish with numerous, darker, tessellate markings. Periostracum bright green.
Similar to *M. viridulus* H. Adams, 1871 (Red Sea) and *M. mirandus* Smith, 1884 (Australia) but altogether heavier. These species live in rafts embedded in sand flats but the habits of the eastern Arabian form is unknown. **Habitat:** ? **Distribution:** GO, Mas.

Gregariella simplicifilis Barnard, 1964: **959**
To 15mm. Modioliform, distinctly wedge shaped with a relatively long posterior margin, ventral margin straight or slightly concave. Feeble, sculpture of about 10 anterior riblets, a narrow smooth median zone, about 15 narrow riblets on the posterior slope. Inner margin crenulated along the anterior and posterior edges. Large dysodont teeth present behind ligament and below umbo. White. Periostracum shiny, chestnut brown with long robust simple hairs on the posterior slope. **Habitat:** attached in crevices and old borings of date mussels. **Distribution:** NWG, GO.

Subfamily LITHOPHAGINAE
Botula cinnamomea (Gmelin, 1791): **960**
To 35mm. Beaks terminal, somewhat coiled, prominent. Subcylindrical, ventral margin slightly concave. Internal dorsal margin finely serrated beneath the ligament. White. Periostracum light brown becoming darker with age. **Habitat:** in calcareous rocks. **Distribution:** GO, Mas.

Lithophaga robusta Lamy, 1919: **961**
To 100mm. Cylindrical but posterior slightly expanded. Sculpture of fine vertical riblets; lacking consolidated calcareous deposits. White. Periostracum light yellow-brown. **Habitat:** in calcareous rocks. **Distribution:** NWG.

Leiosolenus lima (Lamy, 1919): **962** NOT ILLUSTRATED
To 100mm. Cylindrical; ligament margin shorter than dorsal slope; anterior margin obtuse. Calcareous deposit variable in extent but usually not worn completely and then smooth ventrally and coarsely reticulate on the posterior area, never projecting beyond the margins. White. Periostracum dark reddish brown with a paler ray often visible from the umbones towards the posterior. **Habitat:** in calcareous rocks and corals. **Distribution:** GO.

Musculus coenobitus 957

Musculus cf *costulatus* 958

Botula cinnamomea 960

Musculus cumingianus 956

Gregariella simplicifilis 959

Lithophaga robusta 961

Superfamily: **MYTILOIDEA, PTERIOIDEA**

Leiosolenus hanleyanus (Reeve, 1857): **963**
To 40mm. Calcareous deposits consolidated over the postero-dorsal area, usually smooth but some with traces of a reticulated surface, overlapping the posterior margin and forming a prominent, chisel-shaped extension. White. Periostracum yellowish brown. **Habitat:** in calcareous rocks and corals. **Distribution:** GO.

Leiosolenus obesus (Philippi, 1847): **964**
To 120mm. Cylindrical, elongate, prominently expanded just posterior to the ligament. Thin calcareous deposits covering most areas, smooth ventrally, granular posteriorly and dorsally, never projecting beyond the margins. White. Periostracum yellowish brown. **Habitat:** in calcareous rocks and dead coral. **Distribution:** GO.

Leiosolenus tripartitus (Jousseaume, 1888): **965**
To 60mm. Cylindrical, elongate, dorsal margin sloping sharply down from ligament margin. Encrustation extensive, dense, smooth, posteriorly extended into a long point which may be simple, or divided into three with a median extension which may almost cross that of the opposing valve. White. Periostracum chestnut to pale brown. **Habitat:** in dead coral, rock and large shells. **Distribution:** SEG, GO, Mas, SO.

Superfamily PTERIOIDEA
The pearl, wing and hammer oysters are the most familiar of the Pterioidea. They are byssally attached and the shell has a deep

Leiosolenus hanleyanus **963**

Leiosolenus tripartitus **965**

Pteria macroptera **966**

Leiosolenus obesus **964**

anterior notch with the beaks close to the anterior. The posterior adductor scar is much larger and placed almost centrally. The hinge bears no true teeth and the ligament is external, set in a single or multiple series of shallow triangular pits. Sculpture is limited to scaly growths; lined internally with the typical, lustrous mother of pearl. Three families are represented in eastern Arabian waters: the Pteriidae, Malleidae and Isognomonidae.

Of the Pteriidae, the wing oysters are generally obliquely ovate in outline with prominent extensions of the dorsal margin. They all live in association with colonial coelenterates such as hydroids, gorgonians and soft corals. Pearl oysters are also obliquely ovate to subquadrate but lack prominent wing like projections. They live attached to hard substrates among rocks or in crevices.

The most familiar of the Malleidae are the hammer oysters where the body of the shell is greatly elongate and spathulate, the dorsal margins being drawn out into long narrow projections. Most of the Malleidae lack the long dorsal wings but all are spathulate. They live partially embedded in soft substrates or in crevices, with the exception of *Vulsella* which is commensal with sponges.

The Isognomonidae differ from the others in that the ligament is in a multiple series of small pits along the hinge line. Otherwise they resemble the Malleidae in form and habits.

Family PTERIIDAE

Pteria macroptera (Lamarck, 1819): **966**
To 100mm. Obliquely ovate, broadly expanding ventrally so that height becomes greater than wing length. Hinge teeth well developed. Sculpture virtually smooth. Adductor and pedal retractor scars joined. Externally dark chestnut brown, uniform or with darker or black, irregular, subconcentric bands; internally white to lustrous copper. **Habitat:** attached to sea whips, offshore. **Distribution:** SEG, GO.

Pteria tortirostris (Dunker, 1848): **967**
To 80mm. Very obliquely ovate, posterior wing usually long, anterior wing prominent, twisted. Sculpture of very fine, closely spaced, concentric lines. Posterior adductor scar and pedal retractor scar not joined. Reddish-buff to dark brown with sparse, pale, radial markings; internally white. **Habitat:** attached to sea whips, offshore. **Distribution:** SEG, GO, Mas.

Pteria tortirostris **967**

Pteria penguin (Röding, 1798): **968**
To 250mm. At first obliquely ovate but becoming expanded ventrally. Sculpture weak. Black with narrow, radial, pale lines in juveniles, internally lustrous bluish with a narrow black margin. **Habitat:** attached to sea whips, offshore. **Distribution:** SEG, GO.

■ *Pteria* shells were locally called "Mussel" and were extensively collected to provide bluish mother of pearl for decorative uses. Most were collected off Ras al Khaymah and the Iranian islands of Hinderabi, Sheyk Shoeyb and Kais. Annual exports in 1901 totalled 400 - 500 tons. F. W. Townsend noted that in all his dredgings he rarely obtained this species but sackfuls were commonly seen at Dubai awaiting export. Today it is rarely seen.

Pteria penguin **968**

Superfamily: **PTERIOIDEA**

Pterelectroma zebra 969

Pterelectroma zebra (Reeve, 1857): **969**
To 20mm. Very fragile, translucent. Obliquely wedge shaped with a weak posterior wing. White with narrow black zigzag lines. **Habitat:** attached to hydroids, offshore. **Distribution:** SEG.

Pinctada margaritifera (Linnaeus, 1758): **970**
To 200mm. Subequivalve, thick, heavy. Subcircular to squarish; dorsal margin straight, shorter than diameter; posterior margin almost straight. Sculpture lamellose with radial rows of broad, appressed scales. Externally greyish green with white or yellowish, radial rows of scales; internally vividly lustrous with narrow, greyish green margin. **Habitat:** attached among rocks from lower shore and below. **Distribution:** NG, NWG, SEG, GO, Mas.

Pinctada radiata (Leach, 1814): **971**
To 65mm. Subquadrate usually higher than long; dorsal margin longer than body of shell; posterior margin slightly concave becoming subalate in some. Hinge with persistent teeth. Sculpture lamellose often with radial rows of sharp appressed spines. Muscle scar a more or less regular ellipse with a broad, poorly demarcated, dorsal tail. Externally variable, tan, brown and shades of red often with darker, almost black markings or with radial rays; internally lustrous with a narrow, non-nacreous margin. **Habitat:** attached to rocks from lower shore and below. **Distribution:** NG, NWG, SEG, GO, Mas.

Pinctada* cf *nigra (Gould, 1850): **972**
To 50mm. Subquadrate, often irregular; dorsal margin longer than body of shell; posterior margin slightly concave becoming subalate in some. Hinge without teeth. Sculpture lamellose, often with radial rows of sharp, appressed spines. Muscle scar with a weak basal bulge and a narrow, sharply demarcated, dorsal tail. Externally variable; radially rayed or uniform, yellow, tan, brown, shades of red or almost black; internally lustrous with a narrow non-nacreous margin. **Habitat:** in crevices and under rocks, mid to lower shore. **Distribution:** NWG, GO, Mas.

Family MALLEIDAE
Malleus* cf *albus Lamarck, 1819: **973**
To 150mm. Triangular spathulate with anterior and posterior wings, the anterior shorter and often further shortened by wear, spathulate section often curved posteriorly, margins with large interlocking undulations. Nacreous area an elongate oval, black; non-nacreous area with a distinct median ridge, buff in colour. Externally dirty white or pinkish flecked with black. **Habitat:** attached to rocks, often lying flat, lower shore and below. **Distribution:** GO, Mas.

Pinctada radiata 971

■ *This is the Pearl Oyster of the Arabian Gulf and was an important source of pearls before the development of culture methods in Japan. The Gulf trade centred on Kuwait and Bahrain with other collecting areas around the coast of what is now the United Arab Emirates. The pearl trade at the turn of the century was valued at £750,000 and about 2000 tons of shell were exported to London annually.*

Pinctada cf *nigra* 972

Pinctada margaritifera 970

■ *The shells of this species were once exported to London to provide mother of pearl for inlay and cutlery. Most were collected in the Gulf from off the islands along the Iranian coast. At the peak of the trade, around 1900, about 150 tons of shell were exported annually and this was rather small by comparison with other regions. Locally the shells were called "mother-o'-pearl" and in the trade were termed "Bombay Shell".*

Superfamily: ***PTERIOIDEA***

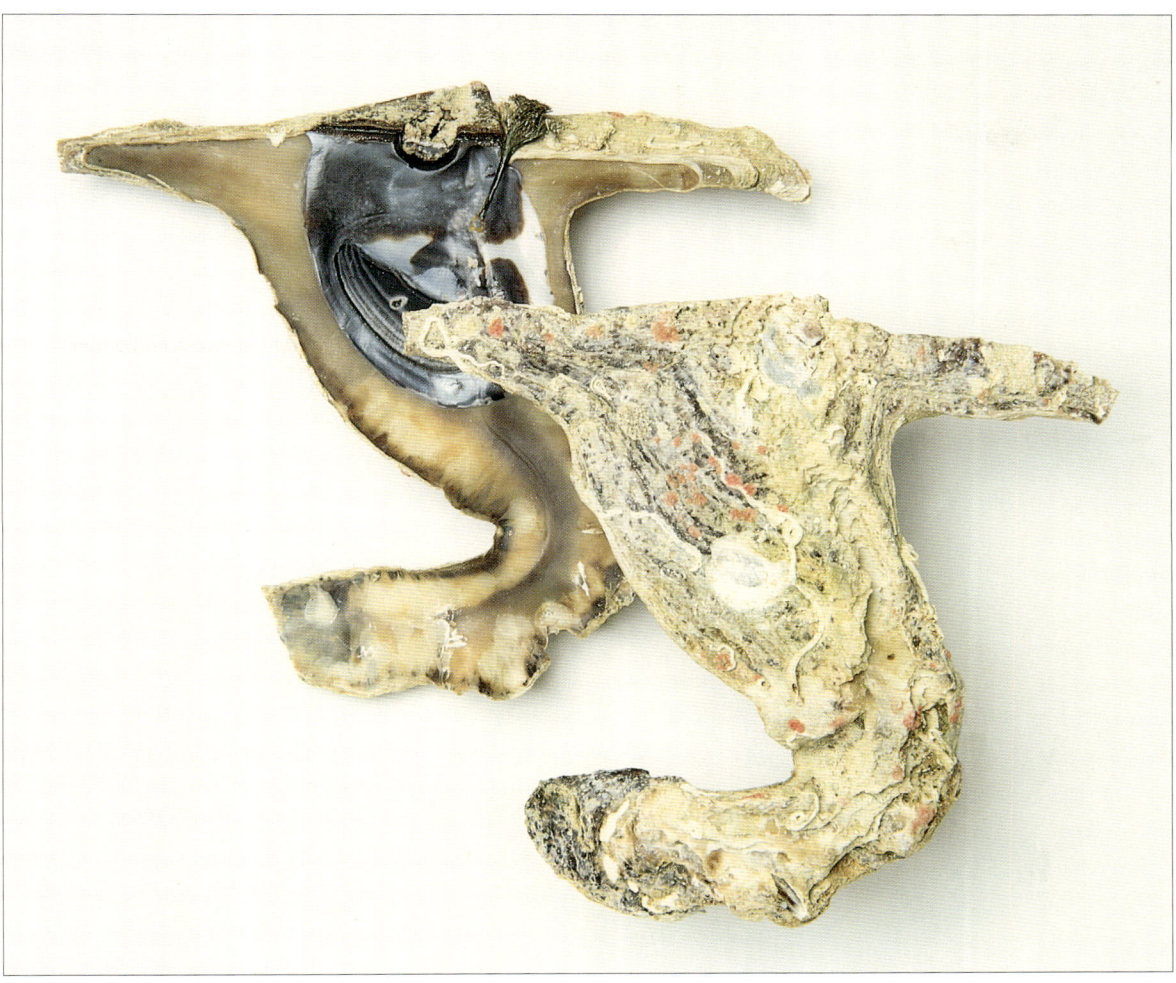

Malleus cf albus 973

Malvufundus normalis 974

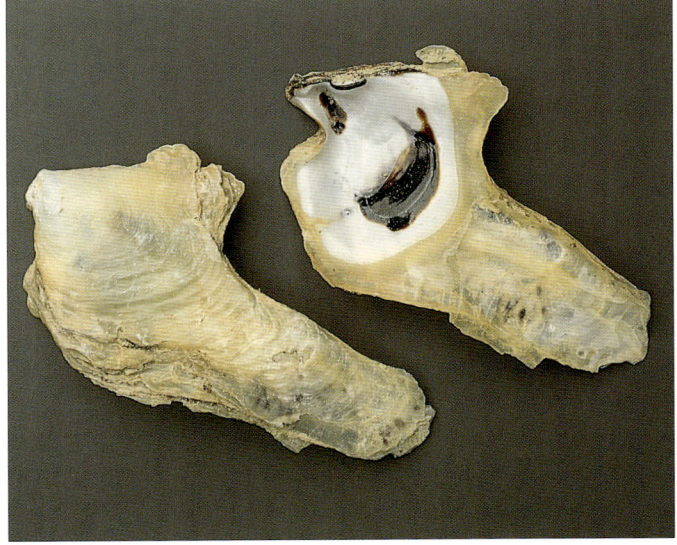

Malvufundus regula 975

Malvufundus normalis Lamarck, 1819: **974**
To 150mm. Elongate portion straight or curved; margins regularly waved; posterior wing short but broad; anterior wing absent. Externally greyish-cream with dense spots or larger streaks of a dark grey-brown colour; internal nacreous area is black, remainder greyish-cream. **Habitat:** among rocks, offshore. **Distribution:** SEG.

Malvufundus regula (Forsskål, 1775): **975**
To 40mm. Outline usually spathulate but irregular, occasionally curved, often sinuous, some ventrally expanded; posterior wing usually absent, if present very small. Nacreous area restricted, suboval; non-nacreous extension may have weak median ridge. Colour pattern variable, brown-purple throughout to greyish-yellow with purple brown spots to greyish yellow. **Habitat:** attached in crevices under rocks, lower shore and below. **Distribution:** NWG, SEG, GO, Mas, SO.

Superfamily: **PTERIOIDEA, PINNOIDEA**

Parviperna nucleus **979**

Vulsella vulsella **976**

Isognomon legumen **978**

Vulsella fornicata **977**

Vulsella vulsella (Linnaeus, 1758): **976**
To 80mm. Outline variable, spathulate to elongate oval, ears absent or very small. Ligament in a deep pit, umbones not divergent or only slightly so. Sculpture of minute, sharp scales on closely spaced, concentric lamellae; overall with very weak radial lines. Initially rather dark brown but becoming paler with dark radial lines or stripes. **Habitat:** embedded in sponges, sublittoral. **Distribution:** SEG.

Vulsella fornicata (Forsskål, 1775): **977**
To 70mm. As *Vulsella vulsella* but beaks point anteriorly, adjacent anterior dorsal margin prominent often projecting beyond the beaks. Ligament in a deep triangular pit, umbones widely divergent. **Habitat:** embedded in sponges, sublittoral. **Distribution:** GO, Mas.

Family ISOGNOMONIDAE
Isognomon legumen (Gmelin, 1791): **978**
To 90mm. Outline variable from narrowly spathulate, irregularly crescent-shaped to tailed. Hinge line usually shorter than widest part of body of shell with multiple ligament pits. Surface flaky often with weak radial lines or threads. Non-nacreous margin of shell usually wider than nacreous area. Externally tan, some with radial lines of darker brown or purple brown; internal nacreous area lustrous white. **Habitat:** attached under rocks from mid-shore to shallow waters. **Distribution:** all.

Parviperna nucleus (Lamarck, 1819): **979**
To 30mm. Subquadrate, beaked, byssal notch deep. Sculpture of raised narrow scales but frequently eroded and smooth. Ligament pits few, usually relatively large. Adductor scar narrow, crescent-shaped and situated close to the posterior margin. Externally greyish to purple-black; internally lustrous, tinged with purple, margin purple-black. **Habitat:** in crevices and among oysters, high in intertidal zone. **Distribution:** NG, NWG, SEG, GO, Mas.

Superfamily PINNOIDEA
The fan shells are large, but thin and brittle, triangular bivalves hinged along the straight dorsal margin. The ligament is similar to that of the mussels and there are no hinge teeth. The anterior adductor is very small and situated at the pointed end whereas the posterior adductor is large and placed subcentrally. The byssus is well developed and consists of numerous very fine threads. Fan shells are semi-infaunal, living embedded in sand and muddy substrates where the byssus is attached to sediment particles. In life only the edge of the posterior margin remains above the sediment and is subject to considerable damage. Shell repair is common and the posterior margin is frequently irregular where replacement has occurred.
The Pinnoidea of the Indo-Pacific were reviewed by Rosewater (1961) and his systematics are followed here. As outline and sculpture are variable the positions of the adductor scars are key characters. Unfortunately in eastern Arabian shells these characters do not always hold true and their use may need revision.

Pinna bicolor Gmelin, 1791: **980**
To 400mm. Ligament margin straight, ventral margin shorter and often curved, posterior margin rounded and oblique. Sculpture of 8 - 17 radiating ribs with sparse spines but often obsolete. Median sulcus short becoming wide and faint. Tan to dark purplish brown occasionally horn coloured, often with irregular alternating bands of dark and light. **Habitat:** in sand and gravel among rocks, lower shore and below. **Distribution:** NWG, SEG, GO, Mas.

Pinna bicolor 980

Superfamily: **PINNOIDEA**

Pinna muricata **981**

Atrina vexillum **982**

Streptopinna saccata **984**

Pinna muricata Linnaeus, 1758: **981**
To 300mm. Ligament and ventral margins straight, diverging equally; posterior margin truncate. Median sulcus long, very narrow and deep. Spines strong to obsolete. Usually pale with discreet greyish to black markings. Although many specimens of *Pinna* match the general appearance of *P. muricata* all the specimens from the Arabian Sea lack the diagnostic character of the adductor scar overlapping onto the ventral nacreous layer. This suggests that all may be variants of *P. bicolor*. (see above) **Habitat:** embedded between rocks or in eel grass beds, lower shore and below. **Distribution:** NWG, SEG, GO.

Atrina vexillum (Born, 1778): **982**
To 200mm. Robust. Triangular, broadly rounded posteriorly to hatchet shaped. Nacreous layer not divided into two lobes. Muscle scar large, subcircular at or protruding beyond the extent of the nacreous layer. Black occasionally reddish or brown.
Masirah shells have 10-17 major radial ribs with large upright spines posteriorly; on Gulf shells sculpture is weak or absent. **Habitat:** among rocks, offshore. **Distribution:** SEG, GO, Mas.

Servatrina pectinata (Linnaeus, 1767): **983**
To 300mm Fragile. Triangular wedge-shaped, posterior margin a little rounded. Radial sculpture of 15-30 weak ribs on posterior slope; very fine riblets or smooth on ventral area; ribs weakly imbricate with a few short spines. Nacreous layer not divided into two lobes. Adductor scar well within the edge of the nacreous layer. Tan to reddish-brown. **Habitat:** in soft sediments, offshore. **Distribution:** GO, Mas.

Superfamily: **PINNOIDEA, LIMOIDEA**

Servatrina pectinata **983**

Lima sowerbyi **985**

Ctenoides annulatus **988**

Limaria fragilis **986**

Limatulella viali **987**

Streptopinna saccata (Linnaeus, 1758): **984**
To 200mm. Roughly triangular, often elongate but usually distorted and twisted. Valves fused along the dorsal margins. Sculpture of 5 to 12 radiating ribs, usually broad and smooth. Greyish white through tan to dark reddish brown. **Habitat:** among rocks, mostly sublittoral. **Distribution:** SEG, GO, Mas.

Superfamily LIMOIDEA

The Limoidea or file shells resemble the scallops in having shells with small ears at the sides of the hinge. However, they are usually oblong or suborbicular in outline and often asymmetrical. The hinge has a central triangular, external ligament and lacks teeth. As in the scallops there is a single adductor scar. The sculpture is often of radial ribs, with erect flattened spines and the colour is white to pale brown but never bright.

The Limoidea live mainly under rocks or in crevices attached by a byssus. These forms often gather byssus threads around themselves like a nest. Free-living forms are less common and these burrow in the surface of gravel and sand. Some of the byssiferous forms are mobile but are not as efficient swimmers as the scallops. The mantle edge is lined with tentacles often orange or yellow in colour, which are sticky and produce acrid secretions. It is believed that they deter predators.

Lima sowerbyi Deshayes, 1863: **985**
To 40mm. Solid, not gaping. Obliquely oval, anterior margin very long, straight; anterior slope vertical; ears small. Prominent sculpture of 18-24 radial ribs, all with erect flat scales.
Often called *L. lima* (Linnaeus, 1758) or *L. l. vulgaris* Link, 1807. **Habitat:** under rocks, lower shore and sublittoral. **Distribution:** NWG, SEG, GO, Mas, SO.

Limaria fragilis (Gmelin, 1791): **986**
To 40mm. Thin, fragile, translucent. Valves almost flat, anterior and posterior gaping widely. Obliquely ovate, hinge line much narrower than length of shell, ears small. Faint sculpture of numerous radial riblets. White. Animal has many reddish tentacles, lives in a byssus nest below rocks and can swim if disturbed. **Habitat:** under rocks, lower shore and below. **Distribution:** NWG, SEG, GO, Mas.

Limatulella viali (Jousseaume in Lamy, 1920): **987**
To 20mm. Fragile, inflated without gapes. Obliquely oval, ears subequal, small. Sculpture of 20-25 raised radial riblets, sparse anteriorly, more dense posteriorly, occasional radial threads between riblets. White. **Habitat:** ? **Distribution:** NWG.

Ctenoides annulatus (Lamarck, 1819: **988**
To 40mm. Flattened. Byssus gape with lip-like reflected edges. Oval, not strongly oblique. Sculpture of weak, rounded radial riblets, divaricating ventrally and with flattened scales overall. Pale brown to white. **Habitat:** ? **Distribution:** Mas, SO.

Superfamily: **LIMOIDEA, OSTREOIDEA**

Limatula leptocarya **989**

Limatula leptocarya (Melvill, 1898): **989**
To 8mm. Not gaping, deeply concave. Oval, much higher than long, ears short, of equal size. Sculpture of 25-28 smooth radial riblets, median ribs strongest with a weak median sulcus. White. **Habitat:** in sand or muddy sand, offshore. **Distribution:** NWG, SEG, GO.

Limea juglandula Melvill & Standen, 1907: **990** NOT ILLUSTRATED
To 4mm. Solid, not gaping, deeply concave. Oval, hinge line relatively long. Sculpture of about 20 radial ribs, imbricated by concentric lines. Hinge plate thick with about 10 denticles on either side. White. **Habitat:** in shell sand, deep water. **Distribution:** GO.

Superfamily OSTREOIDEA

The Ostreoidea or true oysters are epifaunal bivalves living cemented to a variety of substrates by the left valve. Shell shape is varied, often influenced by and conforming to the substrate. In general they are subcircular to elongate oval. They are more or less equilateral but are usually inequivalve with the cemented left valve the more inflated. Sculpture is primarily of thin, foliaceous lamellae which are often eroded but occasionally develop into hollow (hyote) spines. The valves may be flat or variously folded, most noticeably at the margins. All oysters have a single large adductor situated off centre towards the posterior ventral edge. The hinge is without teeth but a variety of marginal ridges or pustules known as chomata may be present and are of systematic importance. Vermiculate chomata are long, narrow ridges present in dense patches just beyond the ligament. Nodular chomata are circular to oval, rounded denticles which often have corresponding pits in the opposite valve; they occur singly or in small groups. Pustulose chomata are like tiny pin heads which occur in multiples over the inner margin. The ligament is simple, forming a three-part triangle beneath the beaks. Oyster species vary in colour; most are of a greyish white ground colour with shades of red to purple-black occurring as radial or irregular markings. Some are more brightly coloured in shades of orange-red or even blue. Internally they are usually white but again washes of red-purple, green or brown are not unusual.

Oysters are not easy to identify and this is compounded here by the lack of studies on eastern Arabian forms. General shape and sculpture cannot be trusted and confirmation must be based on details of the chomata, adductor scar, shell structure and sculpture. Colour may be consistent in some but in others it is highly variable. Even then the taxonomy of some species is uncertain and will remain so until anatomical or even genetic studies are undertaken.

Family OSTREIDAE

Lopha cristagalli (Linnaeus, 1758): **991**
To 90mm. Small attachment area with clasper spines. Roughly circular to oval, with 4-8 large, acute folds on both valves. Outer surface at least in part with pustulose striations. Multiple pinhead (lophine) chomata present on both valves but more prominent on RV. Externally shades of brown to lilac, internally bronze to white. **Habitat:** attached to corals and rocks, mainly sublittoral. **Distribution:** NWG, SEG, GO, SO.

Lopha cristagalli **991**

Dendrostrea frons **992**

Dendrostrea frons (Linnaeus, 1758): **992**
To 60mm. Attachment area variable, often with short clasper spines. Irregularly subcircular to elongate. Both valves convex with up to 20 rounded folds, which are usually

Superfamily: **OSTREOIDEA**

Striostrea margaritacea **995**

Alectryonella crenulifera **994**

strongly developed at the margins where they interlock. Sculpture smooth or weakly lamellose, without pustules. Nodular chomata present near the hinge, pustulose chomata often present on RV. Externally whitish but often shades of pink to blue, internally lustrous white to reflections of the outer colour. Those attached to whip coral stems have a longitudinal ridge corresponding in size to the stem. **Habitat:** attached to rocks, corals and whip corals, mostly sublittoral. **Distribution:** NWG, SEG, GO.

Alectryonella plicatula (Gmelin, 1791): **993**
To 70mm. Thin but strong. Attachment area variable, lower valve very shallow conforming to substrate, with clasper spines; upper valve flat to convex. Outline very variable as shell conforms to space available on rock. Sculpture weakly lamellose, radial folds irregular but some usually interlocking at margins. Chomata sparse, nodular, often in small clumps close to ligament but may occur all round. Adductor scar kidney shaped. Externally purple-red, some with radial paler lines or stripes; internally mottled malachite green, purple-red at margins. **Habitat:** attached to rocks, lower shore and sublittoral. **Distribution:** GO, Mas.

Alectryonella crenulifera (Sowerby, 1871): **994**
To 40mm. Roughly circular. Lower valve extensively cemented with many small marginal folds. Upper valve almost flat with indented margins, surface irregular, lacking spines and rather smooth. Chomata sparse.
Externally off-white with weak reddish lines, internally flushed green. **Habitat:** attached to rocks, middle shore. **Distribution:** NWG.

Striostrea margaritacea (Lamarck, 1819): **995**
To 80mm. Lower valve almost completely cemented. Elongate oval, narrowing towards umbo. Sculpture of weak lamellae and irregular folds, outer layers with dense, irregular, radial striations, often worn smooth. Nodular chomata at edges below umbones. Adductor scar kidney shaped. Externally dull purple brown, white if worn; internally lustrous white with weak purple-reddish brown areas, especially at margins. **Habitat:** attached to rocks in dense patches, lower shore and sublittoral. **Distribution:** GO, Mas, SO.

Alectryonella plicatula **993**

SEASHELLS OF EASTERN ARABIA 227

Superfamily: **OSTREOIDEA**

Saccostrea cuccullata (Born, 1778): **996**
To 70mm. Irregularly circular to oval. Lower valve with large area of attachment, upturned margin with numerous folds. Upper valve flat with marginal lobes fitting folds of lower valve, sculpture of appressed lamellae, some becoming spiny, others worn smooth. Muscle scar kidney shaped. Nodular chomata usually present around all margins. Externally shades of purple-black often with pale or white radial streaks. Internally white with a broad purple-black border, muscle scar may be darker than surrounding shell area. **Habitat:** covering rocks or in clumps in mangroves, upper middle shore. **Distribution:** all.

"Ostrea" subucula Lamy, 1925: **997**
To 40mm. Thin and brittle. Shallow lower valve conforming to substrate often irregular. Flat upper valve translucent with sparse chalky thickenings. Both valves generally elongate, narrowing towards umbo. Weak nodular chomata near umbo visible in some shells. Muscle scar subcircular, usually more flattened dorsally but not kidney shaped. Sculpture, if present, of very weak concentric lamellae. Externally dirty buff, internally lustrous white. This species has been synonymised with ***Neopycnodonte cochlear*** (Poli, 1795) a deep-water oyster from the Mediterranean and Atlantic. Given the habitat of the eastern Arabian shells such a synonymy seems highly unlikely. Until a full anatomical study is done its placement in ***Neopycnodonte*** is doubtful and ***Ostrea*** in its widest sense is used here provisionally. **Habitat:** under rocks, upper middle shore. **Distribution:** NWG, SEG, GO.

Ostrea sp: **998**
To 70mm. Solid shells with large attachment areas. Lower valve shallow, upper valve generally flat. Outline variable and irregular, roughly circular to oval. Weak sculpture of appressed, concentric, irregular lamellae, some with radial folds interlocking at margins. Large nodular chomata restricted to dorsal margins. Adductor scar large, flattened dorsally. Externally from uniformly dirty beige to reddish-purple with radial pale lines and bands; internally usually suffused with dark green to olive but some with rust red or copper tones. This oyster may be related to the South African ***O. atherstonei*** Newton, 1913, but until a proper systematic review is undertaken the Masirah shells remain unnamed. **Habitat:** on and under rocks, mid to lower shore. **Distribution:** Mas.

Family GRYPHAEIDAE

Hyotissa hyotis (Linnaeus, 1758): **999**
To 150mm. Robust. Lower valve partially free. Roughly circular to oval, both valves thick with up to 12 rounded folds, surface with lamellar scales developed as hyote spines. Dense, ridge like (vermiculate) chomata at sides of ligament. Adductor scar subcircular. Variable shades of reddish brown through olive-brown to black, paler internally. **Habitat:** attached to rocks and coral, lower shore and below. **Distribution:** NWG, SEG, GO.

"Ostrea" subucula **997**

Saccostrea cuccullata **996**

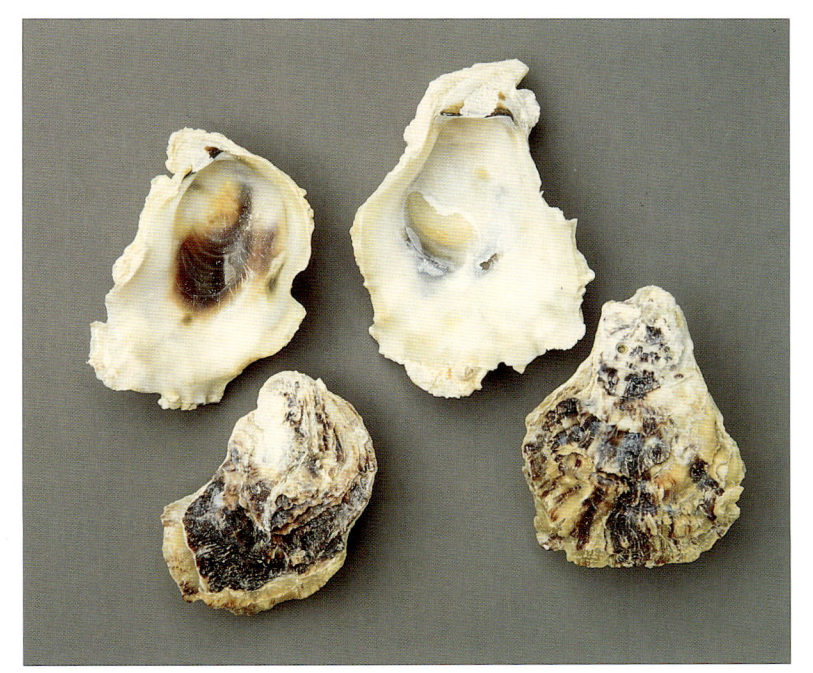

Superfamily: **OSTREOIDEA, PLICATULOIDEA**

Ostrea sp. **998**

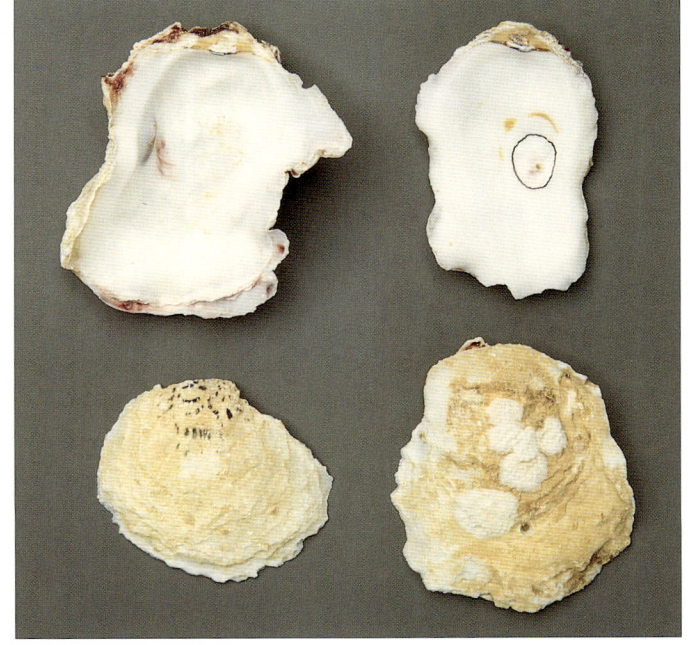

Hyotissa numisma **1000**

Hyotissa numisma (Lamarck, 1819): **1000**
To 70mm. Subcircular. Lower valve extensively cemented to substrate, shallow with short steep sides, upper valve more or less flat unsculptured, marginal folds weak. Dense, ridge like (vermiculate) chomata at sides of ligament. Adductor scar subcircular. Mainly white, purplish brown patches internally. Indistinguishable from *Plicatula australis* without examining the hinge. **Habitat:** attached to rocks, mid-shore and below. **Distribution:** GO, Mas.

Superfamily PLICATULOIDEA
The Plicatulidae resemble the Spondylidae (see below) in that they are cemented and possess an isodont hinge. They are much smaller and sculptured with divaricating, smooth or imbricate, radial ribs. The attachment area ranges widely from being restricted to the umbo to involving the whole of the cemented valve.

Hyotissa hyotis **999**

Plicatula australis Lamarck, 1819: **1001**
To 50mm. Irregularly discoid and compressed, attached by whole or most of lower valve. If perfectly developed then sculpture of 10-20 low, weakly angulate, diverging ribs, some developing short spines, these forming interlocking crenulations at margin. Umbonal area with dense, pale brown lines and sparse, darker to black flecks; crural teeth brown. More frequently shells are worn and devoid of sculpture, marginal crenulations vary in depth, colour lost or that of encrusting algae or epifauna. **Habitat:** cemented to rocks, mid-shore and below. **Distribution:** all.

Plicatula australis **1001**

SEASHELLS OF EASTERN ARABIA 229

Superfamily: PLICATULOIDEA, PECTINOIDEA

Plicatula plicata (Linnaeus, 1767): **1002**
To 20mm. Roughly pear shaped, compressed. Cemented at umbo only. Sculpture, on both valves, of 5-10 diverging, weakly angulate ribs lacking ornamentation. Patterned with dense, reddish brown, irregularly radial, broken lines on a paler background. **Habitat:** cemented to rocks and coral, lower shore and below. **Distribution:** GO, Mas.

Superfamily PECTINOIDEA
The scallops (Pectinidae) with their ribbed, fan-shaped shells are perhaps the easiest of bivalves to recognise. Together with the Propeamussidae and the thorny oysters (Spondylidae) they make up the Pectinoidea. All are roughly discoidal to fan shaped in outline and in life lie on one of the valves. All have a single, large, adductor muscle and all have a simple triangular ligament set on an internal shallow resilifer. They may be free living, byssally attached or cemented, many of the free-living species are able to swim by clapping the valves together.
The Pectinidae living in eastern Arabian waters are either free living or byssally attached. The shells are fan shaped, often unequally convex and have ear like extensions (auricles) of the dorsal margin. Hinge teeth are generally absent except that some have weak simple pairs of teeth (crural teeth) along the auricles. The sculpture is usually of radial ribs but some are smooth. Orientation of scallop shells may be confusing but those with a byssus have the byssal notch anterior. The byssal notch usually has a set of teeth (ctenolium) between which the byssus threads pass. Most scallops are brightly coloured and patterned but as identification characters these are generally unreliable as they can be very variable. Identification is based on the overall shape and details of the sculpture and microsculpture.
Similar in most respects to the Pectinidae is the Propeamussidae but the latter lack a ctenolium, have fragile shells usually strengthened by internal ribs. They are predominantly deep water forms. The Spondylidae are all sessile, living with the right valve cemented to rocks, corals etc. They are usually heavy shelled, have weak auricles and are sculptured with radially arranged rows of spines. The hinge is reinforced by the presence of heavy crural teeth which often interlock so much so that they have to be broken before the valves can be separated. As the lower valve is cemented, growth of the ligament is often asymmetric resulting in the extension of the dorsal area of the attached valve. Due to their cemented life style the shells are subject to great variation in form and development of the sculpture, making them one of the most difficult groups to identify.

Family PECTINIDAE
Chlamys senatoria (Gmelin, 1791): **1003**
To 70mm. Disc almost circular. Sculpture of 22-26 low rounded ribs, secondary riblets developing on margins of primaries; ribs often smooth but scales often persist along edges of ribs. Reddish brown, red, mauve or yellow, either uniform or with mottling or zigzag bands. **Habitat:** among rocks, lower shore and below. **Distribution:** NWG, SEG, GO, Mas.

Chlamys townsendi (Sowerby, 1895): **1004**
To 200mm. Disc almost circular. Sculpture of 18-23 low rounded ribs, almost smooth in large shells but sparsely scaled in young. Colour consistent; externally mottled beige, reddish brown and purple brown, generally dull; internally with purple at least around hinge. **Habitat:** attached to or among rocks, lower shore and below. **Distribution:** GO, Mas, SO.

Chlamys noduliferus (Sowerby, 1842): NOT ILLUSTRATED.
40mm. Broadly fan shaped. Sculpture of 7-9 broad, rounded, striated ribs, nodulose at junctions with deeply stepped, concentric ridges. Mottled, orange and white. **Habitat:** ? **Distribution:** SEG, Mas.

Plicatula plicata 1002

Chlamys senatoria 1003

Superfamily: **PECTINOIDEA**

Chlamys townsendi **1004**

Chlamys livida (Lamarck, 1819): **1006**
To 90mm. Disc narrowing towards umbo. Sculpture of 22-28 low, narrow, primary riblets with varying numbers of secondaries or incised grooves; primaries with partially flattened scales. Auricles with latticed sculpture between riblets, umbonal area with weak divaricating microstriations. Usually purple-red but white, pink and red also occur. The purple-red eastern Arabian shells are often referred to as *C. ruschenbergeri* Tryon, 1869.
Habitat: attached by byssus to rocks, littoral and offshore.
Distribution: all.

Chlamys cf tincta (Reeve, 1853): **1007**
To 25mm. Distinctly higher than wide, posterior auricles almost absent. Disc with about 20 narrow, rounded riblets and rapidly developing secondaries and tertiaries so that adults have 50 - 70 riblets all with small erect scales. Prominent, divaricate microsculpture between riblets, no netted sculpture on auricles. Usually red but some beige and red-brown. The relationships of these shells to *C. multistriata* (Poli, 1795), *C. natalensis* (Smith, 1906) and *C. humilis* (Sowerby, 1904) need clarification. **Habitat:** known from strandline valves only. **Distribution:** Mas, SO.

Chlamys livida **1006**

Chlamys cf tincta **1007**

SEASHELLS OF EASTERN ARABIA

Superfamily: **PECTINOIDEA**

Decatopecten plica (Linnaeus, 1758): **1008**
To 40mm. Auricles small, almost equal. Sculpture of 6-9 broad, rounded ribs, the outer pair rather smaller; with numerous radial raised threads and minutely concentrically scabrous overall. Hinge with prominent crural teeth, byssal gape absent. LV cream to beige with beige to reddish patches; RV paler, often uniformly cream. **Habitat:** on rough ground, offshore. **Distribution:** SEG, GO, Mas.

Pecten dorotheae Melvill in Melvill & Standen, 1907: **1009**
To 60mm. LV flat but umbo concave, RV deeply concave. Sculpture of 14 - 17 broad flat topped ribs all with at least a median groove and often with up to four grooves. Pale shades of brick red over white, flat valve darker.
This species has often been called **P. erythraeensis** Sowerby, 1842 but that has smooth ungrooved ribs. **Habitat:** free living on sand and gravel, offshore. **Distribution:** SEG, GO, SO.

Family PROPEAMUSSIDAE
Propeamussium steindachneri (Sturany, 1899): **1010**
To 15mm. Very thin, almost transparent; valves almost equally inflated. Higher than long with small auricles, byssal notch obsolete. Sculpture of disk smooth, minute ridges on the auricles. Internal ribs reaching about 2/3 the height but as ventral margins are usually broken the ribs often apparently reach the margin. RV with amber patches and fine, subconcentric, interrupted lines; LV white. **Habitat:** on muddy sand, deep water. **Distribution:** GO.

Parvamussium siebenrocki (Sturany, 1899): **1011**
To 7mm. Slightly inequivalve at the ventral margin, LV inside RV and often bent. Subcircular. Auricles small, anterior ones slightly larger with a byssal notch. Sculpture very discrepant, LV with fine concentric raised threads, RV with strong narrow riblets with secondary and tertiary elements developing towards the margins, totalling about 25 in the larger specimens. The ribs and interspaces are crossed by widely spaced, concentric threads making the ribs minutely imbricate. Both valves with 10-13 internal ribs reaching almost to the margins. White. **Habitat:** on muddy sand, deep water. **Distribution:** GO.

Parvamussium thyrideum (Melvill in Melvill and Standen, 1907): **1012**
NOT ILLUSTRATED
To 4mm. Height equal to length, auricles subequal, byssal notch small. Sculpture of RV cancellate, 13-17 primary ribs, occasional secondaries developing marginally, crossed by concentric lamellae almost equal in size to the ribs, nodules at intersections. Internal ribs weak, medially obsolete, 5-8 in total. White. **Habitat:** on sand and mud, deep water. **Distribution:** GO.

Similipecten eous (Melvill in Melvill and Standen, 1907): **1013** NOT ILLUSTRATED
To 5mm. Thin, transparent except for umbonal area. Outline circular, anterior auricles defined by groove, posterior auricles weak. Smooth

Propeamussium steindachneri **1010**

Decatopecten plica **1008**

Parvamussium siebenrocki **1011**

Pecten dorotheae **1009**

232 SEASHELLS OF EASTERN ARABIA

Superfamily: **PECTINOIDEA**

except for concentric ridges on anterior auricles. White. **Habitat:** on sand and mud, deep water. **Distribution:** GO.

Family SPONDYLIDAE
Spiny oyster shells are very variable in shape and sculpture and are difficult to identify. Nearly all the eastern Arabian shells that have been examined have a reddish inner margin with beige colouring around the hinge; externally, if not worn, the sculpture is of flattened spines. Despite the extent of the external sculpture all shells with these internal characters are here referred to a single species.

Spondylus marisrubri Röding, 1798: **1014**
To 100mm. Irregular discoid, attachment area large. Sculpture of rows of flattened spines sometimes worn smooth, some in spaced rows, others crowded. Externally reddish; inner margin reddish to purple-red with beige around the hinge. **Habitat:** cemented to rocks and corals, lower shore and below. **Distribution:** all.

Spondylus hystrix **1015**

Spondylus hystrix Röding, 1798: **1015**
To 60mm. Irregular discoid, attachment area large. Sculpture of numerous radial rows of narrow semi-erect spines. Externally white with sparse purple to black markings; internally white with some beige or yellow around margins. **Habitat:** cemented to corals, offshore. **Distribution:** GO.

Spondylus marisrubri **1014**

SEASHELLS OF EASTERN ARABIA 233

Superfamily: PECTINOIDEA, ANOMIOIDEA, LUCINOIDEA

Spondylus gloriandus Melvill & Standen, 1907: **1016**
To 90mm. Irregularly discoid, attachment area large. Sculpture of numerous rows of long almost erect narrow spines of various lengths. Externally dull orange with weak pink tracery around umbones; internally white with dull orange on margins. **Habitat:** cemented to rocks and cables offshore. **Distribution:** SEG.

Superfamily ANOMIOIDEA
The Anomioidea include the saddle oysters (Anomiidae) and window pane shells (Placunidae).
In life saddle oysters resemble true oysters in that the lower valve closely conforms with the substrate but is not cemented to it. Attachment is achieved by a plug-like, calcified byssus which passes through a large notch in the lower valve. In outline the shells are irregularly circular or oval. True hinge teeth are absent but divergent umbonal ridges (crural teeth) may be present. Distinctive muscle scars are present, best seen in the upper valve, consisting of a subcentral adductor scar with one or more pedal and byssus retractor scars above it. Saddle oysters live attached to rocks and other shells from mid shore and below.
The window pane shells are free living and do not have a byssal notch. They live in sheltered muddy lagoons in shallow water.

Family ANOMIIDAE
Anomia achaeus Gray, 1850: **1017**
To 50mm. Irregularly circular to oval, rather flat to deeply concave; with three retractor scars. Weak upper-valve sculpture of irregular concentric lines and minute radial striae. Lustrous, white, yellow, orange to green. Lower valve very thin with a prominent notch; usually white. **Habitat:** attached to rocks and shells from mid shore and below. **Distribution:** all.

Family PLACUNIDAE
Placuna placenta (Linnaeus, 1758): **1018**
To 150mm. LV a little convex, RV a little concave, very compressed. Subcircular. Hinge line straight but short, with two ventrally diverging crural teeth supporting ligament material. Concentric sculpture of very thin appressed lamellae which are radially vermiculate overall.
In recent years only the occasional dead shell has been found. **Habitat:** free living in shallow muddy lagoons. **Distribution:** NWG, GO.

Subclass HETERODONTA
Superfamily LUCINOIDEA
The Lucinoidea superficially resemble the venerid clams in that they are predominantly lenticular or subovate in form and are burrowers into sand, gravel and mud. Anatomically they are quite distinct as they do not possess paired posterior siphons and the pallial line is always entire. Unlike the venerids the inhalant aperture is anterior and the contact with the surface is maintained by the finger-like foot which builds a mucus-lined tube.
In the Lucinidae and the Thyasiridae the anterior adductor muscle is usually elongate and this is reflected by the scar which is partly free from the pallial line. The Lucininae have a fully developed dentition of cardinal and lateral teeth. The Lucinidae are further divided into a number of subfamilies.
The Lucinidae are lenticular often with an anterior and posterior sulcus. The hinge has two cardinal teeth and anterior and posterior laterals developed to various degrees. The ligament is mostly external but deeply sunken in some. Sculpture is primarily concentric, often with a secondary radial element. The anterior adductor scar is relatively short. The Myrteinae are compressed, roundly squarish in outline and lacking any obvious external sulcus. Sculpture is concentric. Anterior scars are medium sized. The Milthinae are more variable in

Spondylus gloriandus **1016**

Anomia achaeus **1017**

Placuna placenta **1018**

Superfamily: **LUCINOIDEA**

Lucina dentifera **1019**

Codakia tigerina **1021**

Bellucina semperiana **1020**

Ctena divergens **1022**

Cavilucina pamela **1023**

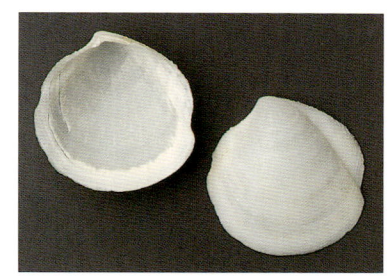

"Lucina" victorialis **1024**

outlines but characteristically have a long anterior adductor scar. Sculpture is weak and always concentric.

The Divaricellinae are lenticular with a divaricate sculpture of lines or ridges.

The Thyasiridae have a poorly developed hinge with either no teeth or a small cardinal peg. They are poorly represented in tropical seas and only a single species is described here.

In the Ungulinidae the anterior adductor is not elongate and species of this family are easily confused with venerids. In eastern Arabia all are related to the genus *Diplodonta*. They are roughly orbicular with a smooth sculpture. The adductor scars are not elongate, the pallial line is thick but thins out over a short area postventrally. The hinge consists of two cardinal teeth in each valve and no laterals. The absence of laterals distinguishes the Ungulinidae from venerids which lack a pallial sinus.

Family LUCINIDAE
Subfamily LUCININAE

Lucina dentifera Jonas, 1846: **1019**
To 30mm. Subcircular-trigonal; posterior margin sloping, sulcus prominent forming a posterior indentation, a small anterior sulcus and indentation also present. Lunule small, sunken. Sculpture of up to 30 widely spaced, fragile lamellae which when well preserved are frilled along their margins. Hinge with 2 cardinals and full laterals in each valve. Inner margin weakly denticulate. White. **Habitat:** in mud and shell sand, offshore. **Distribution:** SEG, GO.

Bellucina semperiana (Issel, 1869): **1020**
To 8mm. Beaks central, spherical. Circular with a prominent posterior sulcus and a slightly demarcated anterior area. Lunule small but deep. Sculpture of heavy concentric ridges crossed but not interrupted by numerous radial riblets. Inner margin crenulate, denticulate on posterior sulcus. **Habitat:** in shell sand, lower shore and below. **Distribution:** NWG, SEG, GO, Mas.

Codakia tigerina (Linnaeus, 1758): **1021**
To 75mm. Lenticular, posterior sulcus indistinct, lunule small. Sculpture of numerous radial riblets intersecting slightly weaker concentric ridges, more or less reticulate. White often with rose tints dorsally; inside flushed yellow. **Habitat:** in sand, offshore. **Distribution:** SEG, GO, Mas, SO.

Ctena divergens (Philippi, 1850): **1022**
To 25mm. Beaks central. Variable, from subelliptical with lateral margins evenly rounded to oval with posterior margin longer and more straight than anterior. Lunule always prominent and excavated. Sculpture of numerous radial riblets, some dividing and diverging on lateral margins; cut by numerous concentric grooves overall. White or tinged pink-yellow. **Habitat:** in sand and gravel, lower shore and below. **Distribution:** all.

Cavilucina pamela (Melvill & Standen, 1907): **1023**
To 8mm. Beaks central, rather tumid. Subcircular, posterior dorsal margin short, sloping gently; posterior margin straightening. Lunule asymmetric, large, cordate, sunken. Sculpture initially of raised concentric threads these developing into low but rather stout, somewhat rugose ridges. Cardinal teeth strong, laterals insignificant. Inner margin smooth. Shell white or brownish. **Habitat:** offshore. **Distribution:** NWG, GO.

"Lucina" victorialis (Melvill, 1899): **1024**
To 20mm. Thin. Beaks central. Oval-subtrigonal, slightly higher than long; umbones small, pointed; posterior sulcus strong, extending well down the posterior margin; anterior sulcus short and weak. Lunule small, cordate and slightly sunken. Sculpture of dense, concentric, raised threads giving a rough feel, rasp-like over lateral slopes. White.
This species was originally placed in the thyasirid genus ***Cryptodon*** but the presence of teeth negates this. Its correct position is uncertain and it is placed here in ***Lucina*** in the widest sense. **Habitat:** in muddy sand, offshore. **Distribution:** SEG, GO.

Superfamily: **LUCINOIDEA**

Pillucina angela (Melvill, 1899): **1025**
To 13mm. Solid, beaks almost central. Almost circular, posterior margin becoming straight. Lunule small. LV with 2 strong cardinals. Sculpture of numerous weak, radial riblets diverging slightly and crossed by concentric threads; weaker medially. Inner margin weakly denticulate. Anterior adductor free for half of its length. White. **Habitat:** offshore. **Distribution:** GO, Mas.

Pillucina fischeriana (Issel, 1869): **1026**
To 14mm. Beaks almost central. Almost circular but straightening both on anterior and posterior. Lunule elongate-cordate. Divergent radial sculpture of low, somewhat broad, riblets, these absent medially; with dense slightly wavy concentric threads overall. Inner margin weakly denticulate. Anterior adductor free for two thirds its length. White, pale yellow or pale orange. **Habitat:** in muddy sand, intertidal, especially in lagoons and on mud flats. **Distribution:** all.

Wallucina erythraea (Issel, 1869): **1027**
To 12mm. Beaks almost central. Subcircular with a distinct posterior sulcus and a much weaker anterior sulcus. Lunule slightly sunken. Sculpture of evenly spaced concentric threads with an indistinct, underlying radial sculpture. Ligament internal, cardinals strong, laterals weak. Inner margin evenly denticulate. White. **Habitat:** in muddy sand, lower shore and below. **Distribution:** NWG, SEG, GO, Mas.

Subfamily MILTHINAE

Anodontia edentula (Linnaeus, 1758): **1028**
To 75mm. Beaks almost central. Circular, globose. Lunule ill defined but lunule area depressed. Sculpture of fine irregular concentric lines with faint radial striae. Hinge lacking teeth except for a minute cardinal peg in the juvenile. Ligament deeply sunken. Anterior adductor scar strongly diverging from pallial line. Inner margin smooth. White. **Habitat:** in mud and muddy sand, lower shore and below. **Distribution:** NWG, SEG, GO, Mas, SO.

Eomiltha voorhoevei (Deshayes, 1857): **1029**
To 75mm. Beaks slightly towards the anterior, umbones small and pointed. Compressed; transversely oval but irregular, anterior narrowly rounded, posterior broad. Lunule small but deeply sunken. Ligament external but sunken. Surface often irregular, sculpture of concentric lines and weak radial striae. Anterior adductor very long extending to below beaks. White.
This very rare shell was originally described from Mozambique. Only three valves have been found in eastern Arabia. **Habitat:** ? **Distribution:** GO, Mas.

Subfamily MYRTEINAE

Myrtea fabula (Reeve, 1850): **1030**
To 10mm. Beaks slightly in front of mid-line. Compressed; transversely oval with prominent but small pointed umbones; posterior dorsal margin long, sloping, straight; lateral margins rounded. Lunule large, narrowly oval and producing a distinct depression in front of the beaks. Sculpture of closely spaced, fragile, concentric lamellae, usually fractured except along the dorsal margin where they form a row of spines; sulci absent. White with rusty deposits anteriorly. **Habitat:** in mud, offshore. **Distribution:** GO.

Subfamily DIVARICELLINAE

Divalinga arabica Dekker & Goud, 1994: **1031**
To 20mm. Beaks central. Circular with straight dorsal margins. Lunule narrow not sunken. Sculpture of widely spaced, incised, divaricate grooves. Cardinal teeth strong, laterals insignificant. Ligament in a narrow submarginal groove. Anterior scar free for half its length but close to pallial line. White to pale orange.

Superfamily: **LUCINOIDEA**

Habitat: in sand, intertidal. **Distribution:** NWG, SEG, GO.

Bourdotia boschorum Dekker & Goud, 1994: **1032**
To 25mm. Beaks central. Outline circular with straight dorsal margins. Lunule cordate, sunken. Sculpture of closely spaced, incised, divaricate grooves. Cardinal teeth weak, anterior lateral prominent. Ligament in a deep submarginal groove. Anterior scar free for half its length but diverging from pallial line. White. **Habitat:** offshore. **Distribution:** GO, Mas, SO.

Divaricella ornatissima (Orbigny, 1846): **1033**
To 20mm. Lenticular. Sculpture of about 50 narrow elevated divaricate riblets (counted down the line of divarication). Riblets weakly rugose close to dorsal margins. Angle of divarication acute. White. **Habitat:** offshore. **Distribution:** GO.

Lucinella sp: **1034**
To 20mm. Beaks almost central. Subcircular a little longer than high with short, sloping, dorsal margins. Lunule cordate, sunken. Sculpture of concentric lines except on posterior area where irregular, incised, divaricate grooves are present. Ligament internal on a deep inclined resilium. Anterior scar free for half its length. Inner margin smooth. White. **Habitat:** ? **Distribution:** SO.

Family THYASIRIDAE
Axinopsida sp: **1035**
To 7mm. Irregularly orbicular, umbones pointing forward over lunular depression. Ligament internal. Hinge teeth of one peg like cardinal in RV and 2 cardinal protuberances in LV. Sculpture of concentric lines and weak ridges, often dented. Adductor scars continuous with pallial line, posterior scar displaced ventrally. White. **Habitat:** in sand, shallow sheltered waters. **Distribution:** NWG, GO.

Family UNGULINIDAE
Diplodonta genethila Melvill, 1898: **1036**
To 13mm. Beaks almost central, tumid. Subspherical-trigonal, umbones prominent; anterior narrower than posterior, anterior dorsal margin sloping steeply. Ligament on a short, steeply inclined, excavated nymph. Posterior cardinal eroded by encroaching nymph. Sculpture of fine concentric lines only, except in some shells which develop 2-3 radial folds on the anterior slope. White. **Habitat:** in muddy gravel, offshore. **Distribution:** NWG, SEG, GO, SO.

Diplodonta holosphaera Melvill, 1899: **1037**
To 10mm. Beaks in front of mid-line, very tumid. Spherical, all margins equally curved. Ligament on a long shallow nymph. Posterior cardinal subhorizontal, parallel to nymph, RV anterior cardinal very small. Sculpture of concentric lines only. Adductor scars proportionately large. White. **Habitat:** offshore. **Distribution:** NWG, Mas.

Diplodonta* cf *globosa (Forsskål, 1775): **1038**
To 26mm. Beaks slightly in front of mid-line, tumid. Subspherical, anterior narrower, anterior ventral margin distinct and sloping less than posterior ventral margin; anterior dorsal margin distinct, almost horizontal. Ligament on deeply excavated nymph, partly concealed by dorsal margin. Bifid cardinal narrow in both valves. Both adductors rather narrow. Periostracum thin, brown-olive. White. **Habitat:** in muddy sand, intertidal eelgrass beds and khors. **Distribution:** NG, NWG, SEG, GO, Mas.

Diplodonta subrotundata Issel, 1869: **1039**
To 28mm. Beaks slightly in front of mid-line, not inflated. Subcircular, anterior narrower, anterior ventral margin distinct and sloping less than posterior ventral margin; anterior dorsal margin distinct, usually sloping. Ligament on shallow nymph, not concealed by dorsal margin. Bifid cardinal narrow in both valves, posterior cardinal separated from nymph by a groove. Both adductors rather narrow. White. **Habitat:** in sand, lower shore and below. **Distribution:** SEG, GO, Mas, SO.

Diplodonta holosphaera **1037**

Diplodonta subrotundata **1039**

Diplodonta cf *globosa* **1038**

Divaricella ornatissima **1033**

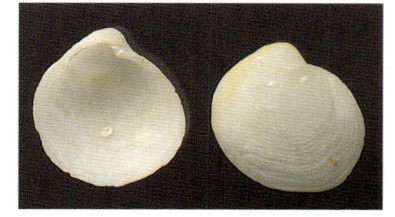

Axinopsida sp. **1035**

Bourdotia boschorum **1032**

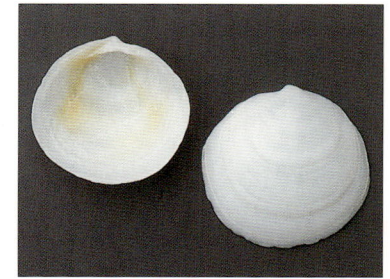

Lucinella sp. **1034**

Diplodonta genethila **1036**

SEASHELLS OF EASTERN ARABIA

Superfamily: LUCINOIDEA, GALEOMMATOIDEA

Diplodonta crebristriata **1040**

"Corbula" mirabilis **1041**

Amphilepida cf *faba* **1042**

Amphilepida aurantia **1043**

Diplodonta crebristriata Sowerby, 1905: **1040**
To 14mm. Beaks almost central, not tumid. Suboval, dorsal margins indistinct; anterior and posterior margins straightening. Ligament in a short excavated nymph. Sculpture of concentric lines only. White, orange or pink with a thin olive periostracum. **Habitat:** in sand, offshore. **Distribution:** NWG, SEG, GO, Mas, SO.

Superfamily GALEOMMATOIDEA
The Galeommatoidea comprises a large group of small bivalves which are rarely collected and consequently poorly known. They are divided into a number of families: Erycinidae, Kelliidae, Leptonidae, Montacutidae and Galeommatidae. For the non-specialist these family divisions are difficult to employ except perhaps for the type genus of each. Many taxa are based on European or N. American forms and the tropical species do not always fall easily into established genera. In this book, therefore, we have chosen not to define the families and present the genera separately. The Galeommatoidea are variable in shell outline and dentition. They are generally equivalve with adductor muscle scars of equal size and a broad pallial line. All lack a pallial sinus but there are three mantle apertures represented by anterior inhalant, pedal and posterior exhalant openings. The hinge usually has small, often tuberculate cardinal teeth with elongate laterals but there is a tendency for these to become obsolete. The ligament is usually set internally on a weak resilium.

The Galeommatoidea are either free living in soft sediments or byssally attached to the undersides of rocks or in crevices. Some species have adopted an almost slug-like existence creeping on the undersides of rocks. In these forms the mantle is reflected over the shell and is often ornamented with fleshy tentacle-like lobes. Other species are commensal with marine invertebrates such as burrowing shrimps, sea urchins, sea cucumbers, anemones and also with tubiculous polychaetes and crevice-dwelling sipunculans. In these circumstances the bivalve may attach to its host or may live free in the host's burrow.
The lack of knowledge of the species inhabiting tropical seas cannot be exaggerated and there is enormous scope for making simple observations of life histories and habitat. Relatively few species are known from eastern Arabia but recent collections from Masirah indicate that many species are present. Few shells are available for study so only the more distinctive species are described below.

"Corbula" mirabilis Lynge, 1909: **1041**
To 12mm. Strongly compressed, almost flat. Beaks well behind the mid-line. Rhomboidal; posterior acute with a slight flexure, posterior margin steeply angled, straight; anterior rounded, anterior dorsal margin long, sloping gently to beak; ventral margin more or less straight. Posterior with a sharp keel. Sculpture of weak growth lines with weak pustulose ornamentation. Ligament on a weakly defined internal resilium. RV hinge consists of a large inwardly projecting, peg-like cardinal and other "teeth" projecting dorsally consisting of two anterior and one posterior protuberances. White.
This distinctive shell is known only from the single valve described by Lynge from the Gulf of Siam and a single valve from off As Sib, Oman. It was tentatively placed in the genus *Corbula* by Lynge but a new generic name is required. **Habitat:** ? **Distribution:** GO.

Amphilepida cf *faba* (Deshayes, 1856): **1042**
To 15mm. Beaks central. Not gaping ventrally. Elliptical, anterior and posterior evenly rounded. Sculpture of growth lines only. Hinge confined to umbonal area, of weak cardinal tubercles and posterior laterals. Weak ligament mostly external but partly sunken. White. In life mantle partly reflected over shell, white with white papillae and tentacles. **Habitat:** creeping on the undersides of rocks, mid and lower shore. **Distribution:** GO, Mas, SO.

Amphilepida aurantia (Deshayes, 1835): **1043**
To 20mm. Beaks central. Gaping ventrally. Narrowly elliptical, anterior and posterior evenly rounded. Sculpture of growth lines only. Hinge and ligament as in *Amphilepida* cf *faba* (see above)

Superfamily: **GALEOMMATOIDEA**

Orange. **Habitat:** creeping on the undersides of rocks on lower shore. **Distribution:** GO.

Amphilepida peilei (Tomlin, 1921): **1044** NOT ILLUSTRATED
To 15mm. Fragile, semi-transparent. Gaping ventrally. Narrowly elliptical, anterior and posterior narrowly rounded; dorsal margins dipping towards umbones. Sculpture of growth lines and appears minutely bubbled overall. Hinge teeth prominent. White. **Habitat:** under rocks on lower shore. **Distribution:** NWG.

Amphilepida callipareia (Melvill, 1899): **1045**
To 12mm. Fragile, semi-transparent. Not gaping ventrally. Elliptical, posterior distinctly wider than anterior. Sculpture of growth lines only. Hinge confined to umbonal area, of weak cardinal tubercles and posterior laterals. Weak ligament mostly external but partly sunken. Tinted bluish-rose. **Habitat:** offshore. **Distribution:** NG.

Scintilla sp: **1046**
To 10mm. Fragile, semi-transparent. Compressed. Rounded, broadly elliptical; posterior a little wider than anterior; dorsal margins dipping towards umbones. Sculpture of growth lines only. Hinge weak as in *Amphilepida*. White tinged with very pale orange. **Habitat:** under rocks, lower shore. **Distribution:** GO.

Kellia leucedra Melvill & Standen, 1907: **1047**
To 12mm. Solid. Inflated. Beaks slightly in front of mid-line. Rounded squarish, posterior broadly rounded to roundly truncated, anterior a little narrower, ventral margin becoming straight. Hinge with peg-like cardinals (1 in RV, 2 in LV) widely separated by internal ligament from weak posterior lateral. White. **Habitat:** under rocks, lower shore and sublittoral. **Distribution:** all.

Marikellia pustula (Deshayes, 1863): **1048**
To 9mm. Solid, translucent. Beaks slightly in front of mid-line. Rounded rectangular; posterior a little broader than anterior; anterior a little truncated; ventral margin becoming straight. Hinge with single laminar cardinal and single posterior laterals; ligament sunken on oblique resilium. White. **Habitat:** under rocks at low tide. **Distribution:** Mas.

Scintillula variabilis (Sturany, 1899): **1049**
To 15mm. Thick, tumid. Beaks slightly in front of mid-line. Rounded elliptical, posterior a little wider than anterior. Hinge with large cardinal peg and single large posterior lateral separated by gap containing ligament. Sculpture smooth. White. **Habitat:** ? **Distribution:** Mas.

Nesobornia sp: **1050** NOT ILLUSTRATED
To 10mm. Thin. Beaks slightly in front of mid-line. Elliptical, posterior a little wider than anterior. Hinge weak with 2 cardinals and 1 posterior lateral in LV. Sculpture apparently smooth but overall micro-punctate. **Habitat:** ? **Distribution:** Mas.

Curvemysella peculiaris (A. Adams, 1856): **1051**
To 5mm. Sickle shaped, slightly twisted with ventral margin deeply indented. Hinge with two small, divergent cardinals in LV, RV with marginal extensions. White. **Habitat:** attaches itself to the insides of shells inhabited by hermit crabs but as yet not found living in eastern Arabia. **Distribution:** GO.

Curvemysella sp: **1052**
NOT ILLUSTRATED
To 5mm. Thin. Translucent. Beaks central. Elliptical, posterior a little broader than anterior; ventral margin indented. Hinge with two large, divergent cardinals in LV; RV with marginal extensions. Sculpture of growth lines only. White. **Habitat:** ? **Distribution:** Mas.

Amphilepida callipareia **1045**

Scintillula variabilis **1049**

Kellia leucedra **1047**

Marikellia pustula **1048**

Curvemysella peculiaris **1051**

Scintilla sp. **1046**

Superfamily: **GALEOMMATOIDEA, CYAMIOIDEA, CARDITOIDEA**

Callomysia revimentalis (Melvill & Standen, 1907): **1053** NOT ILLUSTRATED
To 7mm. Thin, translucent. Beaks almost at mid-line. Transversely oval; lateral margins rounded, posterior a little broader. Hinge with small cardinals in LV these not diverging steeply from hinge plate; RV with corresponding marginal extensions. Sculpture smooth, margin distinctly dentate. White. **Habitat:** offshore. **Distribution:** GO.

***Mysella* sp: 1054**
To 10mm. Solid. Beaks well to the posterior. Subovate; anterior reduced becoming roundly truncate; posterior elongate-rounded. Hinge: LV with strong laminar cardinals diverging steeply from hinge plate; RV with marginal extensions. Sculpture smooth. White. **Habitat:** offshore. **Distribution:** GO, Mas.

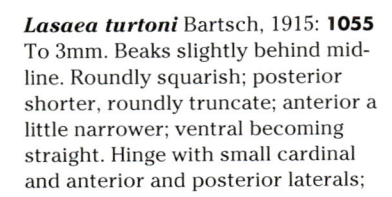

Lasaea turtoni Bartsch, 1915: **1055**
To 3mm. Beaks slightly behind midline. Roundly squarish; posterior shorter, roundly truncate; anterior a little narrower; ventral becoming straight. Hinge with small cardinal and anterior and posterior laterals; ligament internal. Sculpture of growth lines, some with concentric corrugations. White tinged deep red especially around umbones. **Habitat:** nestling among the byssus of mussels and in crevices, mid-shore. **Distribution:** Mas.

Superfamily CYAMIOIDEA

The Cyamioidea comprises four families of which only the Sportellidae is represented in eastern Arabian waters. They are small, reduced anteriorly, with an angular rhomboidal outline. The pallial line is entire and the distinctive hinge bears only a single prominent projecting cardinal. They are thought to be commensal with other invertebrates such as the Polychaeta. Confusion is most likely to occur with some of the Kelliidae in the Galeommatoidea.

Basterotia arcula Melvill, 1898: **1056**
To 12mm. Solid, inflated, beaks well towards the anterior end. Rhomboidal; anterior broadly rounded; posterior slope straight, ventral junction acute; posterior area distinctly keeled. Sculpture of irregular growth lines with coarse pustules on anterior and 2-3 radial ridges on posterior slope. White. **Habitat:** ? **Distribution:** Mas.

Basterotia borbonica (Deshayes, 1863): **1057** NOT ILLUSTRATED
To 10mm. Thin, not inflated, beaks well towards the anterior end. Roundly rhomboidal not unlike a squarish modiolid; anterior narrowly rounded; posterior broad, roundly truncated; posterior slope with a weak rounded keel. Sculpture of growth lines with minute pustules overall. White.
Previously known from Reunion and Gulf of Aden, a single valve has been found at Masirah. **Habitat:** ? **Distribution:** Mas.

Superfamily CARDITOIDEA

The Carditoidea are rather variable in form, some resemble cockles, some mussels and some are minutely trigonal. All groups possess strong radial sculpture. The hinge tends to be very strong with large oblique cardinals but poorly developed laterals. The ligament is external. There is no pallial sinus. The heart-shaped species, like the cockles they resemble, are shallow burrowers in soft sediments. These are free living but some may use a byssus in the early stages. The mussel-like species are epifaunal and live attached by a byssus in crevices on rocky or coral substrates. The minute species have only been collected as dead shells but probably live in clean, coarse sand and gravels.

Cardita ffinchi (Melvill, 1898): **1058**
To 40mm. Elongate trigonal, posterior greatly expanded, length of posterior margin more or less equal to ventral margin. Sculpture of 10-14 ribs, 2-3 on posterior keel very large. All ribs with squamose spines, those on the posterior keel and dorsal margin greatly expanded. Buff to coral pink with very occasional brown spots or blotches. **Habitat:** offshore. **Distribution:** NWG, SEG, GO.

Cardita crassicosta Lamarck, 1819: **1059**
To 70mm. Elongate-oval, modioliform with beaks very close to anterior end, posterior expanded, dorsal and posterior margins continuous. Sculpture of 10-14 ribs, 2-3 on posterior keel very large. All ribs with heavy, flattened spines, those on the posterior keel and dorsal margin greatly expanded. Buff to coral pink with very occasional brown spots or blotches. **Habitat:** offshore. **Distribution:** Mas, SO.

Mysella sp. 1054

Basterotia arcula 1056

Lasaea turtoni 1055

Cardita ffinchi 1058

Cardita crassicosta 1059

Superfamily: **CARDITOIDEA**

Beguina gubernaculum **1061**

Cardita variegata **1060**

Cardites bicolor **1062**

Carditopsis majeeda **1064**

Centrocardita echinaria **1063**

Cardita variegata Bruguière, 1792: **1060**
To 40mm. Beaks subterminal. Elongate-modioliform, not greatly expanded posteriorly. Sculpture of 18-20 acute ribs increasing in size posteriorly, with small spines in early stages, slightly squamose spines later. Uniform white to buff with variable amounts of brown-black spots and blotches. **Habitat:** attached to rocks, offshore. **Distribution:** SEG, GO, Mas.

Beguina gubernaculum (Reeve, 1843): **1061**
To 50mm. Compressed, beaks subterminal. Modioliform, greatly expanded posteriorly, anterior area small, acute. Sculpture of broad rather flat radial ribs incised by radial grooves, the broad ribs along the posterior angle sparsely spined. Rufous-brown-buff with paler anterior area. **Habitat:** attached in crevices of rocks, middle shore and below. **Distribution:** all.

Cardites bicolor (Lamarck, 1819): **1062**
To 55mm. Heavy. Subovate with high inflated umbones. Sculpture of 20-22 broad, flat-topped, radial ribs, some anterior ribs with cross bars, posterior ribs less developed. Hinge with long, arching posterior cardinal. Inner margin deeply crenulate. Pallial line entire. White with various amounts of brown-black blotching. **Habitat:** in coarse sand and gravels, offshore. **Distribution:** all.

Centrocardita echinaria (Melvill & Standen, 1907): **1063**
To 30mm. Tumid, beaks in front of mid-line. Roundly quadrate with prominent inflated umbones; anterior narrowly rounded; posterior expanded, roundly subtruncate. Lunule cordate, impressed. Sculpture of 24-26 narrow radial ribs; anteriorly with cross bars progressively spiny towards posterior angle then becoming smooth on posterior slope. Inner margin crenulate. Buff with few darker markings. **Habitat:** in coral sand, offshore. **Distribution:** NG, SEG.

Carditopsis majeeda (Biggs, 1973): **1064**
To 2mm. Beaks almost central. Trigonal with steeply sloping dorsal margins, posterior margin slightly longer than anterior. Larval shell prominent but rounded. Sculpture of 10-11 radial ribs, slightly scabrous especially on posterior. Inner margin crenulate. Reddish-brown. **Habitat:** in sand, offshore. **Distribution:** NWG, SEG, GO.

SEASHELLS OF EASTERN ARABIA

Superfamily: **CARDITOIDEA, CHAMOIDEA**

Carditella n. sp: **1065**
To 3.5mm. Solid, beaks in front of mid-line. Transversely oval, umbones trigonal, posterior dorsal margin longer than anterior, both excavated; posterior margin more pointed than anterior. Prodissoconch capped. Sculpture of 11 - 13 large rounded radial ribs, interspaces narrower than ribs, weak ornamentation of concentric striations only. White with rust-brown marks on posterior. **Habitat:** in shell gravel, sublittoral. **Distribution:** Mas.

Superfamily CHAMOIDEA

The Chamoidea, jewel box clams, are heterodont bivalves which are epifaunal, living with one valve cemented to a hard substrate such as coral, rock or other shells. They may be confused with oysters and spiny oysters (Spondylidae), but unlike those families have two adductor muscles, a parivincular ligament, a heterodont hinge, and a pair of short siphons. As with other cemented bivalves they are highly variable in form because the lower valve conforms to the substrate and the upper valve becomes very worn losing all trace of colour and sculpture.
The variability of shell form has led to a confused nomenclature and it is often almost impossible to name many shells. Those described here may well not represent separate species but rather may be considered a representative selection of the forms found in the eastern Arabian region.

Chama reflexa Reeve, 1846: **1066**
To 50mm. Circular to oval. Upper valve with dense array of narrow flat spines with a weak or obsolete posterior sulcus. Marginal spines often yellow to orange, ground colour white tinged purple; internal tinged deep purple. **Habitat:** on sides of and under rocks, lower shore and below. **Distribution:** NWG, SEG, GO, Mas.

Chama aspersa Reeve, 1846: **1067**
To 30mm. Circular to oval. Upper valve with short, dense spines and no sulcus. Upper valve dirty white with some orange red tinges, inside of lower valve white with reddish areas. **Habitat:** attached to rocks and shells, offshore. **Distribution:** GO.

Chama douvillei Lamy, 1921: **1068**
To 40mm. Narrowly oval, angulate with upper valve displaced to one side. Externally worn smooth, inside white with red areas. **Habitat:** attached to exposed smooth rocks, lower shore. **Distribution:** GO.

Chama asperella Lamarck, 1819: **1069**
To 20mm. Roughly circular. Upper valve with dense, short, erect, fluted spines these becoming broad and flat on posterior area. Upper valve dirty white to cream, inside off-white. **Habitat:** attached under rocks, intertidal. **Distribution:** GO, Mas.

Chama brassica Reeve, 1847: **1070**
To 80mm. Massive. Roughly circular to oval. Upper valve usually worn but posterior sulcus apparent. Sculpture would be of broad, fluted, interrupted lamellae. Dirty white with pale tinges of mauve internally. **Habitat:** attached to exposed rocks, lower shore. **Distribution:** GO.

Chama asperella **1069**

Chama douvillei **1068**

Chama aspersa **1067**

Carditella n. sp. **1065**

Chama reflexa **1066**

Chama brassica **1070**

242 SEASHELLS OF EASTERN ARABIA

Superfamily: **CRASSATELLOIDEA, CARDIOIDEA**

Bathytormus jousseaumei **1074**

Indocrassatella indica **1071**

Crassatina picta **1072**

Bathytormus radiatus **1073**

Superfamily CRASSATELLOIDEA

The Crassatelloidea are generally quadrangular to trigonal in outline, rounded in front, truncate to rostrate posteriorly. The sculpture is concentric, of lines or broad ridges, and is usually covered by a thick, dark brown periostracum. The ligament is internal and the hinge has prominent, radiating cardinal teeth.

Very little is known about them from warm seas as they tend to live exclusively in deeper water. They probably live in mixed, rather coarse sediments.

Indocrassatella indica (Smith, 1895): **1071**
To 35mm. Beaks in front of mid-line. Subquadrate; posterior dorsal slope shallow, anterior dorsal slope much steeper, posterior roundly truncate, anterior narrowly rounded. Sculpture of narrow concentric ridges, these sometimes fading on the posterior slope. Inner margin denticulate. White. Periostracum thick, brown. **Habitat:** offshore, deep water. **Distribution:** GO.

Crassatina picta (Adams & Reeve, 1850): **1072**
To 20mm. Beaks central. Ovate-trigonal with subtruncate posterior margin; dorsal margins sloping equally. Sculpture of low, narrow, concentric ridges. Inner margin denticulate. White and shades of reddish brown in a tented pattern, internally flushed pink. **Habitat:** offshore. **Distribution:** NG.

Bathytormus radiatus (Sowerby, 1825): **1073**
To 25mm. Compressed, beaks well in front of mid-line. Transversely subtrigonal with rostrate posterior. Posterior dorsal margin concave, rostrum truncate. Sculpture of well spaced concentric ridges. Inner margin denticulate. Purplish black through brown and reds to pale buff. **Habitat:** in muddy sand, offshore. **Distribution:** SEG, GO, Mas.

Bathytormus jousseaumei (Lamy, 1918): **1074**
To 25mm. Very similar to *B. radiatus* (see above) but much deeper with weaker and more closely spaced concentric ridges. **Habitat:** in muddy sand, offshore. **Distribution:** GO, Mas, SO.

Superfamily CARDIOIDEA

The Cardioidea comprises two familiar groups of shells, the cockles (Cardiidae) and the giant clams (Tridacnidae). Cockles have rounded, ovate or ovate-angular shells with prominent umbones. They are usually sculptured with raised radial ribs and the ventral margin is correspondingly crenulate although in a few genera the sculpture is obsolete. The hinge is well developed with two cardinal teeth in each valve, single laterals in the left valve and paired laterals in the right valve. The external ligament is in the form of a highly arched band behind the beaks. The adductor scars are subequal and the pallial line is entire and lacks any trace of a sinus.

Cockles have short siphons and are restricted to a shallow burrowing habit in a variety of soft sediments. The heavy rounded shell of many species enables them to live in mobile sand, while the thinner, smooth shells are typical of those inhabiting tranquil muddy sites. They should not be confused with other bivalves although the rounded forms of some Carditoidea are similar externally. The Cardioidea are divided into a number of subfamilies some of which are useful but others confusing in terms of aiding identification.

The Cardiinae are semicircular to roundly quadrate with a straight hinge line. The ornamentation on the ribs always arises from the middle of each rib. The Trachycardiinae are oval (much higher than long) but not oblique, with an arched hinge line. The rib ornamentation arises from the sides of each rib. The Fraginae are subquadrate, mostly with a strong posterior keel. The Laevicardiinae are obliquely oval, rather thin shells with a relatively weak sculpture. The Protocardiinae are the most variable but generally include those genera with oblique or complex radial sculptures. The Tridacnidae with their massive, ribbed shells are familiar to all and need little introduction. Giant clams are thought not to be living in eastern Arabian waters although subfossil shells are not infrequent around Sur (Oman). When the climate was warmer and reef development was greater it is believed that giant clams were present. The cooling effects of the southern Arabian upwelling and the variable climate of the Gulf may account for their current absence.

Superfamily: **CARDIOIDEA**

Family CARDIIDAE
Subfamily CARDIINAE

Vepricardium exochum (Melvill & Standen, 1907): **1075**
To 9mm. Beaks in front of mid-line, distinctly turned towards the anterior, tumid. Subcircular with margins evenly curving, anterior narrower than posterior. Sculpture of about 37 radial, close-set ribs all bearing rather long, densely spaced, erect, fluted scales. Hinge condensed anteriorly, laterals close to cardinals. Margin crenulate, extending around post-ventral area and interdigitating. White, with a posterior pink area internally. **Habitat:** offshore. **Distribution:** NG.

Bucardium coronatum (Spengler, 1799): **1076**
To 55mm. Subcircular, margins evenly rounded. Sculpture of 36-38 very low radial ribs, these bearing a calcareous raised flange on their posterior margins, this developed most on posterior ribs which interdigitate at the margin. White suffused with weak shades of orange-pink posteriorly. **Habitat:** offshore. **Distribution:** GO.

Papillicardium omanense (Melvill & Standen, 1907): **1077** NOT ILLUSTRATED
To 5mm. Roundly subquadrate, posterior margin a little longer and straighter than anterior. Sculpture of 25 radial ribs all but posterior few bearing large rounded nodules, these widely spaced and not the full width of the ribs, posterior ribs with sharper scales. White. **Habitat:** offshore. **Distribution:** GO.

Parvicardium sueziense (Issel, 1869): **1078**
To 8mm. Roundly subquadrate, posterior margin slightly longer and less rounded than anterior. Sculpture of 20-24, low, rather broad, radial ribs, all but the posterior few bearing large, slightly curved cross bars; posterior ribs with blunt spines. White. **Habitat:** in muddy sand, lower shore and below. **Distribution:** NWG, SEG, GO, Mas, SO.

Plagiocardium pseudolima (Lamarck, 1819): **1079**
To 150mm. Subquadrate, distinctly longer than high, umbones high, posterior margin almost straight. Sculpture of 36 - 40 flat radial ribs these bearing small cup-shaped spines arising from the mid-line of each rib (may be worn to simple nodules). In fresh specimens periostracal bristles arise from the base of each spine. Pink to reddish brown with contrasting white spines. **Habitat:** in muddy sand, lower shore and below. **Distribution:** GO, Mas.

Subfamily FRAGINAE

Afrocardium richardi (Audouin, 1826): **1080**
To 11mm. Beaks well in front of mid-

line. Obliquely oval, almost modioliform; posterior not keeled but roundly angulate; posterior expanded, margin long, almost straight. Sculpture of 35 - 38 narrow ribs, all with thin, imbricate, projecting scales, these becoming longer on posterior ribs. White, yellow, orange or red, some with reddish brown markings. **Habitat:** offshore. **Distribution:** Mas.

Ctenocardia fornicata (Sowerby, 1840): **1081**
To 15mm. Subquadrate, slightly longer than high, anterior margin rounded, posterior margin straight or concave; posterior keel rounded. Sculpture of about 35 ribs, lateral ribs with dense, imbricating scales, median ribs with erect subtubular spines. Pinkish with fulvous irregular bands, internally suffused with vivid pink. **Habitat:** offshore. **Distribution:** GO, Mas, SO.

Fragum hemicardium (Linnaeus, 1758): **1082**
To 35mm. Rhomboidal-trigonal, much higher than long; posterior keel acutely angled, posterior slope almost vertical. Lunule absent. Sculpture of 29-31 flat low radial riblets with sparse scales but generally smooth, interspaces punctate. White. **Habitat:** offshore. **Distribution:** Mas.

Lunulicardia auricula (Niebuhr in Forsskål, 1775): **1083**
To 50mm. Rhomboidal-trigonal, much higher than long; posterior keel acutely angled, posterior slope almost vertical. Lunule large, deeply excavated and separating the cardinal and anterior lateral teeth. Sculpture of 20-25 low ribs with widely separate tubercles. White. **Habitat:** offshore. **Distribution:** SEG.

Subfamily LAEVICARDIINAE
Fulvia australe (Sowerby, 1834): **1084**
To 30mm. Thin. Ovate, slightly higher than long, slightly oblique, not gaping. Appears smooth and shining but sculpture of 40-45 riblets; flattened, smooth, more pronounced on posterior slope; interspaces narrow. Inner margin weakly crenulate. White blotched all over, pale orange, fawn or pink; inside irregularly suffused with deep pink. **Habitat:** in sandy mud, offshore. **Distribution:** NWG, SEG.

Fulvia fragile (Forrskål, 1775): **1085**
To 50mm. Subcircular to obliquely oval, tumid not gaping. Sculpture of 30-35 broad, flat ribs with interspaces almost as wide as ribs; radial periostracal lamellae arising from mid-line of anterior ribs, from posterior edge of others. White to cream, some with transverse, pale fulvous, zigzag bands, others with pink blotches; inside with some pink area usually deeply coloured across posterior. **Habitat:** in muddy sand, lower shore and below. **Distribution:** all.

Subfamily PROTOCARDIINAE
Lyrocardium n. sp: **1086**
To 55mm. Oval, slightly oblique, higher than long, posterior margin a little longer and straighter than anterior. Sculpture of widely spaced oblique ridges on anterior half, of radial raised lines on posterior. Inner margin finely crenulate. Shell white with chestnut patches especially on umbones and down posterior angle, posterior slope flushed vivid pink. This species has been refered to *L. aurantiacum* (Adams & Reeve, 1850) by Bosch & Bosch (1982) but it is most likely distinct from the Asian species. **Habitat:** offshore. **Distribution:** GO, Mas, SO.

Microcardium centumliratum (Melvill & Standen, 1907): **1087**
To 20mm. Roundly subquadrate, longer than high; posterior margin almost straight, anterior broadly rounded, hinge margin long. Sculpture of numerous (about 100) narrow radial riblets with interspaces of equal size; about one third of the interspaces bear widely separated, subtubular spines. The spine-free interspaces show no fixed patterns; the gaps may be of 1-4 ribs. White. **Habitat:** offshore. **Distribution:** GO.

Fragum hemicardium **1082**

Ctenocardia fornicata **1081**

Lyrocardium n. sp. **1086**

Fulvia australe **1084**

Lunulicardia auricula **1083**

Fulvia fragile **1085**

Microcardium centumliratum **1087**

Superfamily: CARDIOIDEA, MACTROIDEA

Subfamily TRACHYCARDIINAE

Acrosterigma maculosa (Wood, 1815): **1088**
To 35mm. Thin. Oval, higher than long. Sculpture of 45-51 narrow, only slightly raised radial ribs; posterior 6-8 ribs with very small raised nodules usually remaining on ribs 1-2 only and as scars on remainder; anterior ribs with faint cross bars; median ribs smooth. Colour pale yellowish brown with areas of pink or deeper red, in some reddish overall with darker areas; internally with deep pink or reddish blotches. **Habitat:** in sand, lower shore and below. **Distribution:** Mas, SO.

Acrosterigma maculosa **1088**

***Acrosterigma* n. sp. (a): 1089**
To 40mm. Ovate, higher than long, narrowing towards umbones, distinctly pointed, posterior dorsal margin long, sloping immediately from beaks. Sculpture of 32-34 low, narrow, radial ribs; posterior 8-9 with smooth front edges and many small, spaced beads behind.

Acrosterigma n. sp. (a) **1089**

Externally off-white with faint spots and zones of beige and pink; internally white. This species is allied to ***A. arenicolum*** (Reeve, 1845). **Habitat:** in sand, offshore. **Distribution:** NWG, SEG.

Acrosterigma lacunosa (Reeve, 1845): **1090**
To 70mm. Solid, moderately thick. Ovate, higher than long. Sculpture of 31-35 radial ribs; posterior 6-8 with oblique, slightly twisted, lamellar spines. Externally cream with zones, patches and spots of ochre, chestnut and mauve; internal margin reddish to purple-brown, umbonal cavity initially with a pair of longitudinal purplish bars and suffused with fawn to orange but this fades in larger shells. **Habitat:** in sand, offshore. **Distribution:** NG, NWG, SEG, GO.

Acrosterigma assimile (Reeve, 1845): **1091**
To 55mm. Solid, moderately thick, tumid. Ovate, higher than long. Sculpture of 28-35 radial ribs; posterior 5-7 with oblique nodulose spines. Externally white with zones, patches and spots of reddish brown often so dense as to appear predominantly brown; internal margin reddish to purple brown, umbonal cavity initially with a pair of longitudinal purplish bars and suffused with fawn to orange remaining visible in most shells. **Habitat:** in sand, lower shore and below. **Distribution:** Mas.

***Acrosterigma* n. sp. (b): 1092**
To 60mm. Solid, rather heavy. Ovate, higher than long. Sculpture of 38 to 41 narrow radial ribs. Posterior 6-8 ribs irregularly rugose but not nodulose, apices overhanging interspaces, in profile resembling breaking waves. Most shells faded, patterned pink, white and beige; inside tinged orange pink along posterior edge and in umbonal cavity. **Habitat:** offshore. **Distribution:** Mas.

Family TRIDACNIDAE

Tridacna maxima (Röding, 1798): **1093** NOT ILLUSTRATED
To 150mm. Massive. Beaks towards the posterior end. Elongate – subtriangular, hinge margin shorter than ventral margin, byssal aperture large. Sculpture variable of 6-7 broad radial folds with narrow interspaces, the folds with closely spaced usually appressed leaf like scales. The scales are very variable, some shells are almost smooth. White. Known from subfossil shells only. **Habitat:** reefs, sublittoral. **Distribution:** GO.

Superfamily MACTROIDEA

The Mactroidea contains a variety of forms united by the presence of an internal ligament set in a deep socket, the resilifer. The eastern Arabian species fall into three families: Mactridae, Mesodesmatidae and Cardiliidae. The Mactridae are oval, oval-trigonal or oblong in outline, equivalve and usually slightly gaping posteriorly. The hinge is well developed, with characteristically inverted, V-shaped, fused cardinals in the left valve. The siphons are fused nearly to their tips.
Species of *Mactra*, trough shells, have oval or oval-trigonal shells, with strong interlocking teeth and a relatively smooth sculpture. They are active burrowers often living in mobile sand. These contrast with the oblong and widely gaping shells of *Lutraria* which are deep burrowers maintaining contact with the surface via long non-retractile siphons. The shells of *Meropesta* are atypical in being radially sculptured and in inhabiting rocky and coral substrates. Some become distorted due to their crevice-dwelling habit. Species of *Raeta* are poorly known but their very thin, fragile shells suggest that they inhabit tranquil offshore mud.
The Mesodesmatidae differ from

Acrosterigma lacunosa **1090**

Acrosterigma assimile **1091**

Acrosterigma n. sp. (b) **1092**

246 SEASHELLS OF EASTERN ARABIA

Superfamily: **MACTROIDEA**

the Mactridae primarily in the form of the siphons which remain unfused. In the Mesodesmatidae the resilifer is narrow and deep, the hinge is contracted and the left-valve cardinals are rarely bifid. For most the habits are poorly known but *Atactodea* lives as a shallow burrower in littoral and sublittoral sand.

The Cardiliidae are odd because the umbones are coiled and the shell is greatly inflated and oval in outline. Only the resilium displays the link with the Mactroidea.

Family MACTRIDAE

Mactra lilacea Lamarck, 1818: **1094**
To 65mm. Subovate-trigonal; posterior dorsal slope convex; anterior margin somewhat narrowed. Weak sculpture of irregularly spaced concentric ridges and lines, most prominent on the anterior area. White to purple-buff, either uniform or with radial rays, umbones always purple; inside white to deep purple. **Habitat:** in sand, offshore. **Distribution:** all.

Mactra aequisulcata Sowerby, 1894: **1095**
To 80mm. Subovate-triangular; posterior dorsal margin sloping steeply; posterior gape large. Sculpture of evenly spaced, concentric, narrow ridges over whole shell. Uniformly white to beige. **Habitat:** offshore. **Distribution:** GO.

Mactra rochebrunei Lamy, 1916: **1096**
To 100mm. Subovate-elliptical; dorsal margins straight; anterior margin subtruncate; posterior margin almost pointed. Posterior area with a keel line but no raised ridge. Rayed in shades of beige and orange; escutcheon and lunule areas white with a brown blotch. **Habitat:** offshore. **Distribution:** GO, SEG.

Mactra ovalina Lamarck, 1818: **1097**
To 40mm. Subovate-elliptical. Posterior area has a narrow raised ridge with a secondary angulation demarcating the escutcheon. Sculpture of fine concentric lines. Pallial sinus almost reaching below the beaks. White. Periostracum grey-brown. **Habitat:** in muddy sand, offshore. **Distribution:** GO, Mas.

Mactra ovalina **1097**

Mactra lilacea **1094**

Mactra rochebrunei **1096**

Mactra aequisulcata **1095**

Superfamily: **MACTROIDEA**

Mactrinula tryphera **1098**

Meropesta nicobarica **1101**

Meropesta pellucida **1100**

Lutraria australis **1099**

Meropesta solanderi **1102**

Mactrinula tryphera Melvill, 1899: **1098**
To 25mm. Fragile. Beaks almost central. Subovate; anterior rounded, anterior dorsal slope concave; posterior acute, posterior slope long, almost straight. Posterior raised ridge set very close to posterior margin. Sculpture of widely spaced, narrow, elevated, concentric ridges. Pallial sinus narrow, short. White. **Habitat:** offshore. **Distribution:** NG, SEG, GO.

Lutraria australis Deshayes, 1855: **1099**
To 130mm. Beaks well in front of mid-line. Oblong, anterior margin roundly pointed, posterior-dorsal margin slightly concave to slightly arched. White. Periostracum thin, beige to brown. **Habitat:** in muddy sand, offshore. **Distribution:** SEG, GO, Mas.

Meropesta pellucida (Gmelin, 1791): **1100**
To 80mm. Beaks in front of mid-line. Subovate, anterior broadly rounded contrasting with narrow, pointed posterior. Sculpture apparently smooth but with microscopic, radial, dendritic lines. Pallial sinus narrow, reaching to below beaks. White with a thin periostracum. **Habitat:** offshore. **Distribution:** GO.

Meropesta nicobarica (Gmelin, 1791): **1101**
To 50mm. Beaks in front of mid-line. Subovate, anterior broadly rounded, posterior narrower. Radial sculpture of many fine, raised riblets except on the posterior margin, dendritic markings also present. Pallial sinus broadly rounded, reaching below beaks. White with a thin periostracum. **Habitat:** in mud and muddy sand, offshore. **Distribution:** NG, NWG, SEG, GO, Mas.

Meropesta solanderi (Gray, 1837): **1102**
To 30mm. Outline variable, oval-elliptical to oval, often distorted, may become higher than long. Sculpture of prominent 15-34 raised, acute riblets, the lower number

248 SEASHELLS OF EASTERN ARABIA

Superfamily: **MACTROIDEA, SOLENOIDEA**

present in distorted shells; last posterior riblet large, forming a keel. White. Periostracum thin, brown. **Habitat:** nestling or boring (?) in corals and rocks. **Distribution:** Mas.

Raeta pellicula (Deshayes, 1855): **1103**
To 40mm. Thin, translucent, rather inflated. Beaks directed posteriorly. Suboval; anterior broad, rounded; posterior narrow, initially acute, later subtruncate. Sculpture of weak concentric plications, irregular posteriorly, plus widely spaced, feeble, incomplete, radial folds. Lateral teeth obsolete, cardinals often irregular. White. **Habitat:** in mud, offshore. **Distribution:** GO.

Family MESODESMATIDAE
Atactodea glabrata (Gmelin, 1791): **1104**
To 30mm. Heavy, beaks behind mid-line. Oval-subtrigonal, posterior roundly pointed. Sculpture of closely spaced weak concentric ridges and lines. Hinge strong, chondrophore deep, narrow but not projecting. Pallial sinus very short. White. **Habitat:** in sand, intertidal. **Distribution:** SEG, GO, Mas.

Caecella horsfieldi Gray, 1853: **1105**
To 35mm. Beaks just behind mid-line. Elliptical with anterior more expanded. Chondrophore projecting beyond hinge plate. Sculpture of growth line only. Pallial sinus narrow, extending to below hinge. White with buff periostracum. **Habitat:** intertidal flats. **Distribution:** GO, Mas.

Caecella qeratensis Morris & Morris, 1993: **1106** NOT ILLUSTRATED
To 25mm. Beaks at mid-line, globose. Rounded with slight anterior extension. Sculpture of growth lines only. Hinge strong, resilifer elongate and spoon-shaped. Internal shelf present in umbonal cavity. Pallial sinus short, U-shaped. White with a straw coloured periostracum. **Habitat:** in soft sand, intertidal. **Distribution:** GO.

Family CARDILIIDAE
Cardilia semisulcata (Lamarck, 1819): **1107**
To 10mm. Very tumid, higher than long, beaks coiled. Ovoid. Posterior adductor on a flange close to the hinge. Posterior half with radial riblets, anterior smooth. White. **Habitat:** offshore. **Distribution:** GO.

Superfamily SOLENOIDEA
The Solenoidea, razor clams, have greatly elongate, cylindrical shells which gape widely at each end. The beaks are more or less terminal; beneath them lies the reduced hinge with only a few cardinal teeth and behind them the external ligament. The shells are smooth, often rather thin and usually covered by a thin, glossy periostracum. The siphons are short and the pallial sinus is correspondingly shallow.

Cardilia semisulcata **1107**

Although this is a typical definition there are exceptions. Some of the Cultellidae, notably the genus *Siliqua*, are less elongate with the beaks close to the mid-line, resembling some tellinids in form. The small-peg like cardinal teeth and lack of cruciform muscle scars are, however, typical solenid characters. In *Siliqua* there is in addition an internal radial rib running from the hinge towards the ventral margin.
Razor clams are highly adapted for rapid burrowing into sand and mud. The elongate shell allows easy penetration of the substrate and this is achieved by the large foot which emerges from the anterior gape. The siphons are short, so the animal is close to the surface when feeding and retreats rapidly if disturbed. Consequently they are difficult to collect alive and are usually broken by collecting equipment such as grabs.

Family SOLENIDAE
Solen sloanii Gray, 1842: **1108**
To 60mm. Slightly curved, posterior margin subtruncate, anterior margin straight, almost vertical, anterior-ventral angle rounded. Anterior adductor scar long, ventral pallial line distinctly sinuous. White with dense pink to reddish brown spots and patches. **Habitat:** offshore. **Distribution:** SEG.

Solen sloanii **1108**

Raeta pellicula **1103**

Atactodea glabrata **1104**

Caecella horsfieldi **1105**

SEASHELLS OF EASTERN ARABIA

Superfamily: **SOLENOIDEA**

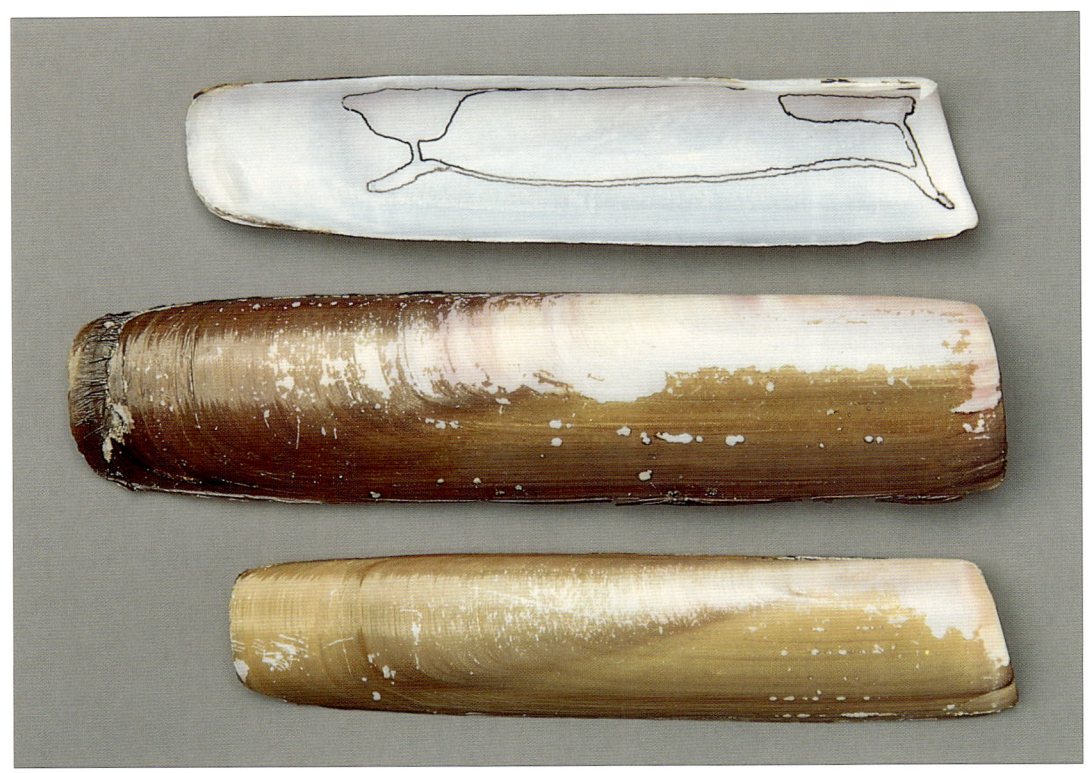

Solen brevis 1109

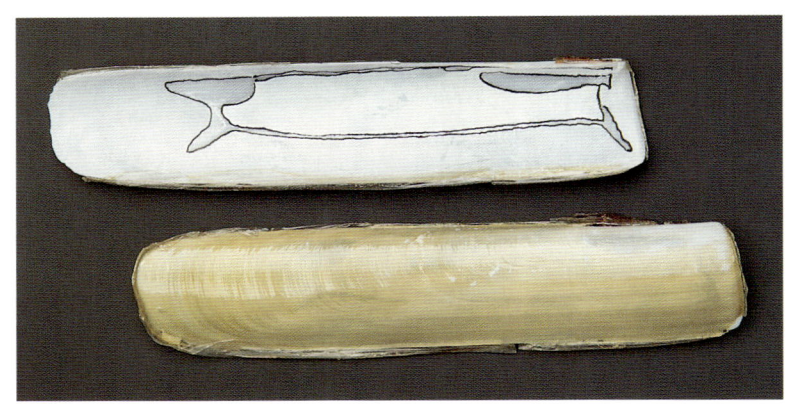

Solen digitalis 1110

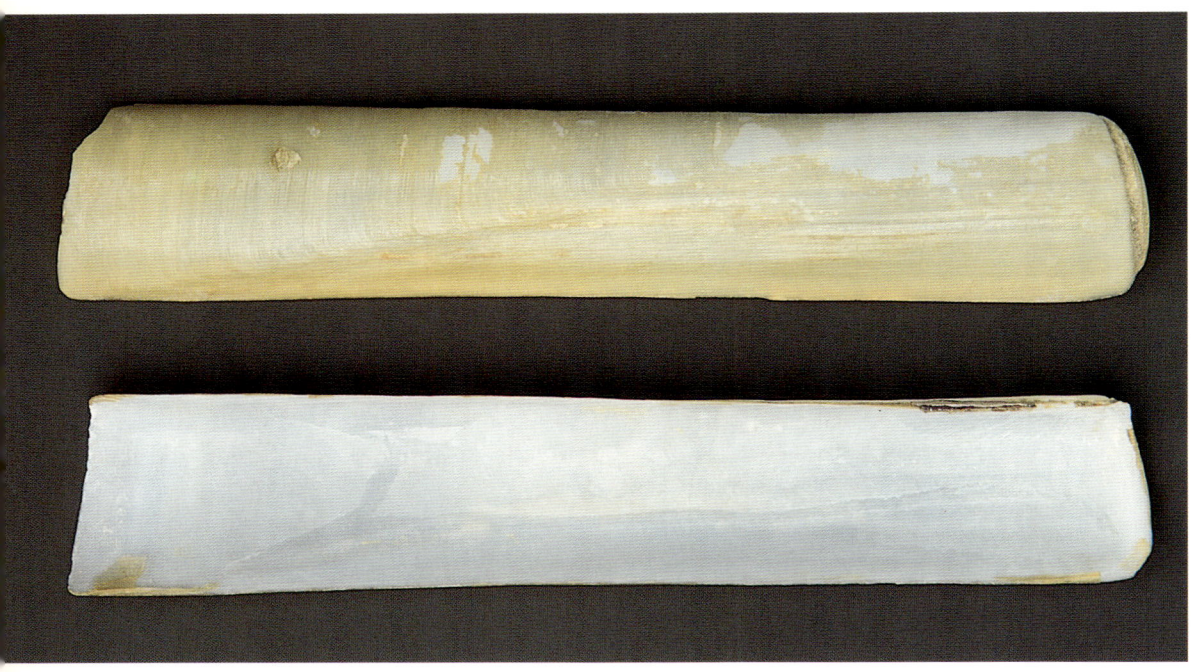

Solen dactylus 1112

Solen brevis Gray, 1842: **1109**
To 100mm. Straight, cylindrical, tapering slightly towards posterior, posterior margin subtruncate, anterior margin distinctly angled. Posterior adductor scar relatively broad. Externally smooth, shining, periostracum persistent, glossy, olive to brown. White, strongly tinged pink posterior to umbones. **Habitat:** offshore. **Distribution:** SEG, GO.

Solen digitalis Jousseaume, 1891: **1110**
To 100mm. Straight, cylindrical, not tapering posteriorly, posterior margin rounded, anterior margin distinctly angled. Posterior adductor scar relatively narrow. Externally smooth, shining, periostracum persistent, glossy, olivaceous. White to pinkish-brown, not banded. **Habitat:** offshore. **Distribution:** GO.

Solen* cf *lischkeanus Dunker, 1865: **1111** NOT ILLUSTRATED
To 100mm. Thin, rather compressed. Straight, cylindrical, not tapering posteriorly, posterior margin rounded, anterior margin weakly angled. Anterior adductor scar narrow, posterior adductor scar relatively narrow. Externally smooth, shining, periostracum pale olive. White.
The identity of these shells is uncertain as the true **S. lischkeanus** has a slightly curved shell and is believed to be endemic to the northern Red Sea. **Habitat:** offshore. **Distribution:** GO, Mas.

Solen dactylus Cosel, 1989: **1112**
To 140mm. Very elongate, straight; anterior weakly angled with rounded ventral junction; posterior straight, almost vertical. Anterior margin with a distinct exterior furrow. White. Periostracum thin, pale olive. **Habitat:** offshore. **Distribution:** NG, NWG.

Solen cylindraceus Reeve, 1843: **1113**
To 70mm. Very narrow, straight, cylindrical; posterior margin rounded, anterior margin almost vertical. Ventral pallial line closer to anterior adductor than to ventral margin. White with an olivaceous periostracum. **Habitat:** offshore. **Distribution:** SEG.

Family CULTELLIDAE
Ensiculus cultellus (Linnaeus, 1758): **1114**
To 40mm. Thin, compressed, beaks close to the anterior end. Narrowly elongate, anterior broadly rounded, posterior acutely rounded. Pale, mottled or spotted, shades of olive and brown. **Habitat:** in muddy sand, offshore. **Distribution:** NWG, SEG, GO.

Siliqua polita (Wood, 1828): **1115**
To 60mm. Thin, compressed, beaks slightly in front of mid-line. Broadly elongate, elliptical; anterior and posterior margins rounded, posterior dorsal margin slightly extended, separated by a shallow groove. Internal rib more or less vertical. Brown to lilac, usually with four white radial rays. **Habitat:** offshore. **Distribution:** all.

Superfamily TELLINOIDEA

The Tellinoidea is a very diverse group of bivalves represented in eastern Arabia by five families: Tellinidae, Psammobiidae, Donacidae, Solecurtidae and Semelidae. Shells tend to be rather thin, flattened, with a narrow posterior gape but in outline are rather variable, from circular to cylindrical. The hinge typically has two small cardinal teeth in each valve but laterals may or may not be present. The ligament is usually external, set on a prominent nymph but the Semelidae also have an internal portion set on a shallow resilifer. The most constant feature of the superfamily is the presence of cruciform muscle scars close to the ventral extremity of the pallial sinus. The siphons are typically very long and the pallial sinus is correspondingly deep. The Tellinoidea are deposit feeders using the inhalant siphon to suck up surface particles. They are found in a wide range of habitats from littoral mobile sand to deep tranquil mud. They are mostly very active burrowers living at relatively deep levels within the substrate.

The Tellinidae represent the more typical form of the superfamily in being compressed, subovate to circular in outline, with a twisting of the posterior end of the shell. The ligament is wholly external. The lateral teeth are variously represented. Many species are strikingly coloured and form a prominent component of the bivalve fauna of most soft-sediment habitats. They are divided into two subfamilies, the Tellininae and Macominae, the latter characteristically lacking lateral teeth. However, given the reduced nature of the lateral teeth in many of the Tellininae, this distinction is of dubious value. The inexact nature of tellinid systematics is even greater at the generic level and many genera and subgenera have been created. While these may be recognisable at a local level we invariably find many species which do not fall easily into these, or they appear to fit a number of subgeneric definitions. Consequently the subgenera are especially confusing and are not used here.

The Psammobiidae are similar to the Tellinidae in many respects and the two families may be difficult to separate. Unlike the Tellinidae the Psammobiidae are not twisted posteriorly, are more oblong in outline, expanded posteriorly, have the ligament set on a raised nymph and usually lack lateral teeth. Many are brightly coloured, especially in shades of pink and purple, but despite their visual appeal are scarce in collections. Most of the species of the most diverse genus, *Gari*, inhabit sand or coarser substrates offshore in deeper water and this may explain their scarcity. The genera *Hiatula* and *Psammosphaerica* inhabit more tranquil places in lagoons and estuaries and shells of *Hiatula* may be very common. *Asaphis* is atypical in that the shell is heavily sculptured and occurs in gravels high in the intertidal zone.

The Donacidae or wedge shells comprise a group of tellinoidean bivalves which are particularly adapted for living on exposed sandy beaches. The shells are wedge shaped, hence the common name, and the posterior area is reduced, its margin being steeply angled. They are mostly small and generally smooth and shining although there is a trend for some to be radially sculptured, especially on the posterior area. They are active burrowers which enables them to withstand the repeated disturbances caused by wave action.

The Solecurtidae appear to be more like razor clams than tellins with their cylindrical, gaping shells. The beaks are, however, placed close to the mid-line, not anteriorly and the pallial sinus is deep. Only two species are described here, both living offshore and are difficult to find.

The Semelidae differ from all other families by having an internal ligament set on a spoon-shaped resilium as well as an external one. In outline they vary from circular to subovate and externally are easily confused with the Tellinidae.

Family TELLINIDAE
Subfamily TELLININAE

Tellina arsinoensis Issel, 1869: **1116**
To 20mm. Thin, fragile. Slight posterior flexure. Beaks just behind mid-line. Subovate, rather triangular; posterior acute, anterior broadly rounded. Sculpture smooth, growth lines only. Hinge weak; RV posterior lateral present but very small. Pallial sinus very large, almost touching anterior adductor; dorsal line strongly angled, very high. White to pink or yellow. **Habitat:** mudflats, intertidal. **Distribution:** all.

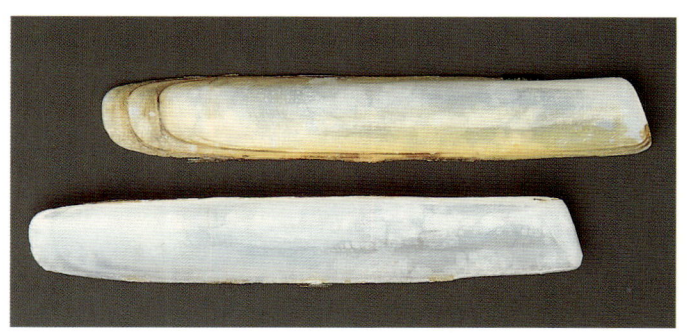

Solen cylindraceus 1113

Ensiculus cultellus 1114

Siliqua polita 1115

Tellina arsinoensis 1116

Superfamily: **TELLINOIDEA**

Tellina asmena Melvill & Standen, 1907: **1117** NOT ILLUSTRATED
To 8mm. Thin, translucent. Beaks central. Subelliptical; anterior rounded, posterior obliquely truncate with a truncate and twisted ventral junction. Sculpture of widely spaced, narrow ridges with radial incisions over lateral margins. Weak laterals present. White. **Habitat:** in shell sand, deep water. **Distribution:** GO.

Tellina claudia **1118**

Tellina claudia Melvill & Standen, 1907: **1118**
To 12mm. Very thin, translucent, compressed. Beaks behind the mid-line. Subovate but broadly rounded anteriorly and subacute posteriorly, with prominent, small, projecting umbones; posterior area set off by an incised line. Hinge very weak, laterals present in RV. Weak sculpture of concentric lines, these stopping at posterior, radial, incised line, beyond this sculpture is indistinct. White to tinged ochre. **Habitat:** offshore. **Distribution:** GO.

Tellina methoria Melvill, 1897: **1119**
To 10mm. Rather solid. Beaks a little behind mid-line. Subovate, rather triangular; posterior dorsal margin long, sloping at about 45° with a slight sulcus, posterior subacute; anterior dorsal margin only slightly less steep than posterior, anterior rounded. RV with very large anterior lateral and weak posterior lateral. Weak sculpture of concentric lines and some with weak anterior or ventral lirae. White. **Habitat:** offshore. **Distribution:** NWG, SEG, GO.

Tellina nitens Deshayes, 1854: **1120**
To 15mm. Thin, translucent, compressed. Narrowly subovate; posterior narrower than anterior but subtruncate. Anterior lateral close but clearly separate from cardinals. Pallial sinus touching adductor. White, some weakly suffused pink or orange. **Habitat:** in muddy sand, offshore. **Distribution:** GO, SO.

Tellina n. sp: **1121**
To 15mm. Solid and shiny. Beaks well behind the mid-line. Narrowly subovate: posterior margin short, steep, posterior area defined by strong posterior angle and with a narrow sulcus close to the post-dorsal margin. Hinge with strong laterals. Sculpture of concentric smooth lirae, these irregularly acentric Pallial sinus almost reaching anterior scar, mostly confluent. White, some tinged yellow internally. **Habitat:** in muddy sand and shell gravels, offshore. **Distribution:** SEG, GO, Mas, SO.

Tellina valtonis Hanley, 1844: **1122**
To 15mm. Thin, compressed. Beaks behind mid-line. Narrowly subovate; posterior narrowly rounded, anterior broader. Hinge teeth weak; RV anterior lateral very close to cardinals. Pallial sinus clearly separated from adductor. White to deep pink with one or two radial rays on posterior. **Habitat:** in muddy sand, intertidal and offshore. **Distribution:** NWG, SEG, GO, Mas.

Tellina vernalis Hanley, 1844: **1123**
To 20mm. Thin, compressed. Broadly subovate; posterior only slightly narrower than anterior. Anterior lateral close but clearly separated from cardinals. Pallial sinus clearly separated from adductor. Pale pink, some with weak, paler radial rays on posterior. **Habitat:** in muddy shell gravels, offshore. **Distribution:** GO, Mas.

Tellinides adenensis (Smith, 1891): **1124**
To 40mm. Beaks central. Subovate, posterior obliquely subtruncate; anterior narrowly rounded. Posterior area set off by a distinct angulation on the LV but less so on the RV. Sculpture virtually smooth, of concentric lines only. Peg-like

Tellina nitens **1120**

Tellina methoria **1119**

Tellina valtonis **1122**

Tellina vernalis **1123**

Tellina n. sp. **1121**

252   SEASHELLS OF EASTERN ARABIA

Superfamily: **TELLINOIDEA**

Tellinides emarginatus 1125

Tellinides adenensis 1124

Cadella obtusalis 1127

Cadella semen 1126

anterior lateral in the RV. Weak internal, raised ridge running radially past the inner edge of the posterior adductor. White with umbonal area suffused yellow-orange. **Habitat:** offshore. **Distribution:** Mas, SO.

Tellinides emarginatus Sowerby, 1825: **1125**
To 40mm. Thin, compressed and posterior strongly flexed to the left. Beaks just behind mid-line. Subelliptical, posterior a little shorter, posterior margin truncate and indented; posterior area sulcate, set off by a low rounded keel. Sculpture smooth and shiny with fine incremental lines only. Hinge weak, RV with a small anterior lateral close to the cardinals. Off-white, some tinged dull yellow over umbones. **Habitat:** offshore. **Distribution:** NG, SEG, GO.

Cadella semen (Hanley, 1845): **1126**
To 11mm. Beaks behind the mid-line. Subovate; posterior slightly narrower than anterior, rounded with a weak flexure and both dorsal margins of similar slope. Hinge with strong laterals in RV. Sculpture of prominent concentric lirae, these becoming raised posteriorly but also anastomosing so that there are about half the number on the posterior area. Dirty white. **Habitat:** offshore. **Distribution:** NWG, SEG, GO, Mas.

Cadella obtusalis (Deshayes, 1854): **1127**
To 9mm. RV slightly larger than LV. Beaks well behind mid-line. Subovate, somewhat trigonal; posterior slope very steep, posterior angle indistinct and posterior area very small; anterior dorsal slope gentle, RV with post umbonal ridge; anterior rounded. Hinge with strong laterals in RV. Sculpture of concentric lirae over the posterior and median areas, then abruptly smooth over anterior. White to pink. **Habitat:** offshore. **Distribution:** SEG.

Peronaea manumissa 1128

Peronaea manumissa (Melvill, 1898): **1128**
To 60mm. Elliptical; anterior margin rounded; posterior margin slightly acute. Posterior slope weakly angulate in both valves. Sculpture of concentric lines and growth stops, some traces of very weak radial striations in some valves, appears smooth overall. White, suffused orange-pink. **Habitat:** offshore. **Distribution:** GO, Mas, SO.

Superfamily: **TELLINOIDEA**

Pharaonella wallaceae **1130**

Phylloda foliacea **1129**

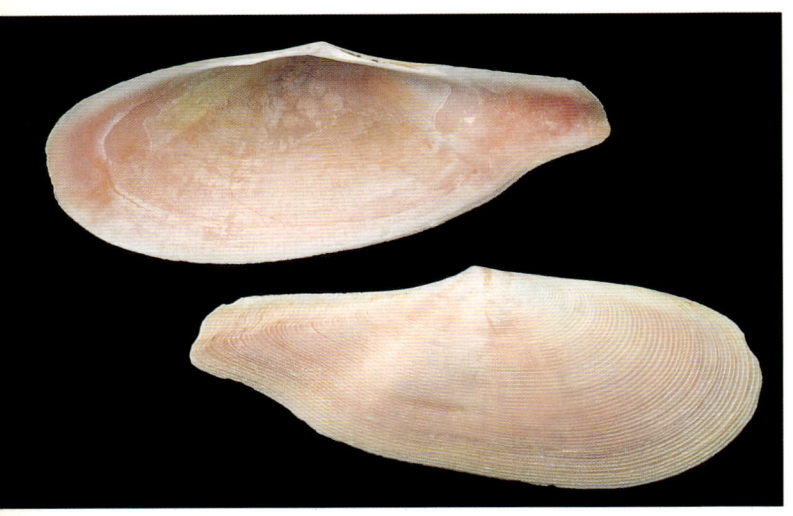

Pharaonella perna **1131**

Laciolina cf *incarnata* **1132**

Tellinella rastella **1133**

Phylloda foliacea (Linnaeus, 1758): **1129**
To 80mm. Very compressed. Beaks almost central. Subrectangular; anterior rounded; posterior slightly expanded, truncate. Sculpture almost smooth except for posterior area which bears fine radial ridges, these minutely spined; a few thorn-like spines are present along the posterior dorsal margin. Deep orange. **Habitat:** offshore. **Distribution:** SEG, GO.

Pharaonella wallaceae (Salisbury, 1934): **1130**
To 30mm. Thin, very compressed. Beaks slightly behind mid-line. Elongate-subovate, posterior subrostrate; posterior dorsal slope slightly concave, posterior area weakly defined by a shallow sulcus. Sculpture of concentric lirae on anterior, these fading and becoming smooth posteriorly. Very weak, anterior lateral in RV, well separated from cardinals. White. **Habitat:** offshore. **Distribution:** SEG, GO.

Pharaonella perna (Spengler, 1797): **1131**
To 75mm. Thin. Beaks almost central. Elongate-oval, posterior rostrate with a strong sulcus, rostrum grooved in LV, angulate in RV. Sculpture of fine concentric raised lines, strongest in RV. RV with distinct lateral teeth. Pallial sinus reaching below anterior lateral. White or pink, occasionally yellow. The eastern Arabian shells all conform with var. *pharaonis* Hanley, 1844. **Habitat:** offshore. **Distribution:** GO.

Laciolina* cf *incarnata (Linnaeus, 1758): **1132**
To 40mm. Close to **P. wallaceae** (see above) but less narrow, posterior shorter. RV anterior lateral prominent. Two posterior radial ridges corresponding to cruciform muscles. Pale orange to deep pink. The few shells known are apparently inseparable from Mediterranean examples. **Habitat:** offshore. **Distribution:** GO, Mas.

Tellinella rastella (Hanley, 1844): **1133**
To 90mm. Beaks almost central. Strong flexure to the right. Elongate-oval, posterior acute, anterior narrowly rounded; RV angulate, LV sulcate. Sculpture of concentric lirae becoming sharper, oblique and rasp-like posteriorly. RV with strong lateral teeth. White or yellowish with strong pink to rust radial rays. **Habitat:** in muddy sand, offshore. **Distribution:** NWG, GO, SO.

Tellinella adamsi (Bertin, 1878): **1134**
To 20mm. Beaks almost central. Subovate - trigonal, just longer than

Superfamily: **TELLINOIDEA**

Serratina capsoides 1138

Quidnipagus palatam 1137

Tellinella cf *pulchella* 1135

Serratina sulcata 1139

Tellinella adamsi 1134

high. Posterior strongly twisted and angulate. Sculpture of dense, concentric, crinkled lamellae. Lateral teeth strong. Rayed brick-red and white. **Habitat:** offshore. **Distribution:** NWG.

Tellinella cf *pulchella* (Lamarck, 1818): **1135**
To 25mm. Beaks just behind midline. Narrowly elongate, posterior narrowed, subrostrate; posterior slope angulate. Hinge with prominent laterals. Sculpture of growth lines and weak lirae becoming raised on posterior area with radial striae. Pallial sinus almost touching anterior adductor. Deep pink with narrow, paler radial rays. Known only from four valves. Apparently identical with Mediterranean examples of *T. pulchella*. **Habitat:** offshore. **Distribution:** GO, Mas.

Elliptotellina erasmia (Melvill, 1898): **1136**
To 7mm. Beaks central. Elongate-elliptical; posterior margin only slightly narrower than anterior. Sculpture of few concentric ridges which become weakly nodulose along the dorsal edges in the anterior third but which, on the posterior third, break down to form a series of robust blunt spines; with faint radial striations between the concentric elements overall. White. **Habitat:** in sandy gravel, offshore. **Distribution:** NG, GO.

Quidnipagus palatam (Iredale, 1929): **1137**
To 60mm. Strongly twisted. Suboval; posterior subacute; anterior broadly rounded. LV with shallow posterior sulcus, RV with posterior angulation. Sculpture of dense, prominent, undulating, raised, concentric ridges, increasing in coarseness marginally and often becoming interrupted forming a rasp-like surface. Radial striations present. Pallial sinus very deep, almost touching the anterior adductor scar. White, tinged yellow. **Habitat:** offshore. **Distribution:** GO, Mas.

Serratina capsoides (Lamarck, 1818): **1138**
To 50mm. Posterior only slightly twisted. Beaks slightly behind midline and rather pointed. Suboval-triangular; posterior margin subacute; anterior margin broadly rounded. Sculpture of dense, evenly spaced, concentric, slightly raised ridges, lightly dissected by fine radial striae, slightly rugose over posterior angle; posterior dorsal margin serrated. White. **Habitat:** in mudflats, low water and below. **Distribution:** NWG, SEG, GO, Mas.

Serratina sulcata (Wood, 1815): **1139**
To 40mm. Beaks slightly behind midline. Distinct posterior twist to the right. Suboval, distinctly longer than high; posterior margin subacute; anterior margin rounded. LV with a distinct posterior sulcus, RV biangulate. Sculpture of well spaced, slightly recurving, concentric ridges, these becoming weak anteriorly, distinctly rugose on posterior area; radial striae dense but indistinct. White, tinged yellow on umbonal region. **Habitat:** offshore. **Distribution:** GO, Mas.

Elliptotellina erasmia 1136

SEASHELLS OF EASTERN ARABIA 255

Superfamily: **TELLINOIDEA**

Tellidora pellyana H. Adams, 1873: **1140**
To 9mm. Beaks almost central. Triangular, dorsal margins steeply sloping, slightly concave. Sculpture of dense, concentric, rugose ridges and raised lamellae posteriorly where they form a series of projections along the dorsal margin. Laterals prominent in the RV. White. **Habitat:** in sand, lower shore and below. **Distribution:** SEG, GO, Mas.

Obtellina sericata (Melvill, 1898): **1141**
To 30mm. Beaks well behind midline. Posterior flexure not apparent. Subovate-truncate; anterior narrowly rounded; posterior straight, almost vertical. Sculpture scissulate, but posterior nodulose, posterior dorsal margin with blunt spines. White, sometimes tinged orange-pink. **Habitat:** offshore. **Distribution:** GO.

Acropaginula inflata (Gmelin, 1791): **1142**
To 40mm. Beaks just behind midline. Subcircular with acute posterior margin. Both valves angulate posteriorly, sulcus lacking but an elongate, flat escutcheon present. Sculpture of growth lines only with very faint radial striations. Single laterals in both valves. White with a straw-coloured periostracum. **Habitat:** in mud, offshore. **Distribution:** NWG, SEG, GO, Mas.

Pinguitellina pinguis (Hanley, 1844): **1143**
To 11mm. Beaks central. Subcircular-trigonal just longer than high; posterior ventral junction angulate; anterior margin rounded. Sculpture smooth, shining with very fine raised concentric threads remaining only at the margins. Pallial sinus extending to a point beyond the anterior lateral tooth but separate from the anterior adductor scar. White, umbonal area often tinged yellow to brick red. **Habitat:** in fine sand, lower shore and below. **Distribution:** NWG, SEG, GO, Mas.

Scutarcopagia scobinata (Linnaeus, 1758): **1144**
To 70mm. Beaks central. Orbicular with a slight posterior twist. Sculpture of concentric ridges broken down into erect crescent-shaped scales. White with faint orange-brown radial rays. **Habitat:** offshore. **Distribution:** SEG, GO.

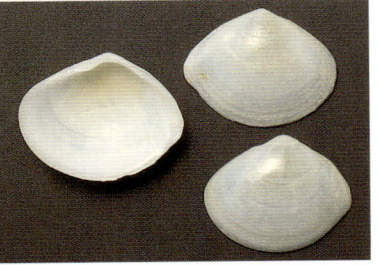

Pinguitellina pinguis 1143

Obtellina sericata 1141

Acropaginula inflata 1142

Tellidora pellyana 1140

Scutarcopagia scobinata 1144

Superfamily: **TELLINOIDEA**

Arcopella isseli (H. Adams, 1871): **1145**
To 12mm. Beaks almost central. Oval-suborbicular; anterior margin broadly rounded; posterior margin less rounded, slightly angulate at post-ventral junction. Sculpture of well spaced, thin, slightly crinkled, raised, concentric lirations with concentric striations between these. Pallial sinus in LV very deep, coalescing with anterior adductor but in RV only extending to the mid-line. White with faint orange-pink radial rays. **Habitat:** in muddy sand, offshore. **Distribution:** NWG, GO.

Subfamily MACOMINAE
Florimetis coarctata (Philippi, 1845): **1146**
To 50mm. Beaks behind mid-line. Moderately inflated. Subquadrate; anterior margin broadly rounded; posterior margin truncate. Posterior slope strongly angulate in RV, smaller specimens distinctly biangulate; LV much less so with a shallow, broad sinus. Sculpture of incremental lines and growth stops only. White. **Habitat:** in mud flats, lower shore and below. **Distribution:** GO, Mas.

Psammotreta praerupta (Salisbury, 1934): **1147**
To 40mm. Fragile. Beaks just behind mid-line. Broadly subovate; posterior weakly angulate, narrower and a little truncate. Sculpture of growth lines only. Pallial sinus narrow, well separated from adductor and only partly confluent. White with a dull greyish periostracum. **Habitat:** in mud flats, lower shore and below. **Distribution:** GO, Mas.

Tellinimactra angulata (Linnaeus, 1767): **1148**
To 60mm. Distinctly twisted. Beaks behind mid-line. Triangular-subovate; anterior margin rounded, continuous with dorsal margin; posterior margin broadly rounded, becoming truncated. Posterior slope in RV angulate, LV sulcate. Sculpture of growth lines only. Hinge with deeply sunken ligament encroaching on cardinal teeth. Pallial sinus deep, almost reaching adductor scar; dorsal line weakly arcuate, scarcely rising above level of adductor scars; anterior adductor long and narrow. White. **Habitat:** offshore. **Distribution:** GO.

Tellinimactra syndesmyoides (Melvill & Standen, 1907): **1149**
To 15mm. Beaks behind the mid-line. Subovate with small prominent umbones; anterior expanded and rounded; posterior reduced and subacute. Sculpture smooth except for weak fragile lirae over posterior area. Ligament partly encroaching on posterior teeth. Pallial sinus deep but well separated from anterior scar, dorsal line rising to a peak and only partly confluent; anterior adductor scar normal. White. **Habitat:** offshore. **Distribution:** SEG, GO.

Loxoglypta rhomboides (Quoy & Gaimard, 1835): **1150**
To 20mm. Beaks well behind mid-line. Narrowly-subovate; anterior margin narrowly rounded; posterior margin subacute. Posterior slope strongly curved but not angular, posterior area steeply sloping. Sculpture scissulate on both valves, most prominent medially, fading completely over posterior area which is concentrically lirate. White, usually with few orange-pink-red radiating rays. **Habitat:** in muddy sand, offshore. **Distribution:** NWG, SEG.

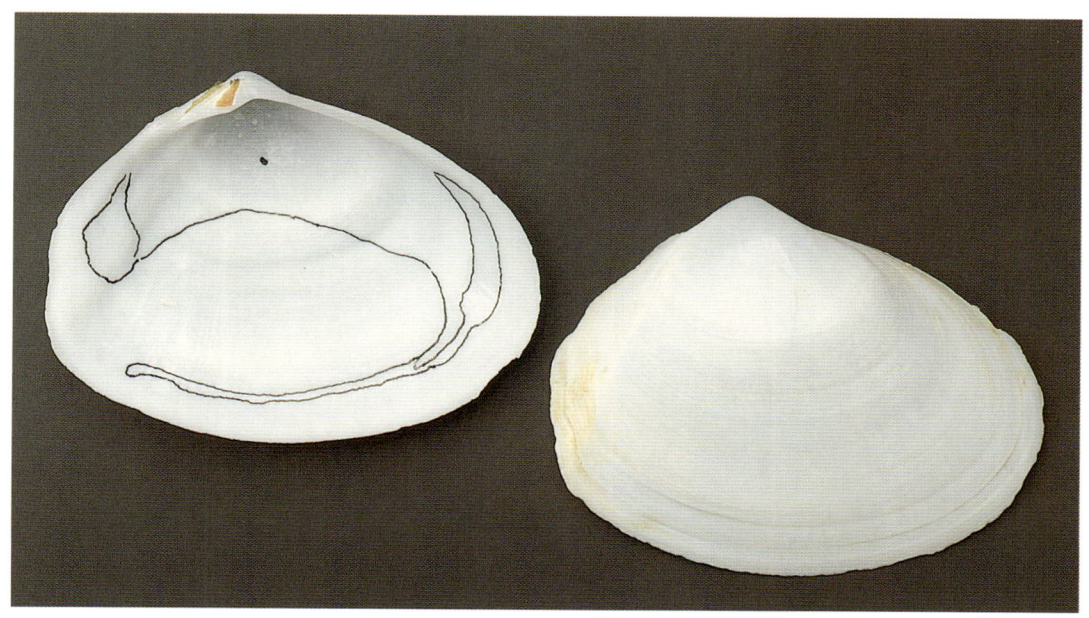

Tellinimactra angulata 1148

Florimetis coarctata 1146

Arcopella isseli 1145

Tellinimactra syndesmyoides 1149

Psammotreta praerupta 1147

Loxoglypta rhomboides 1150

SEASHELLS OF EASTERN ARABIA

Superfamily: **TELLINOIDEA**

Donax erythraeensis 1155

Donax cf *bipartitus* 1152

Donax cuneata 1153

Gastrana multangula 1151

Gastrana multangula (Gmelin, 1791): **1151**
To 40mm. Solid and a little inflated. Beaks at mid-line. Transversely oval; anterior rounded; posterior margin long, curving steeply to angulate posterior-ventral margin, posterior angle equally distinct on both valves. Sculpture of closely spaced, sharp, but little elevated, concentric lamellae with traces of radial striae between. White, some tinged deep orange-yellow. **Habitat:** offshore. **Distribution:** Mas.

Family DONACIDAE
Donax* cf *bipartitus Sowerby, 1892: **1152**
To 15mm. Wedge-shaped but rather narrow, posterior area small, slope about 45°, angle sharp but not keeled. Sculpture of strong, concentric, rugose ridges on posterior slope, as lines on median area ending abruptly before the anterior area. Posterior laterals prominent and very close to cardinal set. Inner margin denticulate. Off-white, inside with purple brown areas. Known from a single specimen collected at Khor Fakkan; needs verification. **Habitat:** ? **Distribution:** GO.

Donax cuneata Linnaeus, 1758: **1153**
To 35mm. Broadly wedge-shaped, almost trigonal. Posterior slope steep, angle distinct but rounded. Sculpture smooth and shining but posterior with dense, wavy, rugose ridges. Lateral teeth present. Shades of greyish mauve and brown. Known from a single specimen collected in the U. A. E.; needs verification. **Habitat:** ? **Distribution:** SEG.

Donax veneriformis Lamarck, 1818: **1154**
To 25mm. Oval-trigonal; posterior area weakly demarcated. Sculpture primarily concentric, anterior area almost smooth, developing towards the posterior end into slightly rugose ridges, posterior area also with radial microsculpture. Hinge heavy, cardinal teeth large, triangular, only anterior laterals present. Inner margins finely serrate but some partially smooth. Variable from white with rose-red radial rays to deep purple with paler rays. Known from a single specimen collected on the Batinah Coast. **Habitat:** ? **Distribution:** GO.

Donax erythraeensis Bertin, 1881: **1155**
To 15mm. Wedge-shaped; posterior area very short, posterior margin steeply sloping. Sculpture smooth anteriorly becoming progressively radially striate posteriorly; has nodulose radial riblets on the posterior area. Anterior laterals lacking. Inner margins coarsely denticulate. Generally pale with brown, reddish or violet markings. **Habitat:** sandy beaches. **Distribution:** SEG, GO.

Donax scalpellum Gray, 1823: **1156**
To 45mm. Beaks well behind mid-line. Very elongate, wedge-shaped: posterior area very short, acute, demarcated by a sharp umbonal ridge. Sculpture of fine concentric lines only. Inner margin weakly crenulate. Pale, lilac to cream with darker brown to purple brown areas, typically two, one median and one posterior. **Habitat:** sandy beaches. **Distribution:** SEG, GO, Mas, SO.

Donax clathratus Deshayes, 1855: **1157**
To 17mm. Beaks behind the mid-line. Wedge-shaped-triangular; posterior slope steep, short; ridge prominent. Sculpture of elevated, concentric

Donax veneriformis 1154

Donax scalpellum 1156

Superfamily: **TELLINOIDEA**

Asaphis violascens 1160

Gari amethystus 1161

Gari bicarinata 1162

Donax townsendi 1159

lamellae intersected by radial riblets and forming closely spaced scales on the intersections which are most developed on the posterior angle and area. White to buff with occasional brown-purple markings. **Habitat:** sandy beaches. **Distribution:** GO, Mas, SO.

Donax nitidus Deshayes, 1855: **1158**
To 10mm. Beaks close to posterior end. Narrowly elongate, wedge-shaped; posterior slope short, slightly steeper than anterior. Sculpture smooth but posterior area with a few oblique lines which turn sharply to continue as radial riblets. Inner margin denticulate. Off-white to violet. Periostracum golden brown. **Habitat:** sandy shores. **Distribution:** GO, Mas.

Donax townsendi Sowerby, 1895: **1159**
To 30mm. Compressed. Ovate-trigonal, only slightly wedge-shaped; posterior area demarcated by a low but distinct posterior angle; posterior margin truncate, steeply sloping. Sculpture smooth but posterior area with fine radial riblets, concentrically ridged towards the posterior margin. Hinge weak, laterals obsolete. Inner margins denticulate. Off-white tinged orange or violet. **Habitat:** sandy beaches. **Distribution:** GO, Mas, SO.

Family PSAMMOBIIDAE
Asaphis violascens (Forsskål, 1775): **1160**
To 60mm. Beaks slightly in front of mid-line. Not gaping. Oblong, anterior rounded, posterior slightly expanded, posterior margin subtruncate. Sculpture of numerous radial riblets, median and anterior acute, those on posterior rounded and irregularly beaded. Off-white to cream tinged with purple; inside shaded entirely or marginally with violet. **Habitat:** gravel, intertidal. **Distribution:** all.

Gari amethystus (Wood, 1815): **1161**
To 80mm. Elongate-subtrapezoidal; posterior margin angulate; posterior area with 1-3 weak ridges and set off by a weak ridge. Sculpture concentric only but complex; anterior with widely spaced smooth ridges merging abruptly into median area with twice as many dorsally elevated lirations; posterior area with low lirations, or if in perfect condition, lirations ventrally elevated into thin lamellae. Shades of lilac to purple, some with radial rays. **Habitat:** offshore. **Distribution:** SEG, GO, Mas.

Gari bicarinata (Deshayes, 1855): **1162**
To 30mm. Beaks almost central. Elongate-subtrapezoidal; posteriorly angulate; posterior ridge prominent with a second smaller ridge ventral to it. Sculpture of widely spaced concentric ridges which are somewhat oblique or irregular close to the first posterior ridge; posterior area with up to 6 scabrous riblets in RV, these are obscure or absent in LV. Buff to shades of red to purple, often rayed. **Habitat:** offshore. **Distribution:** SEG, GO, Mas.

Donax nitidus 1158

Donax clathratus 1157

SEASHELLS OF EASTERN ARABIA

Superfamily: **TELLINOIDEA**

Gari weinkauffi **1164**

Hiatula mirbahensis **1166**

Gari maculosa **1163**

Gari occidens **1165**

Gari maculosa (Lamarck, 1818): **1163**
To 60mm. Elongate-elliptical; posterior margin subtruncate; posterior ventral margin barely angulate; posterior ridge weak, rounded. Sculpture of oblique ridges except on posterior area which bears somewhat irregular concentric ridges crossed by a microsculpture of radial striae. Mottled shades of buff, purple or red. **Habitat:** offshore. **Distribution:** NWG, SEG, GO, Mas.

Gari weinkauffi (Crosse, 1864): **1164**
To 40mm. Fragile. Beaks almost central. Elongate-subtrapezoidal; posterior area expanded, set off by a ridge and change in sculpture, with a secondary median weak ridge; posterior margin subtruncate; posterior ventral margin angulate. Sculpture of fine oblique grooves except on posterior area which is smooth. White to cream, tinged with lilac, with or without darker cream to buff radial rays. **Habitat:** muddy gravels, offshore. **Distribution:** NWG, GO, Mas, SO.

Gari occidens (Gmelin, 1791): **1165**
To 100mm. Elongate-subquadrate; posterior margin subtruncate, posterior angle very weak or absent. Sculpture of irregular incremental lines and ridges often obscured by periostracum. Cream to buff with pink markings. **Habitat:** offshore. **Distribution:** NWG, SEG, GO, Mas, SO.

Hiatula mirbahensis Morris & Morris, 1993: **1166**
To 50mm. Thin. Beaks almost central. Narrowly elongate, subelliptical; posterior slightly narrower, both anterior and posterior narrowly rounded; posterior slope poorly defined, rounded. Sculpture of concentric lines. White, periostracum yellow-tan with weak radial wrinkles and crinkled on posterior dorsal area. **Habitat:** soft sand flats, intertidal. **Distribution:** GO.

Hiatula ruppelliana (Reeve, 1857): **1167**
To 70mm. Oblong, not flared posteriorly; posterior margin obtuse; anterior margin slightly acute. Sculpture smooth with incremental lines only. Off-white, to tinged pink, to deep violet, occasionally with few posterior radial rays. Periostracum thin,

260   SEASHELLS OF EASTERN ARABIA

Superfamily: **TELLINOIDEA**

***Psammosphaerica psammosphaerita* 1168**

olivaceous. **Habitat:** soft muddy sand, lagoons and khors. **Distribution:** all.

Psammosphaerica psammosphaerita Jousseaume, 1894: **1168**
To 20mm. Not gaping. Beaks approximately at mid-line. Subovate; anterior rounded, posterior a little narrower and slightly truncated. Sculpture of concentric lines only. Pale with broad but faint, brownish-purple, radial rays; periostracum thin, yellowish. Syn: *Nanhaia safadensis* Morris and Morris, 1993. **Habitat:** sand flats, lagoons and khors. **Distribution:** GO.

Family SEMELIDAE

Semele carnicolor (Hanley, 1845): **1169**
To 40mm. Beaks central. Subcircular, dorsal margins sloping gently; posterior slightly twisted. Sculpture of well spaced, raised, recurving lamellae and dense radial striations, these cutting the lower edges of the lamellae. White. **Habitat:** sandy mud flats, lower shore and below. **Distribution:** SEG.

Semele zalosa Chesney & Oliver, 1994: **1170**
To 70mm. Subcircular, posterior slightly twisted, anterior broadly rounded, posterior a little truncated. Sculpture up to the 30mm growth stage, smooth with dense radial striae; then progressively developing roundly rugose concentric ridges; the surface over the ridges may appear minutely pustulose or chalky. White except for pink tracery over escutcheon and lunule, hinge plate also with some pink colouring. Periostracum greyish brown. **Habitat:** offshore. **Distribution:** Mas, SO.

Semele sinensis A. Adams, 1854: **1171**
To 40mm. Beaks slightly behind mid-line. Subcircular; posterior subtruncate, anterior rounded, posterior flexure weak. Lunule indistinct. Sculpture of dense radial striae and narrow, weak, usually worn, concentric ridges. Beige with faint orange or reddish rays, inside spotted red. **Habitat:** offshore. **Distribution:** NG, SEG, GO, Mas, SO.

Semele zalosa **1170**

Cumingia striata (Reeve, 1853): **1172**
To 20mm. Beaks slightly in front of mid-line. Subcircular to subquadrate often irregular. Sculpture of well-separated concentric lamellae, these irregularly fractured or worn with traces of radial striae between. Lateral teeth very large. Ligament in a very large deeply inset resilifer. Dirty white. **Habitat:** in crevices on rocky shores. **Distribution:** SEG, GO, Mas.

Hiatula ruppelliana **1167**

Cumingia striata **1172**

Semele sinensis **1171**

Semele carnicolor **1169**

SEASHELLS OF EASTERN ARABIA 261

Superfamily: **TELLINOIDEA**

Semelangulus rosamunda 1173

Ervilia purpurea 1174

Ervilia scaliola 1175

Iacra seychellarum 1177

Rochefortina cf sandwichensis 1176

Leptomya cochlearis 1179

Iacra trotteriana 1178

Semelangulus rosamunda (Melvill & Standen, 1907): **1173**
To 9mm. Beaks well behind mid-line. Narrowly subovate; posterior margin sloping steeply, posterior angle strong in RV, posterior ventral angle subacute with a slight flexure; anterior margin rounded. Sculpture of dense, slightly raised, sharp lirae. Hinge with strong anterior and posterior laterals in RV. Small internal ligament. Pallial sinus very deep almost touching anterior adductor scar. White tinged pink to orange. **Habitat:** muddy sand, offshore. **Distribution:** NWG, SEG, GO.

Ervilia purpurea (Lamy, 1914): **1174**
To 15mm. Beaks slightly in front of mid-line. Elongate-oval; posterior slightly tapering but rounded; anterior rounded. Lateral teeth very short, one cardinal in each valve, that in the LV large, projecting, not bifid. Resilium small, external ligament weak. Sculpture of fine, somewhat irregular, concentric lines and growth stops. Cream to purple-brown, some with broad radial rays. **Habitat:** in sand, offshore. **Distribution:** NWG, SEG, GO, Mas.

Ervilia scaliola (Issel, 1869): **1175**
To 8mm. Beaks at mid-line. Subovate; posteriorly tapering, almost acute; anterior margin narrowly rounded. Sculpture of evenly spaced weak concentric ridges, most with radial striae on the posterior area. Pale cream to rose-pink. **Habitat:** in muddy sand, offshore. **Distribution:** NWG, SEG, GO, Mas.

Rochefortina cf sandwichensis (Smith, 1885): **1176**
To 5mm. Beaks prominent pointing posteriorly. Larval shell distinct and dark. Transversely ovate; anterior rounded, posterior a little truncated. Sculpture of fine, dense, radial, weakly cancellate lines. Ligament internal, between cardinal teeth, laterals weak. White. **Habitat:** valves in shell sand only. **Distribution:** Mas.

Iacra seychellarum (A. Adams, 1856): **1177**
To 20mm. Beaks slightly behind mid-line. Subovate-trigonal; anterior broadly rounded, posterior roundly subacute. Sculpture scissulate, in three parts: anterior concentric; median - posterior strong, oblique; posterior faint, steeply oblique. External portion of ligament small, internal portion in a large oblique resilifer. White. **Habitat:** in sand, offshore. **Distribution:** SEG, GO, Mas, SO.

Iacra trotteriana (Sowerby, 1894): **1178**
As *I. seychellarum* (see above) but external portion of ligament large, internal in a small oblique resilifer. Syn: *Tellina speciosa* Deshayes, 1856, preoccupied by *T. speciosa* Edwards and replaced by *T. kallima* Salisbury, 1934. **Habitat:** in sand, offshore. **Distribution:** GO, Mas.

Leptomya cochlearis (Hinds, 1844): **1179**
To 40mm. Fragile. Subovate-rostrate; anterior margin broadly rounded; posterior acute; posterior dorsal margin long, more or less straight; posterior ventral margin sinuous. Sculpture of feeble, well-spaced,

Superfamily: **TELLINOIDEA**

Azorinus coarctatus **1183**

Leptomya subrostrata **1180**

Theora cadabra **1181**

Solecurtus australis **1184**

concentric lirations. Ligament in a deep oblique chondrophore. White. **Habitat:** offshore. **Distribution:** NG, NWG, GO.

Leptomya subrostrata (Issel, 1869): **1180**
To 30mm. Thin. Subovate; posterior shortly rostrate, subtruncate to subacute: anterior broadly rounded; posterior dorsal margin more or less straight; posterior ventral margin weakly sinuous. May be irregular. Sculpture of closely spaced, thin, raised, concentric lirations, fine radial striations overall causing lirations to be a little crinkled. Ligament in a deep oblique chondrophore, plus a minute sunken external band. Lateral teeth absent (compare with ***Cumingia***). White. **Habitat:** ? **Distribution:** SEG.

Theora cadabra (Eames & Wilkins, 1957): **1181**
To 8mm. Thin, translucent. Transversely oval; posterior bluntly rostrate. Sculpture smooth. Resilium prominent. Cardinals indistinct, laterals distinct in RV. White. **Habitat:** in mud, lower shore and below. **Distribution:** NG, NWG. This species was originally described as an ***Abra***, based on long-dead material, hence the conjuring trick with the name. Unfortunately the shape of the shell clearly allies it with the Indo-Pacific genus ***Theora*** rather than the ***Abra*** of temperate regions.

Syndosmya cistula Melvill & Standen, 1907: **1182**
To 9mm. Thin, semitransparent. Beaks slightly behind mid-line. Subovate, rather triangular; posterior roundly pointed; anterior broadly rounded. Internal ligament on a shallow oblique resilium. White. **Habitat:** in shell sand, deep water. **Distribution:** GO.

Family SOLECURTIDAE
Azorinus coarctatus (Gmelin, 1791): **1183**
To 40mm. Thin and fragile. Gaping at both ends. Beaks in front of mid-line. Elongate-oblong; posterior margin more acute than anterior; posterior-dorsal margin distinctly concave behind the beaks. Sculpture smooth. Hinge without lateral teeth. Pallial sinus deep. White. **Habitat:** in muddy sediments, offshore. **Distribution:** GO.

Solecurtus australis (Dunker, 1861): **1184**
To 50mm. Beaks just in front of mid-line. Gaping at both ends. Oblong, lateral margins roundly truncate. Sculpture of widely spaced, oblique grooves becoming V-shaped along posterior slope and fading anteriorly. Pinkish-buff to deep pink with two white radial rays medially. **Habitat:** offshore. **Distribution:** NWG, SEG, GO, Mas.

Syndosmya cistula **1182**

Superfamily: **ARCTICOIDEA, GLOSSOIDEA, VENEROIDEA**

Trapezium sublaevigatum **1185**

Coralliophaga coralliophaga **1186**

Superfamily ARCTICOIDEA
The Arcticoidea are a diverse superfamily which takes its name from the cold temperate genus *Arctica*. In tropical and subtropical seas the group is most commonly represented by the Trapeziidae which on first sight seem quite unlike *Arctica*. These are elongate, almost modioliform shells, most closely resembling some of the epifaunal Carditoidea such as *Cardita* and *Beguina*. The essential characters are primarily of the hinge which is well developed with two cardinal teeth in each valve and anterior and posterior laterals. The ligament is external and the pallial line is usually entire or has a shallow sinus.

The Trapeziidae are primarily byssate nestling forms, living in crevices or beneath coral debris and rocks. *Coralliophaga* is more specific, inhabiting empty burrows made by date mussels and flask shells.

Trapezium sublaevigatum (Lamarck, 1819): **1185**
To 45mm. Beaks subterminal, close to anterior end. Subrectangular almost modioliform, anterior narrowly rounded, posterior subtruncate or broadly rounded. Sculpture of irregular, coarse, concentric lines and incremental grooves; umbonal region with traces of radial striae. Ligament long, set on a partly sunken nymph. Dirty white with faint brown markings, internal with tinges of brown or purple. **Habitat:** attached to rocks usually in the vicinity of mangroves, intertidal. **Distribution:** NWG, SEG, GO.

Coralliophaga coralliophaga (Gmelin, 1791): **1186**
To 50mm. Beaks close to the anterior end. Elongate-oblong; expanded posteriorly and acute anteriorly but often distorted. Hinge with two cardinals and a single posterior lateral in each valve. Sculpture of radiating lines and elevated concentric lamellae posteriorly. Pallial sinus shallow. White to creamy white, umbo invariably with reddish brown lines. **Habitat:** nestling in empty burrows of date mussels and flask shells. **Distribution:** SEG.

Superfamily GLOSSOIDEA
The Glossoidea are most closely related to the Arcticoidea and share similarities in the hinge structure. Of the two families with living representatives, Glossidae and Vesicomyidae, only the former are found in offshore waters of the eastern Arabian region although the latter are recorded from the deep sea further south. The Glossidae typically have inflated shells in which the umbones are spirally twisted. They are infaunal, shallow burrowers living in soft offshore sediments.

Meiocardia moltkiana (Gmelin, 1791): **1187**
To 45mm. Inflated. Beaks in front of the mid-line, umbones coiled. Trigonal-rhomboidal; almost as high as long in juveniles, becoming more rectangular; posterior margin subtruncate, posterior keel angulate. Sculpture of rounded, concentric ridges but smooth behind the keel. Inner margins smooth. Pallial line entire. White to buff. **Habitat:** offshore. **Distribution:** GO.

Superfamily VENEROIDEA
The Veneroidea or Venus clams are familiar as food items in most parts of the world and there are many regional vernacular names. They are, in general, oval to triangular in outline, tumid and with solid, often thick shells. The escutcheon and lunule are usually well developed. The ligament is always external and set on a prominent nymph. The hinge is strong with well developed teeth, typically comprising three cardinals in each valve with anterior laterals in many, but not all groups. Sculptural patterns are primarily concentric although many display some radial elements as well, resulting in a variety of decussate, reticulate and cancellate forms. The adductor muscle scars are equal or subequal in size. The pallial sinus is very variable, from absent to reaching beyond the mid-line. The inner margins may be smooth or denticulate.

Venus clams are common members of the intertidal and inshore faunas. They inhabit a wide variety of substrates but are most often an important part of the fauna of sand flats and the more consolidated sediments offshore. They are efficient burrowers and possess a large hatchet-shaped foot. They live at a variety of depths and so the siphons vary from very short to moderately long. A few species live among or in hard substrates, adopting a byssate habit. Their shells do not differ greatly except that they are frequently distorted. The Petricolidae are the major exception to the above typical venerid plan. The shells in this family are ovate to elongate or distorted, have only two cardinal teeth in the right valve and their sculpture is primarily radial. Most are borers and use an acid secretion to tunnel into coral and calcareous rocks.

Only the Veneridae and Petricolidae are present in the eastern Arabian region although the Glauconomidae are common around Karachi and sparsely distributed along the Mekran coast. The Veneridae, which have three cardinal teeth in each valve, are divided into a number of subfamilies which are useful groupings.

The Venerinae have three cardinals and an anterior lateral in each valve. They are roughly circular or broadly ovate with a strong sculpture of both radial and concentric elements. The Chioninae have three cardinals in each valve but no laterals. They are roughly circular to ovate with both radial and concentric sculpture. The inner margin is serrated or denticulate. The pallial sinus is short. The Pitarinae have three cardinals and anterior laterals in each valve. They are subovate with a primarily concentric sculpture of lines or low ridges. The pallial sinus is rather deep.

The Circinae have three cardinals and anterior laterals in each valve. They are roundly triangular to subovate with a primarily concentric sculpture. Some have a divaricate sculpture especially on the umbones. The pallial sinus is no more than a slight indentation. The Meretricinae have three radiating cardinals and an anterior lateral in each valve. They are rather triangular in outline with a smooth sculpture. The pallial sinus is moderate. The Sunettinae have a large deeply excavated escutcheon.

The Clementiinae have three cardinals in each valve and no laterals. They are rather thin, lack an escutcheon and have a weak or obsolete sculpture. The pallial sinus is deep. The Dosiniinae have three cardinals and anterior laterals in each valve. They are discoid, generally compressed with concentric sculpture only. The pallial sinus ranges from moderately deep to very deep. The Tapetinae have three cardinals in each valve but no laterals. They range from elongate to subovate with primarily a concentric sculpture with weaker radial elements. The pallial sinus is moderately deep.

Family VENERIDAE
Subfamily VENERINAE

Antigona lamellaris (Schumacher, 1817): **1188**
To 35mm. Beaks just in front of midline. Subovate, slightly projecting anteriorly. Sculpture of raised, crinkled, concentric lamellae with underlying radial riblets. Hinge with small but prominent anterior laterals. Pallial sinus small, V-shaped. Inner margin crenulate. White-cream with areas of darker brown, internally flushed pink. **Habitat:** offshore. **Distribution:** NWG, SEG, GO, SO.

Periglypta puerpera (Linnaeus, 1771): **1189**
To 90mm. Roundly subquadrate, without any angulation of the posterior margin. Sculpture of low, raised, concentric lamellae, made crinkly by well-spaced underlying radial riblets. Hinge with bifid anterior cardinal. Pallial sinus short, just reaching the posterior end of the nymph. Cream with buff and brown radial but interrupted rays, hinge often tinged rose or violet. **Habitat:** in sandy gravels, lower shore and below. **Distribution:** SEG, GO, Mas, SO.

Periglypta reticulata (Linnaeus, 1758): **1190**
To 80mm. Similar to **P. puerpera** (see above) but posterior is more rectangular. Sculpture more delicate, radial ribs more prominent and more numerous. Anterior cardinal tooth not bifid. Pallial sinus wide, U-shaped extending beyond the posterior end of the nymph. White to cream with numerous brown patches or spots, some coalescing into large areas, hinge always white. Known only from a few shells found at Hormuz. **Habitat:** ? **Distribution:** GO.

Antigona lamellaris 1188

Meiocardia moltkiana 1187

Periglypta puerpera 1189

Periglypta reticulata 1190

Superfamily: **VENEROIDEA**

Bassina calophylla 1191

Bassina foliacea 1192

Timoclea arakana 1193

Timoclea costellifera 1194

Subfamily CHIONINAE

Bassina calophylla (Philippi, 1846): **1191**
To 25mm. Beaks distinctly in front of mid-line. Trigonal-subovate, only a little longer than high. Lunule small, cordate, impressed. Escutcheon broad, rather flat and smooth. Sculpture of few strongly elevated, thin, concentric lamellae, these lowered laterally to form a distinct anterior sulcus and a much weaker posterior sulcus. White tinged with brownish radial markings. **Habitat:** in muddy sand and shell gravels, offshore. **Distribution:** NWG, SEG, GO, Mas.

Bassina foliacea (Philippi, 1846): **1192**
To 20mm. As ***B. calophylla*** (see above) but heavier; concentric lamellae thicker almost overlapping; anterior sulcus absent. Pale with reddish markings. **Habitat:** offshore. **Distribution:** GO, Mas.

Timoclea arakana (Nevill, 1871): **1193**
To 15mm. Beaks almost central. Subovate, anterior rounded, posterior subacute. Lunule narrowly cordate, impressed but raised along valve junction. Escutcheon large, raised posteriorly, almost smooth. Sculpture of radial riblets intersecting concentric lamellae; in small shells lamellae are prominent and wavy; with growth the riblets become stronger and can be scaly or smooth especially on posterior. Inner margin denticulate. Beige to pinkish with darker markings. There are several species of ***Timoclea*** described from the region including *T. farsiana* Biggs, 1973; *T. macfadyeni* Dance & Eames, 1966 and *T. mekranica* (Melvill & Standen, 1907). Given the range of form seen in ***T. arakana*** all are probably variations of that species. Records of *T. layardi* (Reeve, 1863) are also ***T. arakana***. **Habitat:** in sand and muddy sand, offshore. **Distribution:** all.

Timoclea costellifera (Adams & Reeve, 1850): **1194**
To 22mm. Beaks almost central. Subovate; anterior rounded, posterior more quadrate. Lunule narrowly cordate. Escutcheon short and narrow. Sculpture of radial ribs regularly dissected into rectangular blocks except posteriorly where they are nodulose. Inner margin denticulate. Off-white to beige with darker markings, inside has mauve-brown areas. **Habitat:** offshore. **Distribution:** GO.

Subfamily CIRCINAE

Circe rugifera (Lamarck, 1818): **1195**
To 45mm. Very heavy. Beaks almost central. Subovate-trigonal, anterior more broadly rounded than posterior. Umbonal disc large with divaricate ribs. Sculpture of narrow concentric ridges, some with a few stepped ridges. Inner margin finely denticulate. Most shells beige with dense darker zigzags, some with broad, radial, brown rays. Syn: More commonly known under the name *C. corrugata* (Dillwyn, 1817) but that name is preoccupied by ***Venus corrugata*** Gmelin, 1791, which is a ***Tapes***. **Habitat:** in sand, lower shore and below. **Distribution:** NWG, SEG, GO, Mas.

Superfamily: **VENEROIDEA**

Circe rugifera **1195**

Circe intermedia **1196**

Circenita callipyga **1198**

Circe intermedia Reeve, 1863: **1196**
To 45mm. As ***C. rugifera*** (see above) but more triangular. Sculpture of a few wide, stepped ridges. Cream with reddish brown tent and zigzag markings. **Habitat:** offshore. **Distribution:** NWG, SEG, GO, Mas, SO.

Circe scripta (Linnaeus, 1758): **1197**
To 40mm. Compressed. Beaks just in front of mid-line. Subovate to subcircular, umbones trigonal. Umbonal disc indistinct, weak divaricate sculpture of narrow ridges, most visible at margins. Lunule narrow, indistinct. Sculpture of numerous fine raised ridges and lines. Inner margin smooth. Beige to brown with darker zigzag lines and radial rays. **Habitat:** in sand, offshore. **Distribution:** NWG, SEG, Mas.

Circenita callipyga (Born, 1780): **1198**
To 50mm. Beaks almost central. Outline variable, from subovate (length distinctly greater than height) to trigonal-subovate (length just greater than height). Posterior slope correspondingly variable from moderate to steep. Sculpture of low narrow ridges and growth lines, often worn medially and almost smooth. Inner margin smooth. Very variable from pale with sparse spots, to radially interrupted rays, to zigzag lines to bicoloured; all in shades of off-white, grey, orange, brown, chocolate and black. **Habitat:** in sand and sandy mud, mid-tide levels. **Distribution:** all.

Comus platyaulax (Tomlin, 1924): **1199**
To 35mm. Heavy. Beaks slightly in front of mid-line. Trigonal, posterior slope longer than anterior, straight. Umbonal disc small with divaricate ridges. Escutcheon excavated. Lunule cordate, outlined by an incised groove. Sculpture of a few very bold, stepped ridges. Inner margin denticulate. Highly patterned with tents, rays and zigzags in shades of brown through pinks to purple. **Habitat:** offshore. **Distribution:** Mas, SO.

Circe scripta **1197**

Comus platyaulax **1199**

SEASHELLS OF EASTERN ARABIA

Superfamily: **VENEROIDEA**

Gafrarium pectinatum **1200**

Tivela mulawana **1202**

Tivela rejecta **1204**

Redicirce sulcata **1201**

Tivela ponderosa **1203**

Gafrarium pectinatum (Linnaeus, 1758): **1200**
To 25mm. Beaks well in front of midline. Elongate-oval, umbones low, lateral margins broadly rounded. Sculpture of 5-8 large, diverging ribs on posterior area, finer ribs elsewhere. Pallial sinus a shallow indentation. Greyish white with brownish spots or blotches, some tinged pink; inside often has brown posterior blotch. **Habitat:** in sandy gravel, intertidal and sublittoral. **Distribution:** NWG, SEG, GO, Mas, SO.

Redicirce sulcata (Gray, 1838): **1201**
To 15mm. Beaks almost central. Subcircular-trigonal, beaks distinctly pointed, lunule margin concave, umbones not flattened. Sculpture of evenly spaced, low, narrow, ridges with traces of fine radial elements, fine divaricate ribbing present across lunule and on escutcheon edges. Pallial sinus a shallow indentation. White with fulvous or brown rays or tents. **Habitat:** offshore. **Distribution:** NWG, SEG, GO, SO.

Subfamily MERETRICINAE
Tivela mulawana Biggs, 1969: **1202**
To 50mm. Compressed. Beaks well to the posterior. Wedge-shaped like a Donax, post-dorsal slope steep, anterior-dorsal slope much longer and more gentle. Smooth, shiny with growth lines. Pallial sinus narrow and deep reaching below anterior lateral teeth, accessory scar below adductor. White with various amounts of radial, fulvous markings. **Habitat:** offshore. **Distribution:** Mas, SO.

Tivela ponderosa (Koch, 1844): **1203**
To 85mm. Beaks almost central. Trigonal, only slightly longer than high. Anterior a little more rounded than posterior. Lunule weakly defined. Smooth, growth lines only, shining. Typically straw coloured with brownish radial rays but may be uniform and darker; inside has some brown or mauve but not rayed. **Habitat:** in clean sand, intertidal and offshore. **Distribution:** SEG, GO, Mas, SO.

Tivela rejecta Smith, 1914: **1204**
To 30mm. Beaks just behind midline. Subovate-trigonal, distinctly longer than high, post-dorsal slope slightly concave with posterior substrate. Smooth with growth lines only. Pallial sinus deep, narrow, reaching below the beaks, accessory scar coalesced with adductor. Off-white with various amounts of zigzag, fulvous to brown lines; inside with patches or rays of mauve to violet. **Habitat:** in muddy sand, lower shore and below. **Distribution:** Mas, SO.

Superfamily: **VENEROIDEA**

Amiantis umbonella **1205**

Callista erycina **1206**

Subfamily PITARINAE
Amiantis umbonella (Lamarck, 1818): **1205**
To 50mm. Beaks just in front of mid-line. Subovate, almost as high as long, umbones prominent, posterior slope long and steep, anterior rounded. Lunule cordate. Sculpture of weak concentric ridges anteriorly, elsewhere almost smooth. Nymph vertically striated. Pallial sinus broad but short. White with mauve-brown umbones to completely mauve. **Habitat:** sand flats, lower shore and below. **Distribution:** NWG, SEG, GO, Mas.

Callista erycina (Linnaeus, 1758): **1206**
To 100mm. Beaks distinctly in front of mid-line. Subovate approaching oblong but post-dorsal margin moderately sloping, posterior margin rounded, anterior narrower and rounded. Initial sculpture almost smooth, soon developing widely spaced but weak, incised, concentric lines giving rise to broad, weakly elevated, flat ridges most prominent anteriorly. Externally with dense radial rays and lines usually in shades of orange but also brown on a pale background; inside white, margins orange. **Habitat:** offshore. **Distribution:** NWG, SEG, GO, Mas, SO.

Callista florida (Lamarck, 1818): **1207**
To 45mm. Very similar to *C. erycina* but less oblong, post-dorsal slope steeper and posterior margin more acute. Sculpture of distinct, regularly spaced, flat, concentric ridges, most prominent anteriorly but occasionally absent from the posterior slope. Externally variable, cream to beige, overlain by sparse

Callista florida **1207**

or dense radial rays, blotches, lines, or interrupted rays in shades of brown or mauve-brown; inside white, usually with some violet colouring. **Habitat:** in muddy sand and gravel, offshore. **Distribution:** all.

Lioconcha castrensis (Linnaeus, 1758): **1208**
To 40mm. Heavy. Beaks central. Subovate, almost as high as long, rounded with prominent umbones, anterior slightly narrowed. Sculpture smooth with growth lines. Pallial sinus very shallow. White with bold brown tents and zigzag lines. Although widely found throughout the Indian Ocean it is recorded only from Tiwi, Oman, in the eastern Arabian region. **Habitat:** in sand, offshore. **Distribution:** GO

Lioconcha castrensis **1208**

SEASHELLS OF EASTERN ARABIA 269

Superfamily: **VENEROIDEA**

Lioconcha ornata 1209

Pitar hebraea var. 1212

Pitar yerburyi 1210

Pitar tumida 1214

Pitar hebraea 1211

Pitar pudicissma 1213

Lioconcha ornata (Dillwyn, 1817): **1209**
To 30mm. Beaks almost central. Subovate, distinctly longer than high, posterior dorsal margin gently sloping, lateral margins narrowly rounded. Sculpture smooth with weak growth lines. Pallial sinus very shallow. White with brown or beige zigzag lines. **Habitat:** in sand, offshore. **Distribution:** NWG, SEG, GO, SO.

Pitar yerburyi (Smith, 1891): **1210**
To 30mm. Thin. Beaks just in front of mid-line. Subovate, umbones prominent, post-dorsal margin sloping moderately, posterior and anterior margins equally rounded. Lunule large, narrowly cordate, not depressed, incised line weak. Sculpture of concentric threads and growth lines, irregularly pitted, often encrusted with sediment particles. Pallial sinus deep, rounded, ascending but not quite reaching below cardinal teeth. White. **Habitat:** in muddy sand, offshore. **Distribution:** NWG, GO.

Pitar hebraea (Lamarck, 1818): **1211**
To 30mm. Beaks almost central. Subovate, weakly umbonate, post-dorsal slope gentle, anterior narrowly rounded and anterior slope gentle. Lunule slightly impressed. Sculpture smooth, slightly shiny. Pallial sinus less ascending and narrower than in **P. yerburyi** (see above). White with fulvous to brown tents. **Habitat:** offshore. **Distribution:** NWG, SEG, GO.

Pitar hebraea (Lamarck, 1818) var: **1212**
To 20mm. Beaks slightly in front of mid-line. Subovate, anterior shortened often slightly distorted, posterior margin narrower than anterior. Weak sculpture of closely spaced lines and narrow ridges, especially where distorted, pitting sparse. Pallial sinus short, almost horizontal. White with sparse fulvous spots and broken zigzags, some in radial rows. The shells from Masirah may be a separate species but the slight distortion may account for all the differences. More specimens will be needed to solve the problem. **Habitat:** offshore. **Distribution:** Mas.

Pitar pudicissma (Smith, 1894): **1213**
To 15mm. Beaks just in front of mid-line. Subovate, longer than high, post-dorsal margin sloping gently, posterior margin subtruncate, anterior narrower, rounded. Escutcheon indistinct, ligament deeply sunken. Lunule not depressed and defined by a weak incised line. Sculpture of weak, raised, concentric threads and growth lines, pitting lacking. Pallial sinus shallow, rounded, anterior line almost vertical. Dirty white. **Habitat:** in mud, deep water. **Distribution:** GO.

270 SEASHELLS OF EASTERN ARABIA

Superfamily: ***VENEROIDEA***

Sunetta donacina **1216**

Sunetta effossa **1215**

Dosinia alta **1218**

Dosinia contracta **1220**

Dosinia ceylonica **1219**

Cyclosunetta contempta **1217**

Pitar tumida (Sowerby, 1895): **1214**
To 18mm. Solid. Beaks just in front of mid-line. Subovate-trigonal, post-dorsal margin sloping steeply, posterior margin subacute, post-ventral margin often sinuous. Sculpture of weak concentric threads and growth lines, pitting present, often encrusted with sediment particles. Pallial sinus large, rounded, and extending to mid-line. White. Periostracum thin, yellowish with agglutinated sand grains. **Habitat:** offshore. **Distribution:** GO, Mas, SO.

Subfamily SUNETTINAE
Sunetta effossa (Hanley, 1843): **1215**
To 45mm. Beaks slightly in front of mid-line. Subovate, anterior roundly pointed, posterior broadly rounded or subtruncate. Sculpture of up to 25 prominent, concentric, smooth ridges; large shells may be smooth at margins. Escutcheon deeply excavated. Beige to lilac with faint, brown, zigzag lines and blotches. **Habitat:** in sand, offshore. **Distribution:** NWG, SEG, GO, Mas, SO.

Sunetta donacina (Gmelin, 1791): **1216**
To 45mm. Form as ***S. effossa***. Sculpture with concentric, smooth ridges on umbo only or on dorsal margins only, or completely smooth. Beige to lilac with faint, brown, zigzag lines and blotches or uniformly beige. **Habitat:** in sand, offshore. **Distribution:** SEG.

Cyclosunetta contempta (Smith, 1891): **1217**
To 20mm. Beaks central. Subovate, anterior broadly rounded, posterior subtruncate. Escutcheon shallow, very long almost to the posterior margin. Sculpture smooth. Pallial sinus reaching the umbonal line. Inner margin finely denticulate. Cream tinged lilac, often with zigzag lines; inside white, some flushed violet. **Habitat:** offshore. **Distribution:** Mas, SO.

Subfamily DOSINIINAE
Dosinia alta (Dunker, 1848): **1218**
To 25mm. Compressed. Orbicular, height equal to or slightly greater than length, escutcheon margin long and sloping very steeply. Escutcheon shallow, lunule short, sunken and narrowly cordate. Sculpture of dense, low, concentric ridges, virtually smooth medially but expressed as weakly raised ridges at the escutcheon and anterior margins. Pallial sinus very deep, narrow, almost reaching anterior adductor. White, some tinged pink to yellow. **Habitat:** in sand, lower shore and below. **Distribution:** all.

Dosinia ceylonica Dunker, 1865: **1219**
To 35mm. Heavy. Orbicular, height equal to or just greater than length, escutcheon margin long and sloping steeply. Escutcheon shallow. Lunule large, not sunken and marked only by a weak incised line. Sculpture of narrow, sharp, raised, concentric ridges, initially well spaced with fine lines between but later becoming more crowded. Pallial sinus narrow, steeply ascending but not extending beyond line of beak. White. **Habitat:** offshore. **Distribution:** NWG, SEG, GO, Mas.

Dosinia contracta (Philippi, 1844): **1220**
To 15mm. Orbicular, height equal to or a little greater than length, escutcheon margin sloping steeply. Escutcheon shallow, ligament deeply inset. Lunule cordate, impressed. Sculpture of dense, evenly spaced, recurved lamellae, later becoming more irregular; in small shells some lamellae project as spines along the escutcheon margin but these are often rapidly worn. Pallial sinus just reaching to below middle cardinal. White, some tinged pink. **Habitat:** mud flats, lower shore and below. **Distribution:** NWG, SEG, GO.

Superfamily: **VENEROIDEA**

Dosinia erythraea 1221

Dosinia histrio 1222

Dosinia tumida 1223

Marcia marmorata 1224

Dosinia erythraea Römer, 1860: **1221**
To 50mm. Solid. Orbicular and more or less circular, escutcheon margin indistinct and hardly sloping. Escutcheon absent. Lunule short, cordate, impressed. Concentric sculpture of dense, narrow, slightly sharp, raised ridges giving a rough feel but not as distinct as in *D. histrio* (see below). Pallial sinus extending to below the median cardinal. White with brownish radial rays or disrupted blotches. **Habitat:** in sand, offshore. **Distribution:** NWG, SEG, GO, Mas.

Dosinia histrio (Gmelin, 1791): **1222**
To 35mm. Heavy. Orbicular, height equal to length, escutcheon margin sloping steeply. Escutcheon shallow. Lunule small, cordate and impressed. Concentric sculpture of sharp, low lamellae giving a distinctly rough feel. Pallial sinus short less than half the distance between the adductor scars. White with brown rays or tents, inside white, occasionally rose or violet. **Habitat:** in sand, offshore. **Distribution:** NWG, SEG.

Dosinia tumida (Gray, 1838): **1223**
To 50mm. Solid. Orbicular, height equal to length. Escutcheon margin sloping moderately, posterior margin straightening. Escutcheon narrow, a double groove, ligament margins raised to form a low crest. Lunule short, cordate, very deeply incised into hinge plate. Concentric sculpture of low, dense, narrow ridges and almost smooth to the touch; some with marginal projections along lunule and escutcheon. Pallial sinus deep but not beyond line of middle cardinal. Uniformly white, some suffused pink over umbones. The shape of this species is rather variable. In the Arabian Gulf the sculpture is weak, except for a series of small projections along the escutcheon margin, and the shell is often tinged pink; these could be called *D. exasperata* Philippi, 1847. Elsewhere the shells are white and the sculpture more even overall, typical *D. tumida* (Gray, 1838). Some shells may be exceptionally heavy and rather angulate, *D. labiosa* Römer, 1862. **Habitat:** in sand, offshore. **Distribution:** all.

Subfamily TAPETINAE

Marcia marmorata (Lamarck, 1818): **1224**
To 55mm. Beaks well in front of to just in front of mid-line. Subovate to roundly trigonal, much to only a little longer than high, anterior rounded, posterior becoming subtruncate, lunule poorly defined, narrowly cordate. Concentric sculpture of numerous narrow ridges often anastomosing medially. Pallial sinus almost reaching under hinge, horizontal, roundly subacute. Beige to cream with a large variety of darker even black, rays, blotches, zigzags or spots. **Habitat:** sand flats, intertidal. **Distribution:** GO, Mas, SO.

Marcia flammea (Gmelin, 1791): **1225**
To 35mm. Very close to trigonal *M. marmorata* but more inflated, with the posterior slope becoming flattened or sulcate and the posterior-ventral margin slightly sinuous. Lunule well defined, depressed and cordate. Escutcheon depressed. White to beige with very weak patterns.
Marcia marmorata and *M. flammea* may be ecological variants of a single species. The salinity and temperature of the Arabian Gulf may affect the growth rate and consequently the form of the shell. **Habitat:** sand flats, intertidal. **Distribution:** NG, NWG, SEG.

Marcia opima (Gmelin, 1791): **1226**
To 45mm. Beaks well in front of mid-line. Subovate, longer than high, anterior rounded, posterior becoming subtruncate, lunule narrowly cordate. Smooth with growth lines only. Pallial sinus reaching under hinge, horizontal, roundly subacute. Beige with darker

272 SEASHELLS OF EASTERN ARABIA

Superfamily: **VENEROIDEA**

zigzag patterned lines. **Habitat:** intertidal mud flats. **Distribution:** GO.

Paphia undulata (Born, 1778): **1227**
To 75mm. Beaks well in front of mid-line. Elongate-subovate, much longer than high, anterior margin a little narrower, slightly pouting. Lunule narrow. Smooth with growth lines only. Pallial sinus shallow, ascending, rectangular. Beige with dense darker zigzag patterned lines, some uniformly brown. **Habitat:** offshore. **Distribution:** NWG, GO, Mas.

Protapes cor (Sowerby, 1853): **1228**
To 80mm. Beaks well in front of mid-line. Globose. Suborbicular, almost as high as long, anterior margin rather narrow, pouting. Lunule not defined and lunule margin horizontal or nearly so. Concentric sculpture initially of very fine lines developing into increasingly coarse but narrow ridges; faint radial lines apparent in most shells. Pallial sinus short, slightly ascending, apex subacute. Off-white to dirty beige, lacking radial colour markings. **Habitat:** in mud, offshore. **Distribution:** NWG, SEG.

Protapes sinuosa (Lamarck, 1818): **1229**
To 70mm. Beaks slightly in front of mid-line. Somewhat inflated. Subovate, much longer than high, anterior margin rather narrow, strongly pouting, lunule margin recurving, posterior slope becoming sulcate and posterior margin becoming truncated and sinuous. Concentric sculpture initially of very fine ridges developing into coarse ridges, occasionally anastomosing on posterior slope. Pallial sinus deep, ascending steeply. Beige to orange-brown with darker zigzag lines, spots and faint radial stripes. **Habitat:** offshore. **Distribution:** NWG, SEG, GO, Mas.

Paphia undulata **1227**

Protapes cor **1228**

Marcia opima **1226**

Protapes sinuosa **1229**

Marcia flammea **1225**

SEASHELLS OF EASTERN ARABIA

Superfamily: **VENEROIDEA**

***Protapes* n. sp: 1230**
To 45mm. Beaks well in front of midline. Subglobose. Subovate, longer than high; anterior margin rather narrow, pouting; lunule margin sloping; posterior slope becoming flat or even sulcate and posterior margin becoming truncated. Concentric sculpture initially of very fine lines developing into fine ridges, faint radial lines apparent in many shells. Pallial sinus deep, ascending but not steeply. Beige to mauve brown with darker radial stripes, lunule and escutcheon paler. **Habitat:** in sandy mud flats, intertidal. **Distribution:** NWG, SEG, GO, Mas.

***Tapes sulcarius* (Lamarck, 1818): 1231**
To 70mm. Beaks well in front of midline. Oblong, post-dorsal margin gently sloping, posterior rounded to subtruncate, anterior narrower and rounded. Concentric sculpture only, of low ridges becoming flatter and broader over the posterior slope. Pallial sinus short, horizontal. Colour variable, shades of beige, rust and brown, tents, zigzags or radially striped; inside white or suffused with yellow. **Habitat:** in muddy sand, lower shore and below. **Distribution:** NWG, SEG, GO, Mas.

***Tapes deshayesi* (Sowerby, 1852): 1232**
To 40mm. Very similar to *T. sulcarius* (see above) but anterior shorter and narrower. Concentric ridges uneven with raised lines. Pale with darker markings in shades of brown, ranging from sparse spots only to radial blotched rays to zigzag lines; inside white, tinged purple over posterior and around hinge. **Habitat:** offshore. **Distribution:** SEG, GO, Mas.

***Tapes bruguierei* (Hanley, 1845): 1233**
To 40mm. Beaks well in front of midline. Oblong, post-dorsal margin sloping gently; posterior rather high, roundly truncated, anterior narrower, rounded. Sculpture of numerous radial riblets interrupted only by growth lines. Uniform cream to beige or with weak rays, umbonal area may be rosy; inside white to deep rose. **Habitat:** offshore. **Distribution:** GO.

***Venerupis rugosa* Sowerby, 1854: 1234**
To 40mm. Beaks well in front of midline. Irregularly oblong to ovate, posterior expanded, rounded or truncate. Young shells (umbonal area) typically tapetid in outline. Umbonal area smooth, then developing irregularly corrugate, concentric lamellae, these becoming foliaceous posteriorly. Beige except for umbonal area which is often purple to fulvous and may be

Tapes sulcarius 1231

Tapes bruguierei 1233

Protapes n. sp. 1230

Tapes deshayesi 1232

274 SEASHELLS OF EASTERN ARABIA

Superfamily: **VENEROIDEA**

Venerupis rugosa **1234**

Irus macrophylla **1235**

patterned. **Habitat:** in crevices in rocks and corals, intertidal and below. **Distribution:** all.

Irus macrophylla (Deshayes, 1853): **1235**
To 35mm. Beaks well in front of midline. Outline variable, oblong to quadrate, posterior usually subtruncate. Sculpture of widely spaced, thin, concentric lamellae increasing in expression posteriorly; with fine radial striae overall. Pallial sinus short, acutely pointed. White often with internal patch of brown posteriorly. **Habitat:** in crevices of rocks and corals, intertidal and below. **Distribution:** all.

Subfamily CLEMENTIINAE
Clementia papyracea (Gray, 1825): **1236**
To 60mm. Fragile. Beaks well in front of mid-line. Subovate, post-dorsal margin long, lateral margins rounded. Lunule absent. Escutcheon narrow, cleft. Sculpture of broad concentric undulations and growth lines. Large hollow in front of cardinal teeth. Pallial sinus acute and reaching almost to below beaks. White. **Habitat:** in mud, shallow water. **Distribution:** NWG, GO, Mas.

Clementia papyracea **1236**

Clementia asiatica (Melvill, 1899): **1237**
To 12mm. Very thin, fragile. Beaks just in front of mid-line. Roundly subovate, anterior a little more expanded than posterior. Escutcheon absent. Lunule very small. RV with 3 cardinals, anterior 2 close and vertical, posterior shelf-like, almost horizontal. Smooth, with growth lines only. Pallial sinus large, slightly ascending, subacute; pallial line well within margin of shell. White. **Habitat:** offshore. **Distribution:** GO.

Clementia asiatica **1237**

"*Petricola*" *ponsonbyi* (Sowerby, 1892): **1238**
To 16mm. Beaks in front of mid-line. Roughly circular to elliptical but rather distorted. Hinge compressed, 3 cardinals in each valve. Sculpture of numerous low, radial, raised lines developing into flat-topped riblets interrupted by irregular concentric ridges; umbonal sculpture of dendritic striae as in *Petricola*. Pallial sinus very large, extending beyond hinge teeth and with an extensive confluent pallial line. White, umbones rosy or fulvous. This species, known previously only from South Africa, has always been

"*Petricola*" *ponsonbyi* **1238**

placed in the genus *Petricola*. It has three cardinal teeth in each valve and therefore is a venerid and may be related to *Kyrina*. **Habitat:** probably nestling in crevices, offshore. **Distribution:** Mas, SO.

SEASHELLS OF EASTERN ARABIA 275

Superfamily: **VENEROIDEA, MYOIDEA**

Kyrina kyrina 1239

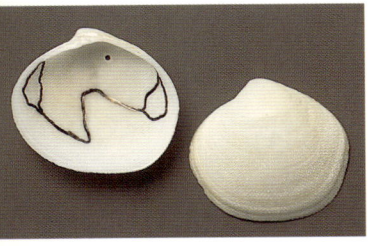

Mysia elegans 1240

Petricola n. sp. 1243

Petricola lapicida 1242

Petricola hemprichi 1241

Asaphinoides madreporicus 1244

Kyrina kyrina Jousseaume, 1894: **1239**
To 15mm. Very thin and fragile, semitransparent. Beaks distinctly in front of mid-line. Subovate, anterior broad and rounded, posterior much narrower, tapering. No escutcheon or lunule. Sculpture of microscopic, divaricate threads. Hinge with 3 delicate cardinals in each valve. Ligament external, on a shallow nymph. Pallial sinus very large, apex rounded and extending to below hinge, pallial line well inside margin and partly confluent. White. **Habitat:** Sand flats. **Distribution:** SEG.

Family PETRICOLIDAE
Mysia elegans (H. Adams, 1870): **1240**
To 16mm. Beaks just in front of mid-line. Roughly circular with prominent umbones, posterior and ventral margins often straightening. Lunule very large, defined by an incised line. Escutcheon not sunken. Sculpture micro-reticulate, forming dense diamond-shaped pits. Pallial sinus deep, ascending, acute. White. **Habitat:** in soft sediments, offshore. **Distribution:** GO.

Petricola hemprichi Issel, 1869: **1241**
To 25mm. Beaks in front of mid-line. Subovate expanded anteriorly and narrow posteriorly but may be distorted. Radial sculpture of numerous raised lines and riblets anteriorly, these becoming sharp ribs on the posterior area. Umbonal area and anterior area initially with dendritic raised lines. White. **Habitat:** Burrowing into limestone and corals, mid shore to sublittoral. **Distribution:** NWG, SEG, GO, Mas.

Petricola lapicida (Gmelin, 1791): **1242**
To 40mm. Beaks very close to the anterior end. Roundly quadrate, umbones prominent. Lunule and escutcheon lacking. Sculpture of a few (about 10) prominent radial riblets on the posterior slope with delicate divaricate striae over the remainder. Pallial sinus very large. White. **Habitat:** boring into soft limestone and corals, subtidal. **Distribution:** NWG.

***Petricola* n. sp**: **1243**
To 17mm. Thin, rather chalky. Beaks close to the anterior end. Roundly rectangular, much longer than high. Sculpture of 25-35 low, well spaced, weakly rugose, radial riblets, most developed over the posterior slope. Hinge with apparently only two cardinal teeth in each valve. Ligament deeply sunken on a strong nymph. Pallial sinus relatively narrow, apex rounded. White. **Habitat:** burrowing in calcareous sandstone. **Distribution:** GO.

Asaphinoides madreporicus (Jousseaume, 1895): **1244**
To 25mm. Outline irregular, subcircular to ovate, valves often twisted and overlapping posteriorly or one valve completely distorted, like an oyster, the other valve normal. Sculpture of well spaced, sharp, radial riblets and a few concentric lamellae. White. **Habitat:** in crevices, subtidal. **Distribution:** Mas, SO.

Superfamily MYOIDEA
Two distinct families, Myidae and Corbulidae, comprise the Myoidea. Externally the shells are dissimilar but both families have a projecting, spoon-shaped chondrophore which supports the internal ligament. The Myidae have elliptical to oval shells, often truncated and mostly with a posterior gape. The siphons are always fused, of varying lengths, but can be very long. Most species burrow into soft sediments but the distorted shells of *Tugonia* and *Sphenia* suggest a nestling habit. The shell sculpture is weak and, if present, of radial lines. The Corbulidae are small with strongly inequivalve shells which do not gape. The pallial sinus is shallow and the siphons correspondingly short. They are shallow burrowers in a variety of sediments, often employing a slender byssus. The shell outline is ovate, rounded or angulate, and often keeled. Sculpture, if present, is concentric and may be heavily ridged.

Family MYIDAE
Tugonia nobilis A. Adams, 1856: **1245**
To 45mm. Inflated. Beaks well in front of mid-line. Widely gaping posteriorly. Transversely oval, often distorted, with a very short, truncate

Superfamily: **MYOIDEA, HIATELLOIDEA**

rostrum. Median sculpture of fine radial riblets, anterior has concentric lines only, rostrum smooth. Chondrophore spoon-shaped. Pallial sinus restricted to the rostrum. White. **Habitat:** probably nestling among rocks, sublittoral. **Distribution:** Mas.

Tugonella decurtata (A. Adams, 1851): **1246**
To 26mm. Fragile. Beaks just behind mid-line. Gaping. Oval-elliptical; anterior margin expanded, rounded; posterior rostrum very short. Sculpture of dense, very fine, radial striae fading anteriorly and absent from the rostrum. Chondrophore shelf-like. Pallial sinus restricted to the rostrum. White. **Habitat:** in sandy mud or mud, offshore. **Distribution:** NWG, GO.

Cryptomya elliptica (A. Adams, 1851): **1247**
To 16mm. Compressed. Gape narrow. Beaks central. Elliptical, posterior slightly angulate. Sculpture of fine radial lines on posterior half of the shell, pustules on anterior. Chondrophore shallow, peg-like anterior tooth in LV. Pallial sinus absent. White. **Habitat:** in sandy mud in sheltered waters, subtidal. **Distribution:** SEG, GO.

Cryptomya n. sp: **1248**
As *C. elliptica* but thinner and lacking any sculpture other than growth lines. Anterior tooth no more than a nodule. White. **Habitat:** in sandy mud in shallow water. **Distribution:** NWG.

Sphenia rueppellii A. Adams, 1851: **1249**
To 12mm. Irregularly elliptical with narrower, subrostrate posterior. Sculpture of growth lines only. Chondrophore variable, no anterior tooth. White with thick periostracum on posterior. **Habitat:** nestling in rocky crevices, lower shore. **Distribution:** SO.

Family CORBULIDAE
Corbula erythraeensis H. Adams, 1871: **1250**
To 8mm. Much smaller LV fitting well inside RV. Inflated. Irregularly subspherical, slightly longer than high; RV sculpture stronger than LV, of rounded concentric ridges but umbonal area smooth. RV with a single cardinal peg, LV with a single dorsally projecting tooth, a narrow projecting chondrophore and a prominent socket. White. **Habitat:** in sand, offshore. **Distribution:** NG, GO, SO.

Corbula taitensis Lamarck, 1818: **1251**
To 14mm. LV only slightly smaller than RV. Beaks slightly in front of mid-line. Transversely subquadrate, anterior rounded, posterior obliquely truncate, posterior ventral junction acute; keel sharp. RV with strong concentric ridges, often stepped, LV ridges narrower. Shades of pink, some with paler radial rays. **Habitat:** in sand, offshore. **Distribution:** NWG, SEG, GO, Mas.

Corbula sulculosa H. Adams, 1870: **1252**
To 8 mm. Form very close to *C. taitensis* but keel less acute. Sculpture much weaker; initially smooth, then developing weak concentric ridges, both valves with faint radial striae, most visible on LV. Uniform white, some tinged pink or yellow. **Habitat:** in muddy sand, lower shore and below. **Distribution:** NWG, SEG, GO, Mas.

Corbula subquadrata Melvill & Standen, 1907: **1253** NOT ILLUSTRATED
To 5mm. Form as *C. sulculosa* but posterior area narrower and posterior ventral junction slightly hooked. Sculpture of prominent concentric ridges developing early; by comparison *C. sulculosa* of equal size will be almost smooth. Radial striae present on both valves. White. **Habitat:** in mud, deep water. **Distribution:** NG, GO.

Superfamily HIATELLOIDEA
The Hiatelloidea are a small group most commonly represented in temperate or cold waters. Only a single genus, *Hiatella*, is recorded and is recognised by the irregular, rectangular shell which has weak keels and eroded hinge. In the eastern Arabian region it is found only within the influence of the Oman upwelling.

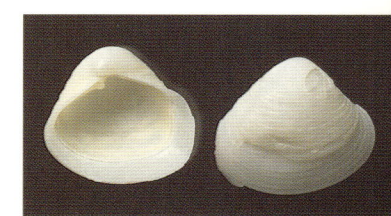

Cryptomya elliptica **1247**

Sphenia rueppellii **1249**

Tugonella decurtata **1246**

Tugonia nobilis **1245**

Corbula erythraeensis **1250**

Corbula taitensis **1251**

Corbula sulculosa **1252**

Cryptomya n. sp. **1248**

SEASHELLS OF EASTERN ARABIA 277

Superfamily: **HIATELLOIDEA, GASTROCHAENOIDEA, PHOLADOIDEA**

Hiatella flaccida (Gould, 1861): **1254**
To 40mm. Beaks in front of mid-line. Transversely quadrate with variable posterior keels; anterior narrow, rounded, posterior broad, subtruncate. Ligament large, extending along dorsal margin. Hinge teeth eroded, single cardinals in each valve usually visible. Sculpture of growth lines and irregular ridges. White with a straw-coloured periostracum **Habitat:** nestling, large shells in sea-squirt tests along with *Musculus cumingianus*, small shells in crevices, intertidal and sublittoral. **Distribution:** GO, Mas, SO.

Superfamily GASTROCHAENOIDEA

Flask shells take their name from the shape of the cavity that they excavate in calcareous rocks and corals. Others build a case or tube around themselves. The end of the tube or flask is very narrow and forms a figure-of-eight outline through which the siphons open. Boring is believed to be mechanical and chemical. The shells are elongate, with a large ventral gape, and lack hinge teeth.

Gastrochaena gigantea Deshayes, 1830: **1255**
To 40mm. Beaks very close to the anterior end. Transversely oval, dorsal margin straight and projects anterior to the beaks. Ventral edge smoothly curved, pedal gape occupies two-thirds of the length of the valves. White. **Habitat:** in dead coral, lower shore and below. **Distribution:** NWG, SEG, GO, SO.

Gastrochaena inaequistriata Lamy, 1923: **1256**
To 35mm. Beaks terminal. Transversely oval, dorsal margin scarcely protruding beyond beaks, ventral margin with an angular curve. Gape occupying only half the length of the valves. Sculpture of weak growth lines, some developed as low sharp ridges on the anterior-ventral part of the shell. **Habitat:** in dead coral, lower shore and below. **Distribution:** GO.

Gastrochaena dentifera Dufo, 1840: **1257**
To 20mm. Beaks terminal. Narrowly subrectangular, dorsal margin slightly curved and projecting just beyond the beaks, ventral edge almost straight. Gape elliptical, occupying most of the length of the ventral side. Sculpture of somewhat irregular growth lines which project as ridges on the anterior-ventral area. White. **Habitat:** in dead corals and shells, intertidal and below. **Distribution:** NWG, SEG, GO.

Spengleria plicatilis (Deshayes, 1855): **1258**
To 35mm. Beaks towards the anterior. Transversely rectangular, dorsal margin long, abruptly meeting subtruncate posterior margin. Gape narrow, very long. Oblique furrow prominent. Anterior with growth lines, posterior with a heavy, wrinkled periostracum and calcareous spicules. Yellowish. **Habitat:** in dead coral, intertidal and below. **Distribution:** GO.

Cucurbitula cymbium (Spengler, 1783): **1259**
To 12mm. Beaks almost terminal. Narrowly rectangular, dorsal and ventral margins almost parallel, posterior margin truncate. Gape wide, occupying most of the length of the valves. Case flask shaped, apparently segmented exteriorly but not internally. White. **Habitat:** attached to dead shells, usually bivalves, offshore. **Distribution:** SEG, GO.

Superfamily PHOLADOIDEA

The Pholadoidea are exclusively adapted for boring into hard substrates such as rock, coral, other shells, peat and wood. The shells and soft parts are greatly modified for this function and exhibit a number of unique features. Two quite distinct families, the Pholadidae and the Teredinidae, are represented in eastern Arabia and are best considered separately.

Family PHOLADIDAE

The Pholadidae are typically subovate to suboblong with a wide pedal gape and a large posterior gape. The shells are frequently divided into two or more areas, the anterior slope usually having a sculpture of toothed concentric lamellae which act as a rasp when making boring movements. The umbonal area bears a reflected ridge, the umbonal reflexion, to which the anterior adductor muscle is attached. The hinge is toothless and the small ligament is supported by a tiny chondrophore. Also

Gastrochaena gigantea **1255**

Hiatella flaccida **1254**

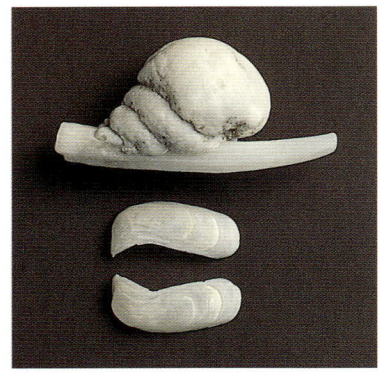

Cucurbitula cymbium **1259**

Spengleria plicatilis **1258**

Gastrochaena dentifera **1257**

Gastrochaena inaequistriata **1256**

Superfamily: **PHOLADOIDEA**

Pholas dactylus 1260

Barnea manillensis 1261

Aspidopholas obtecta 1262

Martesia striata 1263

projecting from the umbonal cavity is a long calcareous process called the apophysis to which the pedal retractors are attached. Along the dorsal area there are accessory shell plates which have the following terminology, dependent on their position in relation to the beaks. The protoplax which is anterior to the beaks lies over and protects the anterior adductor muscle. The mesoplax lies over the beaks themselves and the metaplax posterior to them covers the posterior-dorsal gape. A fourth plate, the hypoplax, may be present covering the posterior-ventral gape. One or any combination of these plates may be present and they are important systematic characters. In some the pedal gape may become completely closed in the adult stages by a calcareous, smooth layer known as a callum. A few genera also possess calcareous tubular lamellae around the siphons but this fragile siphonoplax is often eroded or broken. Shells of this type are typical of most of the Pholadinae almost all of which bore into rock, coral, peat or hard clay. The only exception is *Martesia* which is wood boring.

Pholas dactylus Linnaeus, 1758: **1260**
To 60mm. Beaks close to anterior end. Elongate, tapering posteriorly, anterior area beaked. Anterior pedal gape large, never closed by a callum. Umbonal reflexion septate. Three dorsal accessory plates all calcareous. Sculpture of concentric ridges crossed by radial riblets or threads, these elements combining to form fine teeth over the anterior area. White. Known only from single valves found at Ras Duqm and Masirah. It is equally scarce in the Red Sea. The Arabian shells cannot be distinguished from Mediterranean and Atlantic examples. **Habitat:** in soft rock and peaty sand, lower shore and below. **Distribution:** Mas, SO.

Barnea manillensis (Philippi, 1847): **1261**
To 40mm. Equivalve. Beaks well in front of mid-line. Elongate-subovate, broadest across umbones and tapering posteriorly. Pedal gape relatively small, always open. Dorsal reflexion simple. Sculpture most developed on anterior slope, of widely spaced, concentric, raised lines with radial rows of spines emerging from them, posterior smooth or pustulose. Accessory plates of Oman specimens unknown. White. **Habitat:** in soft rock, lower shore and below. **Distribution:** GO.

Aspidopholas obtecta (Sowerby, 1849): **1262**
To 25mm. Beaks well in front of mid-line. Subovate, broadest across the umbones, tapering posteriorly and irregular here, bulbous anteriorly. Pedal gape very large in juveniles, partially closed by callum in adults. Shell with median furrow, anterior sculpture of serrated concentric ridges, posterior of concentric lines, these fading on posterior calcareous extension (siphonoplax). Mesoplax large, eventually enclosing anterior. **Habitat:** in sandstone, lower shore and below. **Distribution:** NWG, GO, SO.

Martesia striata (Linnaeus, 1758): **1263**
To 20mm. Adults elongate, gape closed by callum, juveniles almost globular, widely gaping anteriorly, beaked, similar to a *Teredo*. Shell divided by a weak umbonal-ventral groove, anterior area with closely spaced, serrated, concentric ridges, posterior area with widely spaced smooth concentric ridges. Mesoplax subcircular without regular sculpture. **Habitat:** in submerged wood. **Distribution:** NWG, GO.

Family TEREDINIDAE
The shells of the Teredinidae are globose but otherwise resemble those of the Pholadidae, except that the anterior is rarely or weakly serrated and there are no accessory plates. The animal is, however, much larger than the shell, greatly elongated, worm like, with a pair of short siphons at the end. Close to the siphons and inserted at their base is a pair of pallets which serve to close the burrow. The pallets are partly periostracal, partly calcareous and may be simple or segmented. They are the prime diagnostic characters as the shells are all very similar and rather variable.

Readers interested in the precise identification of ship-worms should refer to the studies of Dr. Ruth Turner as the details are beyond the scope of this book. Particularly useful is her *Identification of Marine Wood-boring Molluscs* (Turner, 1971).

Subclass ANOMALODESMATA
Superfamily PANDOROIDEA

The superfamily Pandoroidea contains seven families with living species. These families show little immediate similarities in shell form except for the absence of true hinge teeth and for the position of the ligament which, partly anyway, lies on an internal resilium and is often further reinforced by a calcareous lithodesma.

The Pandoridae are a small group of bivalves, typified by the unequal valves in which the dorsal border of one overlaps the other. They are part of the Anomalodesmata and thus lack a heterodont dentition which is substituted by laminar buttresses and an internal resilium, often reinforced by a lithodesma. The shell itself is predominantly nacreous and rather thin. Little is known of their habitats but, typically they live in tranquil muddy situations, burrowing shallowly and leading a rather sedentary existence.

The Laternulidae or lantern shells are typically very thin shelled and oblong in outline. They lack true hinge teeth and the ligament lies on a projecting internal chondrophore supported by two buttresses. There is a distinct split in the shell from the umbo ventrally. Lantern shells burrow deeply in mud or sandy mud from the intertidal zone downwards.

The Periplomatidae are also thin shelled but more ovate in outline with a short, truncated rostrum. Like the Laternulidae the umbones are split. They live in mud and muddy sand, usually offshore.

The Thraciidae have thin, smooth, non-nacreous shells which are inequivalve, the LV fitting into the RV. Hinge teeth are lacking. There is an internal chondrophore directed obliquely backwards but the ligament is mostly external. The sculpture is smooth and often there is a pustulose micro-sculpture, especially posteriorly. They are poorly known in tropical seas but in temperate areas they inhabit muddy sand or mud from the intertidal zone to considerable depths. They burrow but are not mobile like the venerids or tellinids.

Family PANDORIDAE

Pandora ceylanica Sowerby, 1835: **1264**
To 20mm. Very compressed, only 2 - 3 mm deep, almost equivalve. Beaks well in front of mid-line. Subovate, with a short posterior rostrum; posterior-dorsal margin concave; ventral margin equally convex. Sculpture smooth or with weak growth lines; RV with slight posterior-dorsal sulcus corresponding to rostrum. White. **Habitat:** in muddy sand and gravel, offshore. **Distribution:** SEG, GO, Mas.

Pandora flexuosa Sowerby, 1820: **1265**
To 25mm. Strongly inequivalve, LV deeply concave with a strong posterior flexure, RV flat. Beaks well in front of mid-line. Narrowly elongate, posterior rostrate, truncate; posterior dorsal margin concave; anterior margins rounded. Feeble sculpture of concentric lines only. White. **Habitat:** in muddy sand and gravel, offshore. **Distribution:** SEG, GO, Mas.

Family LATERNULIDAE

Laternula anatina (Linnaeus, 1758): **1266**
To 50mm. Elongate-subelliptical, posterior shortly rostrate and upturned. Umbones low with a transverse external slit. Sculpture of weak concentric undulations and lines, covered by evenly spaced, small pustules, these usually worn from the umbones and absent from the rostrum. Hinge without teeth, ligament on a spoon-shaped chondrophore; supporting buttress for chondrophore very thin. White. **Habitat:** in mud and muddy sand, intertidal flats and offshore. **Distribution:** SEG, GO, Mas.

Laternula erythraensis Morris & Morris, 1993: **1267**
To 30mm. Narrowly elongate-subelliptical, posterior rostrate and rarely upturned. Umbones low with a short, transverse, external slit. Sculpture of weak, concentric undulations and lines, rostrum not demarcated, pustules minute and rarely preserved. Hinge without teeth, ligament on a spoon-shaped chondrophore; supporting buttress for chondrophore a low ridge angled more obliquely than in *L. anatina*. White. **Habitat:** in mud and muddy sand, intertidal flats. **Distribution:** SEG, GO.

Pandora flexuosa 1265

Laternula anatina 1266

Pandora ceylanica 1264

Laternula erythraensis 1267

Superfamily: **PANDOROIDEA, CLAVAGELLOIDEA, POROMYOIDEA**

Thracia salsettensis 1270

Periploma indicum 1268

Thracia adenensis 1269

Cardiomya alcocki 1272

Cuspidaria approximata 1273

Family PERIPLOMATIDAE
Periploma indicum Melvill, 1898: **1268**
To 20mm. Fragile. RV overlapping flatter LV, beaks well behind mid-line. Subovate; anterior broadly rounded; posterior narrowly truncate and demarcated by a fine ridge in LV and a double flexure in RV. Sculpture of incremental lines only. Ligament on a fine spoon-shaped chondrophore which appears incomplete on the anterior side, the resulting gap being filled with a lithodesma which is lost in isolated valves; umbonal slit obscure. White. **Habitat:** in muddy substrates, offshore. **Distribution:** GO.

Family THRACIIDAE
Thracia adenensis Melvill, 1901: **1269**
To 30mm. Fragile. Inequivalve, LV a little flatter and overlapped by RV. Beaks well behind mid-line. Subelliptical, posterior broadly truncate and set off by a distinct angled ridge. Sculpture of incremental lines with a microsculpture of dense pustules most visible on posterior area. Ligament external and internal, the latter on a small, shallow, oblique resilifer. Hinge teeth absent. White, often with oxide deposits on posterior area. **Habitat:** in muddy substrates, offshore. **Distribution:** NWG, SEG, GO, Mas.

Thracia salsettensis Melvill, 1893: **1270**
To 50mm. Fragile. Inequivalve, LV compressed, RV tumid, beaks behind mid-line. Oblong, anterior rounded, posterior broadly truncate. Sculpture of prominent concentric undulations, some rather irregular. White. **Habitat:** ? **Distribution:** NWG, SEG.

Superfamily CLAVAGELLOIDEA
The Clavagelloidea are unusual among bivalves because they assume, in part or completely, the form of a tube. The most familiar are the watering-pot shells of the genus *Brechites* where both valves are fused into the wall of the tube which is multi-perforated at one end and open at the other. These forms live embedded in sand with the multi-perforate end downwards. In the genus *Clavagella* only one valve is incorporated into the tube, the other remaining free. The species live in excavated burrows in rock or coral with only the tube projecting. Only the watering-pot shells are recorded from the eastern Arabian region although *Clavagella* is known from Aden and Djibouti. *Clavagella* may have been overlooked because of its rock-boring habit and reduced shell.

Brechites attrahens (Lightfoot, 1786): **1271**
To 150mm. Tubular, both valves fused into the tube and visible just above the multi-perforate end. Siphonal third of tube with pleated ruffles at irregular intervals. Perforate end fringed. Tube variously encrusted with sand grains or shell fragments. **Habitat:** buried in stable sand, offshore. **Distribution:** Mas.

Superfamily POROMYOIDEA
Family CUSPIDARIIDAE
The Cuspidariidae are usually strongly rostrate shells, with hinge teeth and a resilifer. They are characterised by their gill form in which the lamellae are fused to form a muscular septum across the mantle cavity. Movement of this septum causes inhalant and exhalant expulsions of water through the mantle cavity. By these actions the animals trap small passing animals and so can be considered truly carnivorous. The Cuspidariidae are mainly deep-water forms and a number of species have been collected in the Arabian Sea by the exploring expeditions of the "Investigator" and, more recently, the "John Murray". Only two species have been recorded in water of less than 100m depth and these are the only species included here.

Cardiomya alcocki (Smith, 1894): **1272**
To 7mm. Inequivalve, RV fitting into LV. Suboval, with a distinct short rostrum. Sculpture of radial, raised lines developing anteriorly into a few

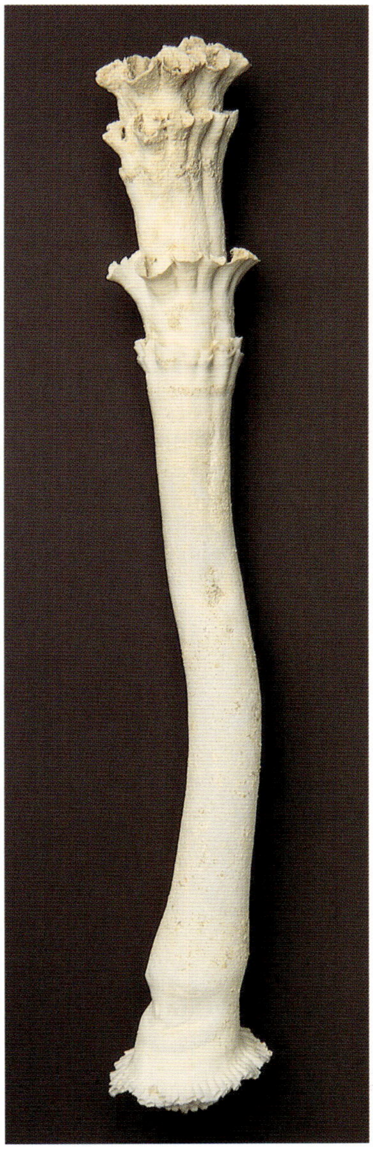

Brechites attrahens 1271

well spaced ribs close to the rostrum. White. **Habitat:** in mud, deep water. **Distribution:** GO.

Cuspidaria approximata Smith, 1896: **1273**
To 10mm. Suboval-subtriangular, posterior rostrum narrow but only a little longer than the body of the shell, set off from rest of shell by a distinct umbonal ridge. Sculpture of numerous fine, concentric threads angled over the rostrum. White. **Habitat:** in mud, offshore. **Distribution:** GO.

GLOSSARY

Sur Masirah, Masirah, Oman: Peter Dance collecting at low tide and (inset) shells he collected at one locality on one visit to Sur Masirah.

GLOSSARY OF TERMS USED IN THIS BOOK

It is almost impossible to avoid using unfamiliar words when describing the physical appearance of molluscs, the places where they live and the circumstances which affect them in their lives. Also it is unforgivable to use such words without saying what they mean. This glossary sets out to explain those which may puzzle the reader. The definitions given here, however, may not be current beyond the confines of this book. They are definitions: they are not meant to be definitive.

ga = gastropods; *ch* = chitons; *sc* = scaphopods; *bv* = bivalves

acentric (*bv*, sculpture): oblique, cutting growth lines at an angle.

adductor muscle scars (*bv*): scars left on inner surface of valves by muscles which close shell. Two or, less often, one present normally.

alate (*bv*, outline): winged or wing-like.

alivincular (*bv*, ligament): flattened and situated on cardinal area.

amphidetic (*bv*, ligament): extending on both sides of beaks.

angulate: with angular profile.

anisomyarian (*bv*): anterior adductor muscle reduced in size.

anterior angle (*bv*): distinct change in curvature separating anterior and median areas.

anterior keel (*bv*): where anterior angle develops into a distinct raised ridge.

anterior sulcus (*bv*): radial depression on anterior area.

antero-dorsal (*bv*): front end of upper side.

apophysis (*bv*): narrow projection from beneath umbones inside valves of some Pholadoidea.

appressed (*bv*): pressed closely against surface. (*ga*) when two adjacent whorls meet closely and produce a weakly defined suture.

arcuate (*bv*, outline): arched like a bow.

articulamentum (*ch*, structure): porcellaneous under layer of shell.

attenuate (outline): slenderly tapering, often to a point.

auricle (*bv*): ear-like lobe.

auriculate (*bv*, outline): with ear-like projections.

axial (*ga*, sculpture): direction of growth parallel with central axis of shell, more or less vertical or longitudinal.

bead: a rounded granule or tubercle.

beak (*bv*): tip of umbo.

biconic (*ga*, structure): like two cones joined at their bases.

bifid: divided into two by a notch or a groove.

bifurcate (*bv*, sculpture): forked.

biotope: a habitat supporting a biological community and classified according to the members of that community.

byssal gape/notch (*bv*): gap or notch between margins of valves through which byssus projects.

GLOSSARY

byssus (*bv*): bunch of chitinous threads anchoring members of some groups.

calcareous: made of lime.

callum (*bv*): accessory shell plate closing anterior gape in Pholadidae.

callus (*ga*, sculpture): calcareous pad, often thick, deposited upon primary sculpture.

cancellate (sculpture): with axial (or radial) ridges intersecting spiral (or concentric) ridges at right angles.

cardinal area (*bv*): region of shell between beaks and hinge margin, usually bearing external ligament.

cardinal teeth (*bv*, hinge): projections about middle of hinge, normally but not always diverging from beaks.

channelled (*ga*, structure): when suture is more or less gutter-like.

chevron: (*bv*): V-shaped ligamental groove on cardinal area; (*ga*): V-shaped colour markings.

chomata (*bv*, margin): small tubercles (and opposing pits) along edges of valves (particularly in oysters).

chondrophore (*bv*, ligament): spoon-like process (which may project) for attachment of internal ligament.

columella (*ga*, structure): lower part of inner lip of shell along central axis.

commensal: living in close association with another species and sometimes sharing its food.

concentric (sculpture): parallel to line of growth.

confluent (*bv*): when pallial line and lower arm of pallial sinus run together.

cord (*ga*, sculpture): a ridge, usually spiral, thicker than a thread.

cordate (*bv*): heart shaped.

coronate (*ga*, sculpture): regularly knobbed or wavy below suture.

crenulate (*bv*, margin; *ga*, apertural edge): regularly notched.

cruciform muscle scars (*bv*): two small scars left by muscles which are used in the Tellinacea to withdraw siphons into the shell, situated below pallial sinus.

crura/crural teeth (*bv*, hinge): ridge-like, symmetrically arranged teeth radiating from apex of ligament pit.

ctenolium (*bv*): comb-like row of denticles lining byssal notch in some Pectinidae.

decussate (*bv*, sculpture): obliquely cancellate.

dendritic (*bv*, sculpture): branched like a tree.

denticle: small tooth.

denticulate (*bv*, margin): with small teeth.

depressed: low.

dextral (*ga*, structure): coiling in a right-handed direction when the shell aperture is on the observer's right and facing.

dimyarian (*bv*): with two adductor muscles in each valve.

discoidal (outline): disc-like.

discrepant (*bv*, sculpture): differing on opposing valves.

divaricate (*bv*, sculpture): diverging or splitting, usually along a line.

domed (*ga*, structure): low and gently curved in profile (protoconch).

dorsal: (*bv*): upper side (i.e. side bearing hinge and umbones); (*ga*): posterior side (i.e. side to rear of aperture).

dorsal area (*bv*): area between beaks.

dorsoventrally: from top to bottom.

dorsum (*ga*, *ch*, structure): rear side (i.e. behind aperture) of gastropod shell; upper surface of chiton.

dysodont (*bv*, hinge): small, often irregular, teeth on margins of some Mytiloidea.

edentulous (*bv*, hinge): toothless.

elliptical (*bv*, outline): compressed circle.

endemic: restricted to a particular geographic area.

entire (*bv*): condition when pallial line is not indented.

epifauna: surface-dwelling animals.

equilateral (*bv*): condition when shell growth either side of beaks is symmetrical or almost so.

equivalve (*bv*): with valves equal.

escutcheon (*bv*): dorsal depression behind umbones, generally set off from rest of shell by a change in sculpture or colour.

exhalant (siphon or current): outgoing tube, or current, for expulsion of water from mantle cavity.

false umbilicus (*ga*, structure): an umbilicus (q.v.) made shallow by secondary shell growth and so not reaching columella.

fasciolar groove (*ga*, structure): spiral groove associated with fasciole, especially in Olividae.

fasciole (*ga*, structure): narrow ridge or broader band encircling base and tracing former positions of siphonal canal.

fluted (*bv*): becoming tubular.

foliaceous (*bv*, sculpture): leaf-like.

forma: scientifically named form of a species.

fossula (*ga*, structure): longitudinal furrow within aperture of cowry shell, situated anteriorly on the columellar side and usually ribbed transversely.

funicle (*ga*, structure): prominent, rounded, spiral ridge entering and sometimes filling umbilicus in Naticidae.

funiculum (*ga*, structure): ridge or warty lump at posterior end of columellar side of shell in Ovulidae.

fusiform (*ga*, structure): spindle shaped, drawn out at each end.

gape (*bv*): space left between valve margins when closed.

girdle (*ch*, structure): flexible band of tissue encircling and retaining valves.

granular (sculpture): with fine, grain-like beads.

growth lines: usually fine axial or concentric lines showing former positions of the growing margin.

head valve (*ch*, structure): first (anterior) valve.

heterodont (*bv*, hinge): with cardinal and lateral teeth well differentiated.

heteromyarian (*bv*): with anterior adductor muscle much reduced.

heterostrophe (*ga*, structure): when whorls of protoconch coil in contrary direction to those of spire.

hinge (*bv*): dorsal region of bivalve along which valves meet.

hinge plate (*bv*): platform on inner dorsal margin of valve bearing hinge teeth and ligament.

holotype: the unique specimen upon which the description of a species is based.

homomyarian (*bv*): bivalve with adductor muscles more or less equal in size.

hypoplax (*bv*): accessory plate blocking posterior end of ventral gape in Pholadidae.

GLOSSARY

imbricate (*bv*, sculpture): overlapping like roof tiles.
inequilateral (*bv*, outline): beaks not central.
inequivalve (*bv*): valves unequal in size and/or shape.
infauna: burrowing animals.
inhalant (siphon or current): tube, or ingoing stream of water passing through it to mantle cavity.
insertion plates (*ch*, structure): projecting margins of valves by which they are attached inside the surrounding girdle.
intermediate valves (*ch*, structure): the six valves between the head and tail valves.
interspace (sculpture): interval between ridges, ribs or other sculptural features.
isodont (*bv*, hinge): with a few symmetrically arranged hinge teeth (crura).

jugal area (*ch*, structure): central portion of median area (q.v.).

keel/keeled: ridge, sometimes prominent and giving shell a distinctive character but occurring in all degrees of angularity.
khor: salt-water lagoon, sometimes well separated from the sea and so relatively tranquil.

lamella (sculpture): a thin, flattened plate.
lamellate/lamellose (sculpture): blade-like, or with lamellae.
lateral area (*ch*, structure): triangular area on each side of an intermediate valve.
lateral teeth (*bv*, hinge): teeth situated at one or both ends of hinge, away from beaks.
lenticular (*bv*, outline): shaped like a convex-sided disc, or biconvex lens.
ligament (*bv*, hinge): chitinous, elastic structure joining hinges of two valves.
lip (*ga*, sculpture): edge of aperture.
lira(e) (sculpture): (*bv*) a narrow concentric ridge; (*ga*) a spiral ridge within the aperture or on the columella.
lirate (sculpture): with lirae.
lithodesma (*bv*, hinge): small calcareous plate strengthening the internal ligament in some bivalves.
littoral: intertidal zone.
lunule (*bv*): depression or delimited area just anterior to beaks.

malleated: as if tapped with a round-ended hammer.
mantle: fleshy fold or skirt lining later developed part of shell and which is responsible for secreting shell-forming material.
mantle cavity: space beneath mantle housing the gills.
median area (*ch*, structure): triangular area of intermediate valve between lateral areas.
median sulcus (*bv*, sculpture): depression on central area of valve.
median, medial: on the mid-line.
median ridge (*ch*, structure): central ridge on valve.
mesoplax (*bv*): calcareous accessory plate lying above umbonal region of shells of Pholadidae.
metaplax (*bv*): narrow, calcareous accessory plate(s) filling postero-dorsal gape of shells of Pholadidae.
modioliform (*bv*, outline): like *Modiolus*, with beaks close to anterior margin.
monomyarian (*bv*): anterior adductor muscle absent, leaving posterior one only.

mucro (*ch*, *ga*, structure): a slight projection on central area of tail valve in chitons; short spike on posterior septum of shell in Caecidae.
multivincular (*bv*, ligament): restricted to a series of conspicuous transverse grooves.
mytiliform (*bv*, outline): mussel shaped, with terminal umbo.

nacre: mother-of-pearl.
nacreous: pearly.
nodose: knobbed, knobbly.
nodule: small knob.
nodulose: with small knobs.
nymph (*bv*, hinge): narrow sometimes raised, platform behind umbo, to which external ligament is attached.

obsolete: weak to almost absent.
obtuse: blunt, forming an angle greater than 90 degrees.
ocelli: (*ga*): colour pattern of eye-like rings; *ch*, pigmented, light-sensitive organs on valves of some chitons (also called aesthetes).
operculum (*ga*, structure): calcareous or chitinous plate, plain or ornamented, attached to foot and usually closing aperture when withdrawn.
opisthodetic (*bv*, ligament): situated entirely behind umbones.
opisthogyrate (*bv*, umbones): directed posteriorly.
orbicular (*bv*, sculpture): circular.
orthogyrate (*bv*, umbones): pointing directly at each other.
ovate: egg shaped.

pallets (*bv*): a pair of spade- or feather-like, calcareous structures which close the burrow opening of species of Teredinidae.
pallial line (*bv*): line near margin on inner surface of valve marking attachment site of mantle edge.
pallial sinus (*bv*): indentation of pallial line marking attachment site of siphonal retractor muscles.
parietal wall (*ga*, structure): region of shell above (i.e. posterior to) columella.
parivincular (*bv*, ligament): long and resembling a split cylinder.
pedal: relevant to the foot.
pedal gape (*bv*): gap between ventral margins in some bivalves.
pedal retractor muscle (*bv*): muscle which withdraws foot.
periostracum: proteinaceous outer layer of shell; sometimes thick, sometimes almost transparently thin, it often flakes off and is seldom retained on dead shells.
peristome (*ga*, structure): edge of aperture.
pipe (*sc*, structure): short, hollow tube projecting from the posterior end of some scaphopods.
plica(e): a fold or ridge.
plicate: ridged.
posterior angle (*bv*): distinct change in curvature separating median and posterior areas from umbo to posterior ventral margin.
posterior canal (*ga*, structure): channel, notch or tube at upper (i.e. posterior) end of apertural lip for reception of exhalant siphon.
posterior keel (*bv*): where a raised ridge represents posterior angle.
posterior sulcus (*bv*): radial depression on posterior area.

GLOSSARY

postero-dorsal (*bv*): upper margin behind umbones.

postero-ventral (*bv*): lower hind margin.

primary (sculpture): sculptural element developing first and becoming strongest.

prodissoconch (*bv*): larval or embryonic shell.

prosodetic (*bv*, ligament): situated in front of umbones.

prosogyrate (*bv*, umbones): anteriorly directed.

protoconch (*ga*): larval or embryonic shell.

protoplax (*bv*): flat, calcareous accessory plate situated in front of umbones of members of the Pholadidae.

punctae: minute pits on shell surface.

punctate: minutely pitted.

pustule: a small tubercle.

pustulose: bearing pustules.

radula(e) (*ch*, *ga*, anatomical): ribbon-like structure beset with rows of rasping teeth (herbivores), or harpoon-like structure functioning as a venomous injecting tool to immobilise prey (carnivores).

recent: the post-Pleistocene period.

recurved: (*bv*), turned back towards umbones; (*ga*), turned back or to the side (referring to orientation of siphonal and posterior canals).

reflected/reflexed: turned backward.

resilifer (*bv*, hinge): hollow in hinge plate containing internal ligament.

resilium (*bv*, hinge): the internal ligament.

reticulate (sculpture): with obliquely intersecting ridges.

rhomboidal: roughly diamond shaped with rounded corners.

rib: (*bv*), prominent radial ridge; (*ch*), (*ga*), prominent ridge, radial, concentric, axial or spiral according to species.

riblet: similar to but less prominent than a rib.

rostrate: with a rostrum.

rostrum: beak-like projection.

rugose: rough or wrinkled.

scabrous (*bv*, sculpture): rough, like a file.

scissulate (*bv*, sculpture): where concentric elements change abruptly to oblique elements.

secondary (*bv*, sculpture): late-developing sculpture, weaker than primary sculpture.

sensu: in the sense of, according to.

septum: partition or shelf.

serrated: (*bv*), where margin is cut by dense, closely spaced grooves; (*ga*), where apertural margin is deeply cut by saw-tooth-like grooves.

sessile: sedentary or fixed to one spot.

sinistral (*ga*, structure): coiling in a left-handed direction when the aperture is on the observer's left and facing.

sinus: notch, often rounded.

siphon: tubular or trough-like extension of mantle through which water enters or leaves mantle cavity.

siphonal canal (*ga*, structure): deep or shallow notch or embayment at anterior end of aperture.

siphonoplax (*bv*): calcareous structure laid down to close posterior gape between valves in some species of Pholadidae.

spathulate (*bv*): elongate, with a flattened end.

spiculation, spicules (*ch*, needle-like or spiky processes on girdle.

spine: more or less sharp protuberance.

spire (*ga*, structure): all the whorls comprising a spirally coiled shell except for the earliest ones (see **protoconch**) and the last one (see **last whorl**); sometimes known as a teleoconch.

spiral (*ga*, structure): coiling about a central axis (see also whorl).

squamose (*bv*): scaly.

stria/striation: (*bv*), fine line (thread or groove); (*ch*), (*ga*), fine groove, sometimes a mere scratch.

striate: with striae.

stromboid notch (*ga*, structure): broad, shallow embayment towards lower (i.e. anterior end of outer lip in Strombidae and some other families.

sub- (*bv*): prefix meaning 'almost' (e.g. subadult, subcylindrical) or 'below' (e.g. subapical).

sublittoral: from below low-tide level.

sulcate (*bv*): grooved.

sulcus (*bv*): a groove.

sutural laminae (*ch*, structure): articulating anterior extensions on all valves but the head valve.

suture (*ga*, structure): where two adjacent whorls meet.

syn: = synonymy; if a taxon has been named more than once the later name(s) is (are) placed in synonymy and called synonym(s).

tabulated: flattened out or flat topped.

tail valve (*ch*, structure): last (posterior) valve.

taxodont (*bv*, hinge): with a row of many small, similar teeth.

taxon (pl. **taxa**): organism or group of organisms which has been given a scientific name; it may not be the name accepted as valid now.

tegmentum (*ch*, structure): surface layer of valves.

tenting: when a series of approximately triangular markings form colour pattern.

tertiary (*bv*, sculpture): third strongest sculptural element.

thread: fine ridge.

transverse: at right angles to main axis.

trapezoidal (*bv*, outline): with four non-parallel sides.

trigonal (*bv*, outline): roughly triangular.

trochoidal: top shaped, resembling a top shell (Trochus).

truncate (*bv*, outline): cut off and generally squared.

tubercle: a knob.

tubercular: knobbed.

umbilicus (*ga*, structure): hole, often within a crater-like depression, at basal end of central axis of shell (see also **false umbilicus**), sometimes at top end of shell.

umbo (pl. **umbones**) (*bv*): surface of valve immediately behind beak.

umbonal reflection (*bv*): when dorsal line turns back in front of and over umbo of each valve.

umbonal ridge (*bv*, sculpture): ridge or keel running from umbo to postero-ventral margin.

variety: distinct form of a species which may have been given a scientific name.

varix (pl. **varices**) (*ga*, sculpture): thickened edge of former outer lip incorporated into shell wall.

ventral: *bv*, on the underside; *ga*, = on the apertural side.

whorl (*ga*, structure): a complete turn of a spirally coiled shell.

wing (*bv*): prominent angular extension of dorsal margin.

REFERENCES

ABBOTT, R. T., 1973. *Acteon eloiseae*, a new opisthobranch from Arabia. *Nautilus* **87**: 91-92.

— & LEWIS, H., 1970. *Cymatium boschi*, new species from the Arabian Sea. *Nautilus* **83**: 86-88.

ADAM, W. & REES, W. J., 1966. A review of the cephalopod family Sepiidae. *John Murray Expedition 1933-34. Scientific Reports* **11**: (1) 1-165, 46 pls.

AHMED, M. M., 1975. *Systematic study on Mollusca from Arabian Gulf and Shatt Al-Arab, Iraq.* pp. 78, 54 figs. Center for Arab Gulf Studies, University of Basrah.

BARRATT, L., 1984. Ecological studies of rocky shores on the south coast of Oman. *Report of IUCN to UNEP Regional Seas Programme, Geneva.* Contract Number KA/0503-82-09 (2362), 104 pp.

BASSON, P. W., BURCHARD, J. E., HARDY, J. T. & PRICE, A. R. G., 1977. *Biotopes of the Western Arabian Gulf.* pp. 284, text figs. many col. Aramco Dept of Loss Prevention & Environmental Affairs, Dhahran, Saudi Arabia.

BEU, A. G., 1986. Taxonomy of gastropods of the families Ranellidae (=Cymatiidae) and Bursidae. Part 2. Descriptions of 14 new modern Indo-West Pacific species and subspecies, with revisions of related taxa. *N. Z. Jl Zool.* **13**: 273-355, 274 figs.

BIELER, R., 1993. Architectonicidae of the Indo-Pacific. (Mollusca, Gastropoda). *Abh. naturw. Ver. Hamburg* **30**: 1-376, text figs.

BIGGS, H. E. J., 1958a. A new species of *Siphonaria* from the Persian Gulf. *J. Conch. Lond.* **24**: 249.

—, 1958b. Littoral collecting in the Persian Gulf. *J. Conch. Lond.* **24**: 270-75.

—, 1969. Marine Mollusca of Masirah I., South Arabia. *Arch. Molluskenk.* **99**: 201-207.

—, 1973. The marine Mollusca of the Trucial Coast, Persian Gulf. *Bull. Brit. Mus. nat. Hist. Zoology* **24**: 343-421, pls 1-6.

— & GRANTIER, L., 1960. A preliminary list of the Mollusca of Ras Tanura, Persian Gulf. *J. Conch. Lond.* **24**: 387-92.

BOSCH, D. & BOSCH, E., 1982. *Seashells of Oman.* pp. 206, illust. col. Longman Group, London.

—, 1989. *Seashells of southern Arabia.* Motivate Publishing, United Arab Emirates. pp. 95, illust. col. (Part of English text reproduced in Arabic.)

CERNOHORSKY, W. O., 1976. The Mitridae of the world. Part 1. The subfamily Mitrinae. *Indo-Pacif. Mollusca* **3** (17): 273-528, col. pls 253-258, text figs.

—, 1984. Systematics of the family Nassariidae (Mollusca: Gastropoda). *Bull. Auckland Inst. Mus.* No. 14, pp. iii + 356, 50 pls, text figs.

—, 1991. The Mitridae of the world. Part 2. The subfamily Mitrinae concluded and subfamilies Imbricariinae and Cylindromitrinae. *Mon. mar. Moll.* No. 4, pp. 164, pls, text figs.

CHARTER, W., 1988. Shell wealth of Mina Qaboos. *Hawaii. Shell News* **36**: 1 & 7.

CHESNEY, H. C. G. & OLIVER, P. G., 1994. Taxonomy of Arabian bivalves. Part 2. A new species of *Semele* (Bivalvia: Tellinoidea). *J. Conch. Lond.* **35**: 33-36, text figs.

CLOVER, P. W., 1972. Description of new species of *Conus* from South East Arabia. *Venus*: **31**: 117-18.

COOMANS, H. E. & MOOLENBEEK, R. G., 1990. Notes on some Conidae from Oman, with description of *Conus stocki* n. sp. (Mollusca; Gastropoda). *Bijdr. Dierk.* **60**: 257-262, text figs.

CURRIE, R. I., FISHER, A. E. & HARGREAVES, P. M., 1973. Arabian Sea upwelling. *Ecol. Stud.* **3**: 37-53.

DANCE, S. P. & EAMES, F. E., 1966. New molluscs from the Recent Hammar formation of south-east Iraq. *Proc. malac. Soc. Lond.* **37**: 35-43, pls 2-5.

DANCE, S. P., MOOLENBEEK, R. G. & DEKKER, H., 1992. *Umbonium eloiseae* (Gastropoda: Trochidae), a new trochid species from Masirah Island, Oman. *J. Conch. Lond.* **34**: 231-235, text figs.

DEKKER, H., & GOUD, J., 1994. Review of the living Indo-West-Pacific species of *Divaricella* sensu auct. with descriptions of two new species and a summary of the species from other regions. (Part 1). *Vita Marina* **42**: 115-136

—, MOOLENBEEK, R. G. & DANCE, S. P., 1992. *Turbo jonathani*, a new turbinid species from the southern coast of Oman (Gastropoda: Turbinidae). *J. Conch. Lond.* **34**: 225-229, text figs.

EAMES, F. E. & WILKINS, G. L., 1957. Six new molluscan species from the alluvium of Lake Hammar near Basrah, Iraq. *Proc. malac. Soc. Lond.* **22**: 198-203, pls 27-28.

EMERSON, W. K. & SAGE, W. E., 1986. A new species of *Lyria* (Gastropoda: Volutidae) from the Arabian Sea. *Nautilus* **100**: 101-104.

FISCHER, P., 1891. Liste de coquilles récueillis par F. Houssay dans le Golfe Persique. *J. Conch. Paris* **31**: 222-30.

GLAYZER, B. A., GLAYZER, D. T. & SMYTHE, K. R., 1984. The marine Mollusca of Kuwait, Arabian Gulf. *J. Conch. Lond.* **31**: 311-30.

HAAS, F., 1952. Shells collected by the Peabody Expedition to the Near East, I. Molluscs from the Persian Gulf. *Nautilus* **65**: 114-18.

—, 1954. Some marine shells from the Persian Gulf collected by Ronald Codrai. *Nautilus* **68**: 46-49.

HARRIS, C. P., 1969. The Persian Gulf submarine telegraph of 1864. *Geogrl J.* **135** (2): 169-90.

HOUBRICK, R. S., 1978. The family Cerithiidae in the Indo-Pacific. Part 1: The genera *Rhinoclavis*, *Pseudovertagus* and *Clavocerithium*. *Monographs of Marine Mollusca* No. 1, pp. 130, 2 col. pls, text figs.

HUDSON, R. G. S., EAMES, F. E. & WILKINS, G. L., 1957. The fauna of some recent marine deposits near Basrah, Iraq. *Geol. Mag.* **94**: 393-401.

ISSEL, A., 1865. Catalogo dei molluschi raccolti dalla missione Italiana in Persia aggiuntavi la descrizione delle specie nuove o poco note. *Memorie Accad. Sci. Torino* (Series 2) **23**: 387-439, 5 pls.

JOUSSEAUME, F., 1912. Faune malacologique de la Mer rouge. Scalidae. *Mem. Soc. zool. Fr.* **24**: 180-246, pls 5-7.

—, 1921. Sur quelques mollusques de la Mer rouge nouveaux ou non figurés. *Mem. Soc. zool. Fr.* **28**: 53-60, pl. 3.

KAAS, P. & VAN BELLE, R. A., 1988. Chitons (Mollusca: Polyplacophora) from the coasts of Oman and the Arabian Gulf. *Am. malac. Bull.* **6**: 115-130, text figs.

KAY, E. A., 1979. *Hawaiian marine shells. Reef and shore fauna of Hawaii. Section 4: Mollusca.* pp. xviii + 653, text figs. B. P. Bishop Mus. Special Publication 64 (4).

KILBURN, R. N., 1980. A new *Ancilla* from the Arabian Sea, and a discussion of two homonyms in the Ancillinae (Mollusca: Gastropoda: Olividae). *Durban Mus. Novit.* **12**: 167-70.

—, 1985. The family Epitoniidae (Mollusca: Gastropoda) in southern Africa and Mozambique. *Ann. Natal Mus.* **27**: 239-337, 171 figs.

— & RIPPEY, E. 1982. *Sea shells of southern Africa.* MacMillan, Johannesburg. pp. 249, 46 col. pls, line drawings.

KNUDSEN, J., 1967. The deep-sea Bivalvia. *John Murray Expedition 1933-34. Scientific Reports* **11**: (3) 237-343, 3 pls, text figs.

MARTENS, E. von, 1874. *Ueber Vorderasiatischen Conchylien nach den Sammlungen des Prof. Haussknecht.* pp. 127, 9 pls. Cassel.

MELVILL, J. C., 1893. Descriptions of twenty-five new species of marine shells from Bombay. Collected by Alexander Abercrombie. *Mem. Proc. Manchr lit. phil. Soc.* (Series 4) 7: 52-67, 1 pl. Reprinted in *J. Bombay nat. Hist. Soc.* **8**: 234-245, 1 pl.

—, 1896. Descriptions of new species of minute marine shells from Bombay. *Proc. malac. soc. Lond.* **2**: 108-116; pl. 8. Reprinted in *J. Bombay nat. Hist. Soc.* **11**: 506-514, 1 pl.

—, 1897a. Description of *Plecotrema sykesii* n. sp. from Karachi. *Proc. malac. Soc. Lond.* **2**: 292, text fig.

—, 1897b. Descriptions of thirty-four species of marine Mollusca from the Arabian Sea, Persian Gulf and Gulf of Oman. (Mostly collected by F. W. Townsend Esq.). *Mem. Proc. Manchr lit. phil. Soc.* **41**, (number 7): 1-26, pls. 6-7.

—, 1898a. Further investigations into the molluscan fauna of the Arabian Sea, Persian Gulf and Gulf of Oman, with descriptions of forty species. (Mostly dredged by F. W. Townsend Esq.). *Mem. Proc. Manchr lit. phil. Soc.* **42**, (number 4): 1-36, 39-40, pls. 1-2.

—, 1898b. Description of a new *Strombus* from the Mekran coast of Beluchistan. *Mem. Proc. Manchr lit. phil. Soc.* **42**, (number 4): 37-38, text fig. (Addendum to previous article.)

—, 1899a. Note on *Scalaria fimbriolata* Melv. *J. Conch. Lond.* **9**: 181.

—, 1899b. Notes on the Mollusca of the Arabian Sea, Persian Gulf and Gulf of Oman, mostly dredged by Mr. F. W. Townsend, with descriptions of twenty-seven species. *Ann. Mag. nat. Hist.* (Series 7), **4**: 81-101, pls. 1-2.

—, 1901. A few further remarks upon the Erythraean molluscan fauna, with descriptions of seven species from Aden, in the collection of Commander E. R. Shopland, R.I.M. *Ann. Mag. nat. Hist.* (Series 7), **7**: 550-556, pl. 9.

—, 1903. A revision of the Columbellidae of the Persian Gulf and North Arabian Sea, with description of *C. calliope* n.sp. *J. Malac.* **10**: 27-31, text fig.

—, 1904a. Descriptions of twenty-three species of gastropoda from the Persian Gulf, Gulf of Oman and Arabian Sea, dredged by Mr. F. W. Townsend of the Indo-European Telegraph Service in 1903. *Proc. malac. Soc. Lond.* **6**: 51-60 pl. 5.

—, 1904b. On *Berthaïs*, a proposed new genus of marine gastropoda from the Gulf of Oman. *Proc. malac. Soc. Lond.* **6**: 61-63, text figs.

—, 1904c. Descriptions of twenty-eight species of gastropoda from the Persian Gulf, Gulf of Oman and Arabian Sea, dredged by Mr. F. W. Townsend of the Indo-European Telegraph Service, 1900-1904. *Proc. malac. Soc. Lond.* **6**: 158-169, plate 10.

—, 1904d. *Conus coromandelicus* Smith, its probable affinities and systematic position in the family Conidae. *Proc. malac. Soc. Lond.* **6**: 170-173, text figs.

—, 1904e. Descriptions of twelve new species and one variety of marine Gastropoda from the Persian Gulf, Gulf of Oman and Arabian Sea, collected by Mr. F. W. Townsend, 1902-1904. *J. Malac.* **11**: 79-85, pl. 8.

—, 1904f. Note on *Mitra stephanucha* Melv., with description of a proposed new varietyy. *J. Malac.* **11**: 86, pl. 8.

—, 1906a. A revision of the species of Cyclostrematidae and Liotiidae occurring in the Persian Gulf and North Arabian Sea. *Proc. malac. Soc. Lond.* **7**: 20-28, pl. 3.

—, 1906b. Descriptions of thirty-one Gastropoda and one scaphopod from the Persian Gulf and Gulf of Oman, dredged by Mr. F. W. Townsend, 1902-1904. *Proc. malac. Soc. Lond.* **7**: 69-80, pls 7-8.

—, 1906c. *Capulus lissus* Smith as type of proposed new subgenus (*Malluvium*) of *Amalthea* Scumacher. *Proc. malac. Soc. Lond.* **7**: 81-84, text fig.

—, 1910a. A revision of the species of the family Pyramidellidae occurring in the Persian Gulf, Gulf of Oman and North Arabian Sea as exemplified mostly in the collections made by Mr. F. W. Townsend (1893-1900), with descriptions of new species. *Proc. malac. Soc. Lond.* **9**: 171-207, pls 4-6.

—, 1910b. Descriptions of twenty-nine species of marine Mollusca from the Persian Gulf, Gulf of Oman and North Arabian Sea, mostly collected by Mr. F. W. Townsend of the Indo-European Telegraph Service. *Ann. Mag. nat. Hist.* (Series 8), **6**: 1-17, pls 1-2.

—, 1912. Descriptions of thirty-three new species of Gastropoda from the Persian Gulf, Gulf of Oman and North Arabian Sea. *Proc. malac. Soc. Lond.* **10**, 240-254, pls 11-12.

—, 1917a. A revision of the Turridae (Pleurotomidae) occurring in the Persian Gulf, Gulf of Oman and North Arabian Sea as evidenced mostly through the results of dredgings carried out by Mr. F. W. Townsend, 1893-1914. *Proc. malac. Soc. Lond.* **12**: 140-201, pls 8-10.

—, 1917b. Description of a new species of *Terebra* from the Mekran coast, Arabian Sea. *J. Conch. Lond.* **15**:188-189, text fig.

—, 1917c. Note on *Conus melvilli* Sowerby. *J. Conch. Lond.* **15**: 222.

—, 1918. Descriptions of thirty-four species of marine Mollusca from the Persian Gulf, Gulf of Oman and Arabian Sea, collected by Mr. F. W. Townsend. *Ann. Mag. nat. Hist.* (Series 9), **1**: 137-158, pls 4-5.

REFERENCES

—, 1928. The marine Mollusca of the Persian Gulf, Gulf of Oman and North Arabian Sea as evidenced mainly through the collections of Captain F. W. Townsend, 1893-1914. - Addenda, corrigenda and emendanda. *Proc. malac. Soc. Lond.* **18**: 93-117.

— & ABERCROMBIE, A., 1893. The marine Mollusca of Bombay. *Mem. Proc. Manchr lit. phil. Soc.* (Series 4), **7**: 17-51.

— & STANDEN, R., 1898. Description of *Conus (Cylinder) clytospira* sp.n. from the Arabian Sea. *Ann. Mag. nat. Hist.* (Series 7), **4**: 461-463.

—, 1901. The Mollusca of the Persian Gulf, Gulf of Oman and Arabian Seas as evidenced mainly through the collections of Mr. F. W. Townsend, 1893-1900, with descriptions of new species. Part 1. Cephalopoda, Gastropoda and Scaphopoda. *Proc. zool. Soc. Lond.* for 1901: 327-460, pls 21-24.

—, 1903a. The genus *Scala* (Klein) Humphrey, as represented in the Persian Gulf, Gulf of Oman and North Arabian Sea with descriptions of new species. *J. Conch. Lond.* **10**: 340-351, pl. 7.

—, 1903b. Descriptions of sixty-eight new Gastropoda from the Persian Gulf, Gulf of Oman and North Arabian Sea, dredged by Mr. F. W. Townsend of the Indo-European Telegraph Service 1901-1903. *Ann. Mag. nat. Hist.* (Series 7), **12**: 289-324, pls 20-23. Reprinted in *J. Bombay nat. Hist. Soc.* **16**: 86-98, 217-234, pls A-D.

—, 1904. The Cypraeidae of the Persian Gulf, Gulf of Oman and North Arabian Sea, as exhibited in Mr. F. W. Townsend's Collections, 1893-1904. *J. Conch. Lond.* **11**: 117-122.

—, 1905. *Rostellaria delicatula* Nevill. Notes upon its distribution and limits of variation. *J. Conch. Lond.* **11**: 161-163, pl. 2.

—, 1907. The Mollusca of the Persian Gulf, Gulf of Oman and Arabian Sea as evidenced mainly through the collections of Mr. F. W. Townsend, 1893-1906, with descriptions of new species. Part 2, Pelecypoda. *Proc. zool. Soc. Lond.* (for 1906): 783-848, pls 53-56.

—, 1917. A revision of the species of *Terebra* occurring in the Persian Gulf, Gulf of Oman and Arabian Sea as evidenced in the collection formed by Mr. F. W. Townsend, 1893-1914. *J. Conch. Lond.* **15**: 204-216.

MIENIS, H. K., 1978. Notes on recent and fossil Neritidae: 8 *Nerita adenensis* a new species from the Arabian Peninsula. *Argamon* **6**: 30-36.

MOOLENBEEK, R. G., 1994. The Orbitestellidae (Gastropoda: Heterobranchia) of the Sultanate of Oman with description of a new genus and two new species. *Apex* **9**: 5-10, text figs.

— & COOMANS, H. E., 1982. Studies on Conidae (Mollusca, Gastropoda). 2. *Conus pusio* Sowerby I (non Hwass) and *C. melvilli* Sowerby III. *Bull. zool. Mus. Univ. Amsterdam* **8**: 145-148, text figs.

—, 1993. New cones from Oman and the status of *Conus boschi* (Gastropoda; Conidae). *Apex* **8**: 19-26, text figs.

— & DANCE, S. P., 1994. *Anachis donnae* (Gastropoda: Columbellidae) a new columbellid species from Masirah Island, Oman. *J. Conch. Lond.* **35**: 119-122, text figs.

— & DEKKER, H., 1992. A new species of *Priotrochus* (Mollusca: Gastropoda: Trochidae) from Oman. *Bull. zool. Mus. Univ. Amsterdam* **13**: 171-174, text figs.

—, 1993a. On the identity of *Strombus decorus* and *Strombus persicus*, with the description of *Strombus decorus masirensis* n. ssp. and a note on *Strombus fasciatus*. *Vita Marina* **42** (1): 3-10, text figs, some col.

—, 1993b. The "Pheasant Shells" of Oman (Gastropoda: Turbinidae). *Venus* **52**: 141-148, text figs.

—, 1994. New nassariids from Oman and Somalia (Neogastropoda: Prosobranchia). *J. Conch. Lond.* **35**: 9-15, text figs.

MORRIS, S. & MORRIS, N., 1993. New shells from the UAE's east coast. *Tribulus* **3** (1) 5-8, 18-19 (figs).

OLIVER, P. G., 1992. *Bivalved seashells of the Red Sea.* pp. 330, 46 col. pls, line drawings. Hemmen, Wiesbaden.

— & CHESNEY, H. C. G., 1994. Taxonomy of Arabian bivalves. Part 1. Arcoidea. *J. Conch. Lond.* **35**: 17-31, text figs.

ROBERTSON, R., 1983. Axial shell rib counts as systematic characters in *Epitonium*. *Nautilus* **97**: 116-18.

ROSEWATER, J., 1961. The family Pinnidae in the Indo-Pacific. *Indo-Pacific Mollusca* **1**(4): 175-226, pls, text figs.

SHARABATI, D., 1981. *Saudi Arabian Seashells.* pp. 119, col. text figs. VNU Books International.

—, 1984. *Red Sea Shells.* pp. 127, 49 col. pls. KPI, London.

SHEPPARD, C., PRICE, A. & ROBERTS, C., 1992. *Marine ecology of the Arabian region.* pp. 359. Academic Press.

SMITH, E. A., 1872. Remarks on several species of Bullidae, with descriptions of some hitherto undescribed forms, and of a new species of *Planaxis*. *Ann. Mag. nat. Hist.* (Series 4) **9**: 344-355.

—, 1877. Descriptions of new species of Conidae and Terebridae. *Ann. Mag. nat. Hist.* (Series 4) **19**: 222-231.

—, 1894. Natural history notes from H. M. Indian Marine Survey Steamer 'Investigator', Commander C. F. Oldham, R. N.- Series 2, No. 10. Report upon some Mollusca dredged in the Bay of Bengal and the Arabian Sea. *Ann. Mag. nat. Hist.* (Series 6), **14**: 157-174, pls 3-5 (and Appendix, 366-368).

—, 1895a. Natural history notes from H. M. Indian Marine Survey Steamer 'Investigator', Commander C. F. Oldham, R. N.- Series 2, No. 19. Report upon Mollusca dredged in the Bay of Bengal and the Arabian Sea during the season 1893-94. *Ann. Mag. nat. Hist.* (Series 6), **16**: 1-19, pls 1-2.

—, 1895b. Natural history notes from H. M. Indian Marine Survey Steamer 'Investigator', Commander C. F. Oldham, R. N.- Series 2, No. 20. Report upon some Mollusca dredged in the Arabian Sea during the season 1894-5. *Ann. Mag. nat. Hist.* (Series 6), **16**: 262-265.

—, 1896. Natural history notes from H. M. Indian Marine Survey Steamer 'Investigator', Commander C. F. Oldham, R. N.- Series 2, No. 22. Descriptions of new deep sea Mollusca. *Ann. Mag. nat. Hist.* (Series 6), **18**: 367-375.

—, 1899. Natural history notes from H. M. Indian Marine Survey Steamer 'Investigator', Commander T. H. Hemming, R. N.- Series 3, No. 1. On Mollusca from the Bay of Bengal and the Arabian Sea. *Ann. Mag. nat. Hist.* (Series 7), **4**: 237-251.

—, 1904. Natural history notes from H. M. Indian

Marine Survey Steamer 'Investigator', Commander T. H. Hemming, R. N.- Series 3, No. 1. On Mollusca from the Bay of Bengal and the Arabian Sea. *Ann. Mag. nat. Hist.* (Series 7), **13**: 453-473. (Continued in *Ann. Mag. nat. Hist.* (Series 7), **14**: 1-14.)

—, 1906. Natural history notes from H. M. Indian Marine Survey Steamer 'Investigator'.- Series 3, No. 10. On Mollusca from the Bay of Bengal and the Arabian Sea. *Ann. Mag. nat. Hist.* (Series 7), **18**: 157-175. (Continued in *Ann. Mag. nat. Hist.* (Series 7), **18**: 245-264.)

SMYTHE, K. R., 1972. Marine Mollusca from Bahrain Island, Persian Gulf. *J. Conch. Lond.,* **27**: 491-96.

—, 1975a. On the occurrence of *Salinator fragilis* (Lamarck) in the Arabian Gulf. *J. Conch. Lond.* **28**: 335-38.

—, 1975b. *Salinator fragilis* (Lamarck) - habitat and behaviour. *J. Conch. Lond.* **28**: 339-342.

—, 1979a. The marine Mollusca of the United Arab Emirates, Arabian Gulf. *J. Conch. Lond.* **30**: 57-80.

—, 1979b. The Tornatinidae and Retusidae of the Arabian Gulf. *J. Conch. Lond.* **30**: 93-98, pl. 4.

—, 1980. Marine Mollusca of Dhofar. *J. Oman Stud. Spec. Rep.* **2**: 90-96.

—, 1982. *Seashells of the Arabian Gulf.* pp. 123, 20 pls, some col. Allen & Unwin, London.

—, 1983. *Seashells of the Sultan Qaboos Nature Reserve at Qurm.* pp. 64, illust. col. Diwan of Royal Court Affairs, Muscat.

—, 1985. Three new buccinids from Oman and notes on *Anachis fauroti* (Jousseaume) (Prosobranchia: Buccinacea). *J. Conch. Lond.* **32**: 25-35, pls 4-5, text figs.

—, 1988. A new species of *Medusafissurella* (Gastropoda: Fissurellidae), a keyhole limpet, from Oman. *J. Conch. Lond.* **33**: 97-101, pls 8-9, text figs.

— & CHATFIELD, J. E., 1981. New species of *Fusinus (Sinistralia)* and *Bullia* from Masirah, Oman (Prosobranchia: Buccinacea). *J. Conch. Lond.* **30**: 373-77, pl. 17.

—, 1982. Living specimens of *Fusinus (Sinistralia) gallagheri* Smythe & Chatfield 1981 from Masirah, Oman. *J. Conch. Lond.* **31**: 95-99, text figs.

— & HOUART, R., 1984. *Favartia (Favartia) paulboschi* (Muricidae: Muricopsinae): a new muricid from Oman. *Inf. Soc. belg. Malac.* (Series 12), **1**: 5-8, pl. 2.

— & OLIVER, P. G., 1986. A new species of *Muricopsis* from Oman (Prosobranchia: Muricacea). *J. Conch. Lond.* **32**: 181-83, pls 19-20.

SOWERBY, G. B. (3rd), 1895a. Descriptions of four new shells from the Persian Gulf and Bay of Zaila. *Proc. malac. Soc. Lond.* **1**: 160-161, pl. 12.

—, 1895b. Descriptions of nine new species of shells from the Persian Gulf. *Proc. malac. Soc. Lond.* **1**: 214-217, pl. 13.

—, 1895c. New species of shells from Kurachi and the Mekran coast, collected by Mr. F. W. Townsend. *Proc. malac. Soc. Lond.* **1**: 278-280, pl. 18.

TADJALLI-POUR, M., 1974. *Contribution à l'étude systématique de la répartition des mollusques des côtes iraniennes du Golfe Persique.* pp. xv + 224, 25 pls. Doctoral thesis, Université des Sciences et Techniques du Languedoc, Montpellier.

TAYLOR, J. D. & SMYTHE, K. R., 1985. A new species of *Trochita* (Gastropoda: Calyptraeidae) from Oman: a relict distribution and association with upwelling areas. *J. Conch. Lond.* **32**: 39-48, pls 6-7.

TOWNSEND, F. W., 1928. Notes on shell collecting in the northern parts of the Arabian Sea, including the Gulf of Oman and Persian Gulf in the years 1890-1914. *Proc. malac. Soc. Lond.* **18**: 118-126.

TREW, A., 1987. *James Cosmo Melvill's new molluscan names.* pp. 84, text figs. National Museum of Wales, Cardiff.

TURNER, H., 1992. Revival of *Mitra sacerdotalis*. *Hawaii Shell News* **40** (12): 1, 4.

TURNER, R.D., 1971. Identification of marine wood-boring molluscs. pp. 17-64 in: E.B.G. Jones & S.K. Eltringham (eds), *Marine borers, fungi and fouling organisms of wood*. OECD, Paris.

VAUGHT, K. C., 1989. *A classification of the living Mollusca.* (Edited by R. T. Abbott and K. J. Boss.) pp. xii + 189. American Malacologists Inc. Melbourne, Florida.

YARON, I., 1983. A review of the Scissurellidae (Mollusca: Gastropoda) of the Red Sea. *Annln naturh. Mus. Wien* **84**: 263-279.

PHOTOGRAPHIC CREDITS

With the exception of the photographs listed below, all photography for *Seashells of Eastern Arabia* was carried out by Neil Fletcher. The authors would like to express their gratitude to him for his professional execution of the photography for this book.

S. Peter Dance; 13, 191.
Una Dance; 12, 18, 19, 20, 21, 28 (bottom two photos).
Horst Kauch; 117 (bottom photo).
Robert G. Moolenbeek; 132 (*Nassarius emilyae emilyae*), 160 (*Conus lischkeanus tropicensis*).
Walter Sage; 141 (*Lyria leslieboschae*).

INDEX OF TAXONOMIC NAMES

This index is arranged on the 'all-through' principle. Current names are shown in roman type, synonyms in italics, and SUPRA-GENERIC names are given in capitals. Where a species is not figured in the book the index entry is followed by '(not illus.)'. The order of terms is 'species, Genus page number(s)' unless the species name is unknown, when it will be listed as 'Genus sp'. Abbreviations used are sp. or spp. for unidentified species, and n.sp. or n.spp. for new species.

abdita, Acar 207
abyssorum, Solarium 173
ACANTHOCHITONIDAE 191
ACANTHOPLEURINAE 190
achaeus, Anomia 234
achatinus, Conus 157
acicula, Creseis 182
acicula, Graptacme 187
acicula, Triphora 105 (not illus.)
ACLIDIDAE 111
aclis [cf.], Syrnola 176
ACTEONIDAE 178
aculeatum, Epitonium 107
acuminata, Amaea 106
acuminata, Costellaria 152
acuminata dayi, Arca 206
acuminata, Nassaria 127
acupicta, Costellaria 152
acus, Pyramidella 177
acuta, Lophiotoma 169
acuta, Ringicula 181
acuta, Triphora 105 (not illus.)
acutissima, Graptacme 187
adamsi, Tellinella 254
adamsianum, Solamen 216
adamsonii, Pseudocypraea 83
adenense, Cerithium 52
adenensis, Nerita 43
adenensis, Tellinides 252
adenensis, Thracia 281
admirandum, Solarium (Torinia) 173
aduncospinosus, Murex 117 (not illus.)
aegle, Triphora 105 (not illus.)
aequisulcata, Mactra 247
aetheria, Anatoma 28
aethiopica, Pusia 153
affinis, Chiton (Rhyssoplax) 190
affinis, Pupa 179
africana, Solemya 203
agagus, Agagus 33
agalma, Cancellaria 156
agatha, Mitrella 129
agnesiana, Columbella 130
alapapilionis, Natica (Naticarius) 87
alatum, Epitonium 107
alauda, Costellaria 153
albatum, Costellaria 153 (not illus.)
albescens gemmuliferus, Nassarius (Niotha) 132
albicilla, Nerita 43
albina, Mitrella 130
albobrunneus, Pterynotus 118
albus [cf], Malleus 220
alchymista, Dentimargo 146
alcocki, Cardiomya 281
alizonae, Epitonium 107
alizonae, Mitrella 130
alouina, Thais (Mancinella) 124
alphesiboei, Pseudonoba 46
alta, Dosinia 271

amabile, Pusia 155
AMATHINIDAE 178
amathusia, Epitonium 107
ambiguus, Fossarus 50, 51
ambrosia, Terebra 170
amethystus, Gari 259
amouretta, Harpa 143
AMPHIBOLIDAE 186
AMPHIBOLOIDEA 186
ampulla, Bulla 179
amydrozona, Volvarina 147
anaclima, Bathyarca 211 (not illus.)
ANADARINAE 210
anatina, Laternula 280
anatomica, Homalocantha 119
anaxares, Morula 122
ANCILLINAE 144
anembatum, Bittium 51 (not illus.)
angela, Pilucina 236
angulata, Tellinmactra 257
angulatus, Vaceuchelus 33
angusta, Pyramidelloides 112
aniesae, Priotrochus 36
anilis, Terebra 170
annellarium, Cyclostrema 39
annulata, Stosicia 48
annulatus, Ctenoides 225
annulus, Cypraea 72
anodonta, Nerita 43
ANOMALODESMATA 280
ANOMIIDAE 234
ANOMIOIDEA 234
antiquata, Anadara 210
antiquata, Colubraria 140
antiquata, Trigonostoma 157
antiquus, Magilus 126
antoniae, Ziba 152
aperta [cf] Philine 180
approximata, Cuspidaria 281
aptus, Fossarus 50
aquatile, Cymatium (Monoplex) 98
arabica, Cypraea 75 (not illus.)
arabica, Divalinga 236
arabica [cf] Glycymeris 213
arabica, Nodilittorina (Nodilittorina) 45
arabica, Notoplax 191 (not illus.)
arabicus, Fusinus 136
arakana, Timoclea 266
ARCHAEOGASTROPODA 28
archeri, "Cyclostrema" 39
ARCHITECTONICIDAE 172
ARCHITECTONICOIDEA 172
ARCIDAE 205
ARCINAE 205
ARCOIDEA 204
ARCTICOIDEA 264
arcula, Basterotia 240
arcularia plicatus Nassarius (Nassarius) 132
ardisiaceus, Conus 158
arenatus, Conus 158
areola, Heliacus (Heliacus) 174 (not illus.)
areolata, Babylonia 126 (not illus.)
argo, Argonauta 193
ARGONAUTIDAE 193
ARGONAUTOIDEA 193
aristaei, Pseudonoba 47
arsinoensis, Tellina 251
articularis, Scalptia 156
articulata, Rhinoclavis (Rhinoclavis) 54 (not illus.)
ashgar, Siphonaria 184
asiatica, Clementia 275
asmena, Tellina 252 (not illus.)
asper, Euchelus 32
aspera, Rhinoclavis (Rhinoclavis) 54 (not illus.)

asperella, Chama 242
aspersa, Chama 242
aspersa, Columbella 129
asperum, Granosolarium 173
assimile, Acrosterigma 246
ATLANTIDAE 84
atramentarium, Bittium 51 (not illus.)
atrimacula [cf], Rissoella 172
attrahens, Brechites 281
aurantia, Amphilepida 238
aurantia aurantia, Mitra 148
aurantia subruppeli, Mitra 149
aurantiacum, Lyrocardium 245 (not illus.)
aureolata, Pusia 155
auricincta, Turritella 58
auricula, Lunulicardia 245
auricula, Stomatella 37
auriculatus, Modiolus 215
aurisjudae, Ellobium 183
auritae, Hypermastus 112
australe, Fulvia 245
australis, Lutraria 248
australis, Plicatula 229
australis, Solecurtus 263
avellana, Arca 205
avellanaria, Barbatia 209 (not illus.)
axicornis, Chicoreus 113
Axinopsida sp. 237
axis, Hemidaphne 167

babylonia, Terebra 171
badius, Capulus 70
bandorensis, Seila 105
banksii, Chicoreus 113
barbatus [cf], Modiolus 215
bardeyi, Tutufa (Tutufa) 102
Barleeia sp. 48
BARLEEIDAE 48
bathyraphe, Terebra 171
belcheri, Siphonaria 185
bellula [cf], Bothropoma 40
bellula, Nuculana 204
beluchiensis, Strombus 61
berghi, Lamellaria 84
bermudezi, Orbitestella 49
betulinus, Conus 158
biangulosa, Pseudominolia 37
bicarinata, Gari 259
bicolor, Cardites 241
bicolor, Pinna 222
bifasciatus bifasciatus, Clypeomorus 53
bifasciatus persicus, Clypeomorus 53
biliosus, Conus 158
bilocularis var. forskali, Septifer 214
bimaculata, Thais 123
bipartitus [cf], Donax 258
biraghii omanensis, Conus 158
birleyana, Anadara 210
bisulcata, Semicassis 94 (not illus.)
bituberculare, Gyrineum (Gyrineum) 95
bitubercularis, Nassaria 127
BIVALVIA 196 et seq.
blanda, Mitrella 130
blanfordi, Pusia 155
bombayana, Diodora 31
bombayanus, Muricopsis 120
bonnieae, Latirus 140
bonum, Epitonium 107
borbonica, Basterotia 240 (not illus.)
boschi, Ancilla (Sparella) 145
boschi, Conus 161
boschi, Cymatium (Ranularia) 99
boschorum, Bourdotia 237
boschorum, Conus 159
boschorum, Hypermastus 112
bovei, Mitra 149

Brachidontes spp. 214
brassica, Chama 242
brevis, Solen 250
brinkae, Persicula 147
bronnii, Allochroa 183
brookei, Nuculana 204
bruguierei, Tapes 274
bruneus, Turbo 41
brunnea [cf], Syrnola 177
brunneomaculata, Tropaeas 177
brunneus, Chicoreus 113
BUCCINIDAE 126
bucciniformis, Litiopa 51
BUCCININAE 126
bufo, Thais 123
bufo, Tutufa (Tutufa) 104
bulbosa, Oliva 144
bulimoides, Limacina (Munthea) 183
BULLIDAE 179
BULLINIDAE 179
BURSIDAE 101
bushirensis, Scaphander 181
bysma, Retusa 181 (not illus.)

cadabra, Theora 263
CAECIDAE 49
Caecum spp. 49
caelatus [cf], Heliacus (Torinista) 174
caerulescens, Terebra 170
caeruleum, Cerithium 51
caillaudii, Pirenella 56
calceatus, Musculus 216 (not illus.)
calculiferus, Phorus 71
calideum, Epitonium 107
caliendrum, Costellaria 153
CALLIOSTOMATINAE 33
callipareia, Amphilepida 239
callipyga, Circenita 267
callista, Pyrgulina 176
CALLISTOPLACINAE 189
CALLOCHITONINAE 189
calophylla, Bassina 266
camelopardalis, Cypraea 72 (not illus.)
camilla, Emarginula 30
CANCELLARIIDAE 156
CANCELLARIOIDEA 156
cancellata, Vanikoro 67 (not illus.)
canephora, Epitonium 108
capensis [cf], Triphora 105 (not illus.)
caperata, Xenophora 71
capsoides, Serratina 255
CAPULIDAE 70
caputserpentis, Cypraea 72
carbasea, Mumiola 176
carbo, Siphonaria 185 (not illus.)
carbonnieri, Murex 117
CARDIINAE 244
CARDILIIDAE 249
CARDIOIDEA 243
Carditella n.sp. 242
CARDITOIDEA 240
carinatum, "Cyclostrema" 39 (not illus.)
carneola, Cypraea 72
carneolata, Ethalia 35
carneolata var. rubrostrigata, Ethalia 35
carnicolor, Domiporta 151
carnicolor, Semele 261
cartwrighti, Mitrella 130
carystia, Mathilda 175
CASSIDAE 92
CASSINAE 93
castanea, Ancilla (Sparella) 145
castaneus, Melampus 184
castrensis, Lioconcha 269
castus, Nassarius (Zeuxis) 134

catena, Turricula 168
caudata, Linatella (Linatella) 101
caurica, Cypraea 73
CAVOLINIIDAE 182
CAVOLINIINAE 182
centumliratum, Microcardium 245
CEPHALOPODA 191
cerdaleus, Heliacus (Torinista) 174
cerdantum, Epitonium 108
CERITHIIDAE 51
cerithiiformis, Rissoina 47
CERITHIINAE 51
cerithinum, Cerithidium 56
CERITHIOIDEA 50
CERITHIOPSIDAE 104
Cerithiopsis spp. 104, 105
CERITHIOPSOIDEA 104
cernica, Natica 86
ceylanica, Pandora 280
ceylonensis, Colubraria 141
ceylonica, Dosinia 271
chaldeus, Conus 159
CHAMOIDEA 242
charbarensis, Gibberula 146
charmophron, "Cyclostrema" 39
charope, Benthonellania 47
chemnitzianus, Turbo 41
chesneyi, Noetiella 212
chiliarches, Monilea 36
chinensis, Cypraea 73
chinensis, Xenophora (Stellaria) 70
CHIONINAE 266
CHITONIDAE 189
CHITONINAE 190
chrysostoma, Morula 122
cibotina, Barbatia 208 (not illus.)
cicatricosa, Cheilea 66
cicercula, Cypraea 73
cinctella, Terebra 171
cingulata, Cerithidea 56
cingulata, Triphora 105 (not illus.)
cingulatum, "Cyclostrema" 39 (not illus.)
cingulifera, Turritella 57 (not illus.)
cinnamomea, Botula 217
CIRCINAE 266
circula, Neocancilla 151
cistula, Syndosmya 263
clandestina, Cypraea 73
clathrata, Emarginula 30
clathratus, Donax 258
clathrus, Neocancilla 151
CLATHURELLINAE 166
claudia, Tellina 252
CLAVAGELLOIDEA 281
CLAVATULINAE 166
CLEMENTIINAE 275
climacota, Pseudominolia 37
CLIONAE 182
clytospira, Conus 162
coarctata, Florimetis 257
coarctatus, Azorinus 263
cochlea, Turritella 58
cochlearis, Leptomya 262
coenobius, Musculus 217
COLEOIDEA 191
collyra, Cylichna 181
coloba, Cypraea 73
COLUBRARIINAE 140
COLUMBELLIDAE 128
COLUMBELLINAE 129
columen, Pseudonoba 47
columna, Cerithium 52
columnaris, Turritella 58
colus, Fusinus 136
comistea, Zafra 131
compressa [cf], Siphonaria 185
concatenata, Triphora 105
concinna, Primovula (Adamantia) 82 (not illus.)

concinnata, Colubraria 141
concinnum, Epitonium 108 (not illus.)
concinnus, Nassarius (Zeuxis) 134
conicus, Hipponix 66
conicus, Potamides 56
CONIDAE 157
conoidalis, Nassarius (Niotha) 132
CONOIDEA 157
CONORBINAE 166
consentanea, Nucula 203
contabulata, Scalptia 156
contempta, Cyclosunetta 271
continens, Epitonium 108
contracta, Dosinia 271
cophinodes, Acrilla 106
cor, Protapes 273
coralliophaga, Coralliophaga 264
CORALLIOPHILIDAE 125
CORBULIDAE 277
cornuammonis, Spirolaxis 175
cornus, Drupella 121
corolla, Cirsotrema 107
coromandelicus, Conorbis 166
coronata, Lunella 40
coronatum, Bucardium 244
coronatus, Conus 159
coronatus, Nassarius (Nassarius) 132
corrugata, Circe 266
corrugata, Venus 266
corrugata, Viriola 106
corrugata, Xenophora (Xenophora) 71
COSTELLARIIDAE 152
costellifera, Timoclea 266
costifera, Trigonostoma 157
costularis, Coralliophila 125
costulatus [cf], Musculus 217
CRASSATELLOIDEA 243
crassicosta, Cardita 240
crassicostata, Trichotropis 70 (not illus.)
crassilabrum, Opalia 110
CRASSISPIRINAE 167
crebristriata, Diplodonta 238
CRENELLINAE 216
crenifera, Cancellaria 156
crenilabris, Haminoea 180
crenulata, Pterygia 151
crenulifera, Alectryonella 227
CREPIDULIDAE 68
CREPIDULOIDEA 68
cribraria, Cypraea 74
cristagalli, Lopha 226
crossei, Leucorhynchia 38
crumena, Bufonaria (Bufonaria) 101
Cryptomya n.sp. 277
cuccullata, Saccostrea 228
CUCULLAEIDAE 213
CULTELLIDAE 250
cultellus, Ensiculus 250
cumingiana, Bullia 135
cumingianus, Musculus 217, 278
cumingii, Melanella 112
cumingii, Naquetia 118
cumingii [cf], Rimula 30
cumingii, Tonna 89
cuneata, Donax 258
Curvemysella sp. 239 (not illus.)
curvotracheatum [cf], "Laevidentalium" 187
CUSPIDARIIDAE 281
CYAMIOIDEA 240
"CYCLOSTREMATIDAE" 39
cyclostoma, Favartia 119
cylindraceus, Solen 250
cylindrica, Atys 179
CYLINDROMITRINAE 151
CYMATIINAE 96
cymbium, Cucurbitula 278

cynocephalum, Cymatium (Ranularia) 99
CYPRAEIDAE 72
CYPRAEOIDEA 72
CYSTISCINAE 146

dactylus, Pholas 279
dactylus, Solen 250
daedala, Costellaria 153
dancei, Phenacovolva (Turbovula) 82
DAPHNELLINAE 167
daphnelloides, (Acamptochetus) 141
davidboschi, Bursa (Bursa) 102
debilis, Nerita 43
deceptrix, Festilyria 143
declivis, Forskaelena 33
decorus masirensis, Strombus (Conomurex) 61
decurtata, Tugonella 277
decussata, Barbatia 208
degregorii, Ethminolia 35
deificum, Epitonium 108
delessertii, Sinum 88
delicatula, Tibia 65, 67
densilabrum, Lucidinella 46
DENTALIIDAE 186
dentatum, Ovulum 81
dentifera, Diniatys 180 (not illus.)
dentifera, Gastrochaena 278
dentifera, Lucina 235
desetangsii, Scabricola 151
deshayesi, Tapes 274
deshayesianus, Nassarius (Niotha) 133
dhofarensis, Trochita 69
diabolica, Huxleyia 203
diaconalis, Costellaria 153
DIALIDAE 54
DIASTOMATIDAE 56
dictator, Conus 161
didyma, Neverita (Glossaulax) 86
digitalis, Solen 250
DIODORINAE 30
diplax, Cerithidium 56
distinctus [cf], Triphora 105 (not illus.)
DIVARICELLINAE 236
divergens, Ctena 235
divina, Argyropeza 51
djiboutina, Ancilla (Sparella) 145
dolabrata var. terebelloides Pyramidella 177
dolichourus, Haustellum 115
dolium, Tonna 89
DONACIDAE 258
donacina, Sunetta 271
donaldi, Boschitestella 49
donnae, Anachis 129
dorbignyi, Sukasitrochus 28
doriae, Mitrella 130
dorotheae, Pecten 232
DORSANIINAE 134
DOSINIINAE 271
douvillei, Chama 242
DRILLIINAE 167
dubia, Medusafissurella 31
duplicata, Duplicaria 170
duplicata, Stomatella 37
duplilirata, "Ziba" 152

ebraeus, Conus 159
echinacantha, Liotia 39
echinaria, Centrocardita 241
echinata, Bufonaria (Bufonaria) 101
edentula, Anodontia 236
edgariana, Calyptraea 68
edgarii, Terebra 170
effossa, Sunetta 271
egenum, Cerithium 53

ehrenbergi, Anadara 210
elachista, Limopsis 213 (not illus.)
elegans, Conus 159
elegans, Macroschisma 31
elegans, Merica 156 (not illus.)
elegans, Mysia 276
elizae, Mitra 153
elliptica, Cryptomya 277
ELLOBIIDAE 183
ELLOBIINAE 183
ELLOBIOIDEA 183
eloiseae, Boschitestella 49
eloiseae, Punctacteon 179
eloiseae, "Umbonium" 36
elongatus, Plesiothyreus 44 (not illus.)
elongatus, Pterynotus 119
elongella, Obtortio 56
elspethae, Pseudonoba 47
emarginatus, Tellinides 253
EMARGINULINAE 30
emilyae emilyae, Nassarius 132
eous, Similipecten 232 (not illus.)
epiphanes, Hypermastus 112
EPITONIIDAE 106 emilyae
EPITONIOIDEA 106 emilyae, Nassarius 132
equestris, Cheilea 66
eranea, Nassa (Alectryon) 133 (not illus.)
erasmia, Elliptotellina 255
ERATOINAE 84
ERGALATAXINAE 121
erithreus, Trochus (Infundibulops) 34
erosa, Cypraea 74 (not illus.)
erycina, Callista 269
erythraea, Dosinia 272
erythraea, Wallucina 236
erythraeensis, Corbula 277
erythraeensis, Donax 258
erythraeensis, Pecten 232
erythraensis, Laternula 280
erythraeonensis, Anadara 210
erythraeus, Onithochiton 190 (not illus.)
erythrinus, Strombus (Canarium) 60
euchilopteron, "Cyclostrema" 39
EULIMIDAE 112
EULIMOIDEA 111
euloides, Cadulus 188
eumares, "Cyclostrema" 39
eumorpha, Volvarina 148
eupoietum, "Cyclostrema" 39
eutropia, Odostomia 176
eutyches, Minolia 37
evansi, Plesiothyreus 44
exasperata, Costellaria 153
exasperata, Dosinia 272
exigua exigua, Ancilla (Chilotygma) 144
exiguum, "Cyclostrema" 39 (not illus.)
exochum, Vepricardium 244

faba [cf], Amphilepida 238
fabreanus, Punctacteon 179
fabula, Myrtea 236
farsiana, Ancilla (Sparella) 145
farsiana, Timoclea 266
fasciata, Rhinoclavis (Rhinoclavis) 54
FASCIOLARIIDAE 136
FASCIOLARIINAE 136
fasciolaris, Mitra 149
fauroti, Anachis 129
fauroti, Homalocantha 120
faurotis, Semicassis 94
felina fabula, Cypraea 74

INDEX

festiva, Festilyria 143
ffinchi, Cardita 240
FICIDAE 91
filaris, Domiporta 151
fimbriata, Cypraea 74 (not illus.)
fimbriolata, Amaea 106
firmus, Trochus (Infundibulops) 34
fischeri [cf], Chrysallida 176
fischeriana, Pilucina 236
fissilabris, Nassarius
   (Plicarcularia) 133
fissurata, Swainsonia 152
FISSURELLIDAE 30
FISSURELLINAE 31
FISSURELLOIDEA 30
flaccida, Hiatella 278
flammea, Marcia 272
flammea, "Ziba" 152
flammeus, Punctacteon 179
flava, Pyrene 130
flavidulus, Inquisitor 167
flavidus, Conus 159
flexipes, Diacavolinia 182
flexuosa, Pandora 280
flexuosa, Patella 32
floccata, Mitra 149
florida, Callista 269
foliacea, Bassina 266
foliacea, Phyllyda 254
foliaceus, Hipponix 66
foliata, Barbatia 207
foliata, Bufonaria (Bufonaria) 101
   (not illus.)
forceps, Fusinus 138
fordiana, Tricolia 42
fornicata, Crepidula 68
fornicata, Ctenocardia 245
fornicata, Vulsella 222
forskoehlii, Murex 117
fosteri, Chiton 190
foveolatus, Euchelus 33
fragile, Fulvia 245
fragilis, Limaria 225
fragilis, Salinator 186
FRAGINAE 244
fragum, Calliostoma 33
frederici, Nassarius (Zeuxis) 134
frons, Dendrostrea 226
FULGORARIINAE 143
fultoni, Trochus (Infundibulops) 34
fultoni, Turritella 58
fulvescens, Nebularia 150
fumosus, Cantharus 127
funiculare, Calliostoma 33
funiculata, Diodora 30
fusca [cf], Scalptia 156
fuscobasis, Terebra 171 (not illus.)
fusiformis, Strombus (Canarium) 60
FUSININAE 136

GADILIDAE 188
GADILINIDAE 188
galathea, Plesiothyreus 45 (not illus.)
GALEOMMATOIDEA 238
gallagheri, Medusafissurella 31
gallagheri, Sinistralia 138
gallinacea, Erato 84
gangranosa, Cypraea 74 (not illus.)
GASTROCHAENOIDEA 278
GASTROPODA 28
gemmata, Triphora 105
generalis maldivus, Conus 160
genethila, Diplodonta 237
gennesi, Clanculus 34
geoffreyana, Pusia 155
geographus, Conus 160
gibberulus gibberulus, Strombus
   (Gibberulus) 60
GIBBULINAE 33
gigantea, Gastrochaena 278

glabrata, Atactodea 249
glabrata, Littoraria (Littoraria) 45
glabratum, Epitonium 108
gladysiae, Phos 127
glans glans, Nassarius
   (Alectrion) 132
glaucum, Phalium 93
globosa [cf], Diplodonta 237
globosa, Niveria (Cleotrivia) 83
globulus, Cypraea 74 (not illus.)
gloriandus, Spondylus 234
gloriola, Epitonium 108
GLOSSOIDEA 264
GLYCYMERIDIDAE 213
gnorima, Bufonaria (Bufonaria) 102
gondola, Argonauta 193
goniochila, Styliferina 55
goniophora, Epitonium 108
gracilis, Cypraea 74
gracilis, Ficus 91
gracillima, Mathilda 175
gradata, Pseudominolia 37
granatina, Domiporta 151
grandimaculatum, Cymatium
   (Lotoria) 96
granularis, Bursa (Bufonariella) 102
granulata, Morula 122
Graphis sp. 111
gratiosa, Leucotina 178
grayana, Cypraea 75
griffithii, Ptychobela 169
GRYPHAEIDAE 228
gualteriana, Natica 86
gualtierii, Architectonica 172
   (not illus.)
gubernaculum, Beguina 241
gueriniana, Vanikoro 67
guillaini, Marginella 147
gyalum, "Cyclostrema" 39
gyratus, Mipus 126

haddoni, Acanthopleura 190
HALIOTIDAE 28
haliotoideum, Sinum 88
HAMINOEIDAE 179
hanleyanus, Leiosolenus 218
HARPIDAE 143
haustellum longicaudus,
   Haustellum 115
hebraea, Pitar 270
hectica, Impages 170
helena, Costellaria 153
helichrysum, Terebra 171
helicoides, Separatista 70
helvola, Cypraea 75
hemicardium, Fragum 245
hemprichi, Petricola 276
henjamense, "Cyclostrema" 39
HETERODONTA 234
HETEROPODA 84
hians, Argonauta 193
HIATELLOIDEA 277
himeroessa, Nassarius (Niotha)
   133
hinduorum, Seila 105
HIPPONICOIDEA 66
HIPPONICIDAE 66
histrio, Cypraea 75 (not illus.)
histrio, Dosinia 272
holosphaera, Diplodonta 237
homalaxis, Solarium 174
HOMALOPOMATINAE 41
horsfieldi, Caecella 249
humilis, Chlamys 231 (not illus.)
hyalina, Cycloscala 107
hybrida, Philippia 174
HYDATINIDAE 180
hyotis, Hyotissa 228
hystrix, Scalptia 156
hystrix, Spondylus 233

iatricus, Chiton 190
icela, Turbonilla 178
ictriella, Pseudonoba 47
idalia, Epitonium 108
idonea, Triphora 105
idyllius, Nassarius 134
ignea, Pisania 128
illustris, Turritella 58
IMBRICARIINAE 151
implexus, Heliacus (Torinista) 174
inaequistriata, Gastrochaena 278
incarnata [cf], Laciolina 254
incisura, Emarginella 30
incolumnis, Triphora 105 (not illus.)
inconspicua, Tornatina 181
inconstans prunulum, Paradrillia 168
indica [cf] Hinemoa 176
indica, Indocrassatella 243
indica, Lophiotoma 169
indica, Scapharca 211
indicum, Periploma 281
infausta, Costellaria 153
inflata, Acropaginula 256
inflata, Scapharca 211
infracostata, Peasiella 46
infundibuliformis, Heliacus
   (Teretropoma) 174
inornata, Ancilla (Sparella) 146
   (not illus.)
inscriptus, Conus 160
insignis, Zebina 48
insulaechorab curta, Tibia 65
intermedia, Circe 267
intermedia, Littoraria
   (Littorinopsis) 45
interpres, Triphora 105
intertexta, Berthais 67
invicta, Turris 169
involuta, Tornatina 181
ios, Tricolia 42
IRAVADIIDAE 46
iridescens, Broderipia 37
iridifulgens, Ethminolia 35
irregulare, Epitonium 108
isabella, Cypraea 75
ISCHNOCHITONIDAE 189
ISCHNOCHITONINAE 189
ischnus, Nassarius (Aciculina) 132
ISOGNOMONIDAE 222
isseli, Arcopella 257
isseli, Granulina 146
isseli, Peasiella 46
iteina, Mitra 154

jacksoni, Anatoma 28
jactabundus, Nassarius (Niotha) 133
janthina, Janthina 111
JANTHINIDAE 111
japonica, Omalogyra 172
jehennei, Recluzia 111
jocosa [cf], Oscilla 176
jomardi, Epitonium 109
jonathani, Turbo 41
jousseaumei, Bathytormus 243
juglandula, Limea 226 (not illus.)

kallima, Tellina 262
karachiensis, Patella 32
kermadecensis, Conus 160
kieneri, Cypraea 75
kochi, Rhinoclavis (Proclava) 54
kochi, Trochus (Infundibulum) 35
konkanensis [cf], Cronia 121
kotschyi, Osilinus 36
kuesterianus, Hexaplex 116
kurracheensis, Siphonaria 185
kyrina, Kyrina 276

labiata, Cucullaea 213
labiosa, Dosinia 272

labiosum, Cymatium
   (Turritriton) 100
labrella [cf], Cassidula 183
labroguttata, Delonovolva 82
lacera, Thais 123
lacteus, Calpurnus
   (Procalpurnus) 81
lacunosa, Acrosterigma 246
LAEVICARDIINAE 245
LAEVIDENTALIIDAE 187
laevigata, Architectonica 173
laevigatum, Sinum 88
laidlawi, Epitonium 109
lalage, Mitra 152
lamarckii, Cypraea 76
lambis, Lambis 62 (not illus.)
LAMELLARIIDAE 84
lamellaris, Antigona 265
lamellosa, Cancellaria 157
lamellosa, Gyroscala 110
lamyi, Chiton 190
lapicida, Petricola 276
lateralis, Sheldonella 212
LATERNULIDAE 280
laxatum, Epitonium 109 (not illus.)
layardi, Timoclea 266
layardii, Nuculoma 203
legumen, Isognomon 222
lemniscatus, Conus 164
lentiginosa, Cypraea 76
leptocarya, Limatula 226
leptomita, Turritella 110
leslieboschae, Lyria 141
leucedra, Kellia 239
ligneus, Modiolus 215
lilacea, Mactra 247
lima, Leiosolenus 217 (not illus.)
lima, Lima 225
limacina, Cypraea 76
LIMACINIDAE 183
LIMOIDEA 225
LIMOPSIDAE 213
LIMOPSOIDEA 213
Linatella (Gelagna) n. sp. 100
lineata, Bullina 179
lineata, Tanea 87
linjaica, Turbonilla 178
LIOTIINAE 39
lischkeanus [cf], Solen 250
   (not illus.)
lischkeanus tropicensis, Conus 160
lissum, Malluvium 66, 67
listeri, Cypraea 74
LITHOPHAGINAE 217
LITIOPIDAE 55
litterata, Strigatella 150
LITTORINIDAE 45
LITTORININAE 45
LITTORINOIDEA 45
livida, Chlamys 231
livida, Glycymeris 213
lividus, Conus 160
Lodderena n.sp. 38
loisae, Terebra 171
longii, Nerita 43
longitrorsum, Laevidentalium 188
longurionis, Conus 161 (not illus.)
lotorium, Cymatium (Lotoria) 97
LOTTIIDAE 31
lucida, Splendrillia 167
Lucinella sp. 237
LUCINIDAE 235
LUCININAE 235
LUCINOIDEA 234
lucificus [cf], Conus 161
luctuosa *forma* rutila, Nebularia 150
luteostoma, Tonna 91
luzonica, Lepidozona 189 (not illus.)
lynx, Cypraea 76
lyra, Epitonium 109

LYRIINAE 141
Lyrocardium n.sp. 245

macandrewi, Cypraea 76
macandrewi, Terebra 171
macfadyeni, Timoclea 266
MACOMINAE 257
macrophylla, Irus 275
macroptera, Pteria 219
MACTRIDAE 247
MACTROIDEA 246
maculata, Terebra 171
maculata, Turritella 58
maculosa, Acrosterigma 246
maculosa, Gari 260
maculosa, Pyramidella 177
maculosum, Epitonium 109 (not illus.)
madreporara, Coralliophila 125
madreporicus, Asaphinoides 276
mahimensis [cf], Alvania 47
majeeda, Carditopsis 241
malabaricus, Haustellum 115
malcolmensis, Costellaria 154
malcolmensis, Epitonium 109
MALLEIDAE 220
mammilla, Polinices 86
mammilliferus, Nassarius (Zeuxis) 134
mammosa, Opalia 111
manceli, Natica (Naticarius) 87
MANGELIINAE 168
manillensis, Barnea 279
manumissa, Peronaea 253
maraisi, Patelloida 31
margarethae, Arcopsis 212 (not illus.)
margaritacea, Striostrea 227
margariticola [cf], Cronia 121
margaritifera, Pinctada 220
MARGARITINAE 32
marginalis pseudocellata, Cypraea 77
marginata [cf], Pseudosimnia 83
marginata, Truncatella 46
MARGINELLIDAE 146
MARGINELLINAE 146
mariae, Haliotis 28
marisrubri, Spondylus 233
marjoriae, Favartia 119
marmorata, Marcia 272
marmoreus, Nassarius (Telasco) 134
martinii, Amaea 106
martinii, Melanella 112
masirana, Persicula 147
maskatensis, Glycymeris 213
mastalleri, Acanthochitona 191 (not illus.)
MATHILDIDAE 175
mauritiana, Bullia (Bullia) 134
mauritiana, Cypraea 77
mauritianus, Strombus 61
maxillaris, Triphora 106 (not illus.)
maxima, Tridacna 246
mazagonica [cf], Gibberula 146
megatrema, Macroschisma 31
mekranica, Timoclea 266
MELAMPODINAE 184
melanoides, Bullia (Bullia) 134
melanostoma, Litiopa 55
melanostoma, Mammilla 85
melanostoma, Merica 156
MELONGENIDAE 136
melvilli, Conus 161
melvilli, Medusafissurella 31
mendicaria, Engina 128
Merelina sp. 47
MERETRICINAE 268
meroclista, Dendropoma 58
MESODESMATIDAE 249

methoria, Tellina 252
microzonias depexa, Pusia 155
miles, Conus 161
milesi, Conus 161
millegrana, Nodilittorina (Nodilittorina) 46
milneedwardsi, Conus 162
MILTHINAE 236
minima, Phasianella 42
minolina, Ethalia 35
minor, Amaea 106
mirabilis, "Corbula" 238
miranda, Pyramidelloides 112
mirbahensis, Hiatula 260
mitra, Mitra 149
mitralis, Otopleura 177
MITRIDAE 148
MITRINAE 148
MODIOLINAE 215
MODULIDAE 51
mollis, Cantharus 127
moltkiana, Meiocardia 264
moneta, Cypraea 78
monile, Conus 162 (not illus.)
monilifera, Laemodonta 184
monilis, Gibberula 147
MONODONTINAE 33
moreleti, Morchiella 38
moritinctum, Cymatium 99
mulawana, Tivela 268
multangula, Gastrana 258
multistriata, Chlamys 231 (not illus.)
munda [cf], Bothropoma, 40
muricata, Pinna 224
MURICIDAE 112
MURICINAE 113
MURICOIDEA 112
MURICOPSINAE 119
musicus, Conus 163
mutabilis mutabilis, Strombus (Canarium) 61
MYIDAE 276
MYOIDEA 276
MYRTEINAE 236
Mysella sp. 240
MYTILINAE 214
MYTILOIDEA 214

nabateus, Trochus 36
NACELLINAE 32
namocanus badius, Conus 162
nana, Hastula 170
nassa, Scalptia 157
NASSARIIDAE 131
NASSARIINAE 132
nassatula forskalii, Peristernia 140
nassoides, Terebra 172
natalensis, Chlamys 231 (not illus.)
natalensis, Engina 128
natalensis, Nodilittorina (Nodilittorina) 46
natalensis, Scapharca 211
natator, Gyrineum (Gyrineum) 95
NATICIDAE 85
NATICINAE 86
NATICOIDEA 85
navarchus, Clavatula 166
nebrias, Strigatella 150
nebrites, Cypraea 78
nebulosa, Monodonta 33
nebulosa, Terebra 172
nedyma, Pseudominolia 37
nelliae spuria, Turricula 168
NEOLORICATA 189
NERITIDAE 43
NERITINAE 43
NERITOIDEA 43
neritoidea, Coralliophila 125
Nesobornia sp. 239 (not illus.)

nicobarica, Meropesta 248
nicobaricum, Cymatium (Monoplex) 98
niger, Planaxis 50
nigra, Coriocella 84
nigra [cf], Pinctada 220
nigrita, Tutufa (Tutufella) 104
nigropunctatus, Conus 162
nitens, Tellina 252
nitidus, Donax 259
nivea, Nassaria 127
nobilis, Tugonia 276
noduliferus, Chlamys 230
nodulosum adansonii, Cerithium 52
nodulosus, Nassarius (Niotha) 133
NOETIIDAE 212
nomadica, Pyrene 131
normalis, Malvufundus 221
notata, Cypraea 74
Notocochlis n. sp. 87
novemcarinata, Lodderia 38
NUCINELLOIDEA 203
nucleus, Cassidula 183
nucleus, Parviperna 222
NUCULANIDAE 204
NUCULANOIDEA 203
NUCULIDAE 203
NUCULOIDEA 203
numisma, Hyotissa 229
nussatella, Conus 162
nux, Scabricola 152

obeliscus, Costellaria 154
obesus, Leiosolenus 218
obesus, Nassarius (Niotha) 133 (not illus.)
obliquata, Barbatia 207
obliquata, Scalptia 157
oboesum, Cymatium (Ranularia) 100
obolos, Pseudomalaxis 174
obscura [cf], Volvarina 148
obscurus, Conus 163
obscurus, Priotrochus 36
obtecta, Aspidopholas 279
obtusa, Marginella 147
obtusalis, Cadella 253
occidens, Gari 260
ocellata, Cypraea 78
OCINEBRINAE 121
ocrinium, Cyclostrema 40
octangulatum, Dentalium 186
ODOSTOMIINAE 176
oldi, Strombus (Tricornis) 62
olearium, Tonna 91
olivaria, Sulcerato 84
OLIVIDAE 144
OLIVINAE 144
OMALOGYRIDAE 172
omanense, Papillicardium 244 (not illus.)
omanensis, Callistochiton 189
omanensis, Muricopsis 120
omanensis, Plesiothyreus 44
omanensis, Retusa 181
omaria, Conus 163
onyx succincta, Cypraea 78
oodes, Granulina 146
opima, Marcia 272
opisthochetos, Ptychobela 169
OPISTHOBRANCHIA 178
orbignyi, Conus 163 (not illus.)
ORBITESTELLIDAE 48
ornata, Lioconcha 270
ornatissima, Divaricella 237
oryza oryza, Triviostra 83
osculans, Plesiothyreus 45 (not illus.)
osiridis, Pusia 155
Ostrea sp. 228
OSTREIDAE 226

OSTREOIDEA 226
otaheitensis, Bullia (Bullia) 135
ovalina, Mactra 247
ovalis, Ancilla (Sparella) 146
OVULIDAE 81
ovum, Ovula 82
oyamai, Tutufa (Tutufella) 104

pachystoma, Rissoina 47
pacifica, Costellaria 154
pagodaeformis, Latirus 140
palatam, Quidnipagus 255
pallasii, Epitonium 110
pallidula, Xenophora (Xenophora) 72
palustris, Terebralia 57
pamela, Cavilucina 235
panama, Purpura 122
PANDORIDAE 280
PANDOROIDEA 280
panhi, Subemarginula 30
papilla, Eunaticina 87
papyracea, Clementia 275
paradisiaca nodosa, Volema 136
pararabicus, Plesiothyreus 45
pardalis, Pusia 155
parthenopeum, Cymatium (Monoplex) 98
parva, Barbatia 208
parvatus sharmiensis, Conus 163
pasithea, Costellaria 154
PATELLIDAE 32
PATELLINAE 32
PATELLOIDEA 31
PATELLOIDINAE 31
paucicostata, Nipponaphera 156
paulboschi, Favartia 119
pauperus, Nassarius (Hima) 132
peasei, Amygdalum 216
peasei, Emarginula 30
pectinata, Servatrina 224
pectinatum, Gafrarium 268
PECTINIDAE 230
PECTINOIDEA 230
pectunculus, Glycymeris 213
peculiaris, Curvemysella 239
PEDIPEDINAE 184
Pedipes sp. 184
peilei, Amphilepida 239 (not illus.)
pellicula, Raeta 249
pellucida, Calyptraea 68
pellucida, Meropesta 248
pellucidula, Triviostra 83
pellyana, Tellidora 256
pellyi, Atys 180
PELYCIDIIDAE 49
penguin, Pteria 219
penitricinctus [cf], Plesiotrochus 53
pennaceus quasimagnificus, Conus 163
perca, Gyrineum (Biplex) 94
peregrinus, Chiton 190
perfragilis, Musculista 216
pergrandis, Volvarina 148
perinesa, Barbatia 208
perinvolutum [cf], Fissidentalium 186
PERIPLOMATIDAE 281
PERISTERNIINAE 140
perna, Pharaonella 254
peronii, Atlanta 84
perparvulum, Bittium 56
perryi, Cymatium (Lotoria) 97
persiana, Tornatina 182
persica, Bullia (Bullia) 135
persica, Coralliophila 125
persica, Cypraea 78
persica, Splendrilla 168
persica, Terebra 171
persicus, Nassarius (Plicarcularia) 133
persicus, Strombus (Conomurex) 61

INDEX

PERSONIDAE 101
perspectiva, Architectonica 173
pervicax, Cerithium 51
peselephanti, Neverita (Neverita) 86
pesmatacis, Anadara 210
Petricola n.sp. 276
PETRICOLIDAE 276
PHALIINAE 93
pharaonis, Sepia 192
pharaonius, Clanculus 34
PHASIANELLIDAE 42
phaula, Zafra 131
PHENACOLEPADIDAE 44
PHILINIDAE 180
PHILINOIDEA 178
philippinarum, Modiolus 215
philotima, Inquisitor 167
PHOLADIDAE 278
PHOLADOIDEA 278
phormis, Rissoina 48
PHOTINAE 127
phymotis, Stomatia 37
physis, Hydatina 180
pica, Patella 32
picta, Crassatina 243
picta, Perna 214
pileare, Cymatium (Monoplex) 99
pilula [cf], Bothropoma 40
pinguis, Colina 53
pinguis, Pinguitellina 256
PINNOIDEA 222
pintado, Babylonia 126
PISANIINAE 127
PITARINAE 269
placens, "Cyclostrema" 39
placenta, Placuna 234
PLACUNIDAE 234
PLANAXIDAE 50, 67
planiliratus, Conus 67 (not illus.)
platyaulax, Comus 267
platysia [cf], Primovula 82
PLEUROTOMARIOIDEA 28
plica, Decatopecten 232
plicarium, Vexillum 154
plicata, Acar 206
plicata, Plicatula 230
plicata, Vanikoro 67
plicatilis, Spengleria 278
plicatula, Alectryonella 227
PLICATULOIDEA 229
plicatus sibbaldii, Strombus (Dolomena) 61
POLINICINAE 85
polita orbignyana, Nerita 44
polita, Siliqua 251
polygona, Costellaria 154
POLYPLACOPHORA 188
pomatiella, Sigatica 87
pomum, Malea 88
ponderosa, Tivela 268
ponderosa unicolor, Casmaria 93
ponsonbyi, "Petricola" 275
POROMYOIDEA 281
POTAMIDIDAE 56
potensis, Scabricola 152
pouloensis, Tomopleura 168
praerupta, Psammotreta 257
pretiosa, Scala 110
pretiosa, "Ziba" 152
princeps, Babelomurex 125
prionotus, Dischides 188
producta, Aspella 113
profunda, Patelloida 32
prolongata, Janthina 111
prominulum, "Cyclostrema" 39
PROPEAMUSSIDAE 232
propinqua, Pyrene 131
propinquum, Ringicula 181
PROSOBRANCHIA 28
Protapes n.sp. 274

proteus, Bittium (Dahlakia) 51
PROTOBRANCHIA 202
PROTOCARDIINAE 245
PSAMMOBIIDAE 259
psammosphaerita, Psammosphaerica 261
pseudoconcinnus, Nassarius (Zeuxis) 134
pseudolima, Plagiocardium 244
pseustes, Natica 87
PTERIIDAE 219
PTERIOIDEA 218
PTERIOMORPHA 204
pudica, Prionovolva 83
pudicissma, Pitar 270
puerpera, Periglypta 265
pulchella [cf], Rissoina 48
pulchella [cf], Tellinella 255
pulchella pericalles, Cypraea 79
pulchellus, Latirus 140
pulcherrima, Trichotropis 70
pulcherrimus, Rubritrochus 33
pulchra, Cypraea 79
pulchrelineatus, Conus 164
pulchrior, Gurmatia 176
pulicaris, Natica 86
PULMONATA 183
punctostriata, Nebularia 150
pupoides, Obtortio 56
purpurata, Architectonica 173
purpurea, Ervilia 262
pusilla, Nassaria 127
pusillum, Gyrineum (Gyrineum) 95
pustula, Marikellia 239
pustulata, Haliotis 28
PYRAMIDELLIDAE 175
PYRAMIDELLINAE 177
PYRAMIDELLOIDEA 175
PYRENINAE 129

qeratensis, Caecella 249 (not illus.)
quadrapicalis, Tesseracme 187
quadrasi, Iravadia 46
quadricarinatum, "Cyclostrema" 39
quasimodoides, Sinum 88
quercinus, Conus 164
quinquecarinatum, "Cyclostrema" 39

radiata, Pinctada 220
radiatus, Bathytormus 243
radiatus, Psilaxis 174
radiatus, Turbo 41
radula, Coralliophila, 125
ramosus, Chicoreus 115
rana, Bufonaria (Bufonaria) 102
RANELLIDAE 94
ranzanii, Cymatium (Cymatium) 96
RAPANINAE 124
rapax, Laemodonta, 184
rapiformis, Rapana 124
raricostum, Epitonium 109
rassierensis, Anachis 129
rastella, Tellinella 254
rattus, Conus 164
rawsoni, Engina 128
raysutana, Anachis 129
recondita, Sulcerato 84
recurva, Nassaria 127
reevei [cf], Architectonica 173
reevei forma lineolatum, Dentalium 186
reflexa, Chama 242
regula, Malvufundus 221
rejecta, Tivela 268
remanalva, Terebra 170
renovata [cf], Crenavolva 81
replicata, Canalispira 146
replicatum, Epitonium 109
requiescens, Bentharca 209 (not illus.)

resplendens, Splendrillia 168
reticularis, Distorsio 101
reticulata, Fenella 56
reticulata, Periglypta 265
retroversa, Limacina (Limacina) 183
Retusa sp. 181
RETUSIDAE 180
revelata, Mitra 154
revimentalis, Callomysia 240 (not illus.)
rhodostoma [cf], Peristernia 140
rhomboides, Loxoglypta 257
richardi, Afrocardium 244
rileyi, Hexaplex 116
RINGICULIDAE 181
RISSOELLIDAE 172
RISSOELLOIDEA 172
RISSOIDAE 47
Rissoina sp. 48
rissoinaeforme, Epitonium 109
RISSOOIDEA 46
robusta, Lithophaga 217
rochebrunei, Mactra 247
rogersi, Bullia (Bullia) 135
romalea, Liotia 40
roni, Gabrielona 42
rosaceus [cf], Mipus 126
rosamunda, Semelangulus 262
rosea rosea, Phenacovolva (Phenacovolva) 82
rosea, Siphonaria 186
roseatus, Phos 127
rota, Cellana 32
rota, Homalocantha 119
rota, Scissurella 28
rotula, Heliacus (Torinista) 174 (not illus.)
rotulacatharinea, Spirolaxis 175
rubeculum rubeculum, Cymatium (Septa) 100
rubens, Leptothyra 41
rubra [cf], Stomatia 38
rubrococcinea, Coralliophila 125
rueppelli, Cerithium 52
rueppellii, Diodora 31
rueppellii, Sphenia 277
rufa, Cypraecassis 93
rugifera, Circe 266
rugosa, Thais 123
rugosa, Venerupis 274
rugulosa [cf], Nucula 203
ruppelliana, Hiatula 260
ruschenbergeri, Chlamys 231

saccata, Streptopinna 225
sacerdotalis, Nebularia 150 (not illus.)
sacra, Tylotiella 168
salebrosa, Medusafissurella 31
salsettensis, Thracia 281
sandwichensis [cf], Rochefortina 262
sanguinolenta, Nebularia 150
savignyi, Siphonaria 186
savignyi, Thais 123
scabridum, Cerithium 52
scabriuscula, Trivirostra 83 (not illus.)
scabrosus, Trochus (Belangeria) 34
scalare, Epitonium 110
scalarina [cf], Scalptia 157
scaliola, Ervilia 262
SCALIOLIDAE 56
scalpellum, Donax 258
SCAPHANDRIDAE 181
scaphella, Ancilla (Sparella) 146
SCAPHOPODA 186
schepmani, Epitonium 110

Scintilla sp. 239
SCISSURELLIDAE 28
scitula, Costellaria 154 (not illus.)
scobinata, Plesiothyreus 45 (not illus.)
scobinata, Scutarcopagia 256
scolopax, Murex 117
scopulorum, Miralda 176
scorpio, Homalocantha 120 (not illus.)
scripta, Circe 267
sculpta, Nuculana 204
sculptilis, Costellaria 154
sebae, Mammilla 85
selasphora, Zafra 131
SEMELIDAE 261
semen, Cadella 253
semiplicata, Bullia (Bullia) 135
semistriata, Diala 54
semisulcata, Cardilia 249
semperiana, Bellucina 235
senatoria, Chlamys 230
senhousia, Musculista 216
SEPIIDAE 192
sepulchralis, Priotrochus 36
sericata, Obtellina 256
serriale, Maculotriton 121
setigera, Barbatia 208
sewelli, Episiphon 188
seychellarum, Iacra 262
siebenrocki, Parvamussium 232
sieboldii, Acteon 179
simiae, Mammilla 85
simplicifilis, Gregariella 217
sinensis, Inquisitor 167
sinensis, Rhinoclavis (Rhinoclavis) 54
sinensis, Semele 261
singaporensis, Diodora 31
singularis, Primovula 82
sinicum, Umbraculum 182
SININAE 87
sinuosa, Protapes 273
SIPHONARIIDAE 184
SIPHONARIOIDEA 184
SIPHONODENTALIIDAE 188
sirahensis, Modiolus 216
situla, Nassa 122
SKENEIDAE 38
sloanii, Solen 249
SMARAGDIINAE 44
smytheae, Bullia (Bullia) 135
solanderi, Meropesta 248
solariellum, "Cyclostrema" 39
solaris, Xenophora (Stellaria) 71
SOLARIELLINAE 36
SOLECURTIDAE 263
SOLEMYOIDEA 202
SOLENIDAE 249
SOLENOIDEA 249
solida, Phasianella 42
solida, Woodringilla 39
sordidula, Rhinoclavis (Proclava) 54 (not illus.)
soror, Lienardia 166
souverbiana, Smaragdia 44
souverbianus, Plesiotrochus 53
soverbii, Colubraria 141
sowerbyanus, Cantharus 127
sowerbyi, Lima 225
speciosa, Tellina 262
spectabilis, Duplicaria 170
spectrum, Turritella 58 (not illus.)
spinifera, Calyptraea 68
spiralis, Cantharus 127
spirata, Babylonia 126
spirata, Mumiola 176
splendidulus, Nassarius (Niotha) 133
SPONDYLIDAE 233
spurca, Etrema 166
spurca, Glycymeris 213

INDEX

squamosissima, Coralliophila 125
staphylaea, Cypraea 79
steindachneri, Propeamussium 232
stellata, Architectonica 173
stellata, Perrinia 33
stephanucha, Costellaria 155
stocki, Conus 164
STOMATELLINAE 37
strenaria, Bullia 135
striata, Cumingia 261
striata, Martesia 279
striatellus, Conus 164
striatula traillii, Crenavolva 81
striatularis, Glycymeris 213
striatum, Taphon 136
striatus, Conus 164
STROMBIDAE 60
STROMBOIDEA 60
strongyla, Natica 87
sturtiana, Nassarius 134
stylifera, Turbonilla 178
subdisjuncta, Daronia 38
subintermedia, Ficus 92
sublaevigatum, Trapezium 264
subquadrata, Corbula 277 (not illus.)
subreticulatus [cf], Macromphalus 67
subrostrata, Leptomya 263
subrotundata, Diplodonta 237
subrugosa, Subemarginula 30
subucula, "Ostrea" 228
succincta, Linatella (Gelagna) 101 (not illus.)
sueziense, Parvicardium 244
sueziensis, Gibberula 147 (not illus.)
sueziensis, Tonicia (Lucilina) 191 (not illus.)
sulcarius, Tapes 274
sulcata, Redicirce 268
sulcata, Serratina 255
sulcatus, Obeliscus 177
sulcatus, Planaxis 50
sulcifera, Eratoena 84
sulcifera, Granata 32
sulcifera martensi, Diala 55
sulculosa, Corbula 277
sulphuratus, Conus 165
SUNETTINAE 271
supremum, "Cyclostrema" 39
suturalis, Nassaria 127
sykesii, Epitonium 110
sykesii [cf], Laemodonta 184
symmetrica, Striarca 212
symphenacis, Arca 205
syndesmyoides, Tellinimactra 257
syrphetodes, Mammilla 85

taeniatus, Conus 164
tahitensis, Bullia 135
taitensis, Corbula 277
talpa, Cypraea 79
TAPETINAE 272
tapparonei, Homalopoma 41
tarutana, Retusa 181
tayloriana, Funa 167
tayloriana, Peristernia 140
tectum, Modulus 51
telaaraneae, Natica 87
telamonia, Mathilda 175
telescopium, Telescopium 56
Tellina n.sp. 252
TELLINIDAE 251
TELLININAE 251
TELLINOIDEA 251

tenebrica, Didimacar 212
tenthrenois, Bittium 51 (not illus.)
tenuirostrum tenuirostrum, Murex 118
terebellum, Terebellum 65
terebra thomasi, Conus 165
TEREBRIDAE 170
TEREDINIDAE 279
teres, Cypraea 80
terverianum, Prunum 147
tessulatus, Conus 165
testudae, Ancilla (Chilotygma) 144 (not illus.)
testudinaria, Pyrene 130 (not illus.)
teulerei, Cypraea 80
textile, Conus 165
textilis, Nerita 44
THAIDINAE 121
THECOSOMATA 182
thelacme, Macromphalus 67
thelcterium, Epitonium 110
theoreta, Drillia 167
thia, Daphnella 167
THRACIIDAE 281
THYASIRIDAE 237
thyraea, Parviterebra 130
thyrideum, Parvamussium 232 (not illus.)
tiberiana, Voorwindia 48
ticaonicus, Hipponix 66
tigerina, Codakia 235
tigris, Cypraea 80
tincta [cf], Chlamys 231
tinctilis, Margovula 82 (not illus.)
tissoti, Thais 123
tomlini, Dentalium 186
TONICIINAE 190
TONNIDAE 88
TONNOIDEA 88
topaza, Drillia 167
tortirostris, Pteria 219
tortuosa, Trisidos 209
torulosa, Turritella 58
townsendi, Chlamys 230
townsendi, Donax 259
townsendi, Epitonium 110
townsendi, Fusinus 138
townsendi, Gibbula (Enida) 33
townsendi, Citharomangelia 168
townsendi, Nassarius 134
townsendi, Nebularia 150
townsendi, Trichotropis 70
TRACHYCARDIINAE 246
tranquebarica, Bullia (Bullia) 136
tranquilla, Natica 86
trapezium, Pleuroploca 136
tremulina, Oliva 144
tricarinata, Amathina 178
tricarinata, Eglisia 110
tricarinata, Vanikoro 68
TRICHOTROPIDAE 70
tricornis, Strombus (Tricornis) 62
tricostata, Patella 178
tridentata, Zebina 48
trilineatum, Cymatium (Ranularia) 100
tripartitus, Leiosolenus 218
TRIPHORIDAE 105
TRIPHOROIDEA 105
tripum, Cymatium (Ranularia) 100
triseriata, Terebra 172
tristoma, Triphora 106 (not illus.)
tritonoides, Pisania 128

TRIVIIDAE 83
TRIVIINAE 83
TROCHIDAE 32
TROCHINAE 34
trochlearis, Fossarus 51
TROCHOIDEA 32
tropaeum, Lippistes 70
tropica, Primovula 82
tropica, Yoldia 204
trotteriana, Iacra 262
truncata sebae, Lambis 62
TRUNCATELLIDAE 46
trygonina [cf], Sepia 192
tryphera, Mactrinula 248
tuberculatum, Cerithium 53 (not illus.)
tumida, Dosinia 272
tumida, Pitar 271
tumidus, Polinices 86
turbinellus, Vasum 143
TURBINIDAE 39
TURBININAE 40
TURBONILLINAE 177
turdus winckworthi, Cypraea 81
turriculata, Atlanta 84
TURRICULINAE 168
TURRIDAE 166
TURRINAE 169
turrigera, Retusa 181
TURRITELLIDAE 57
turtoni, Lasaea 240

umbilicata, Janthina 111 (not illus.)
umbonella, Amiantis 269
UMBONIINAE 35
UMBRACULIDAE 182
UMBRACULOIDEA 182
umbraculum, Umbraculum 182
uncinata, Cavolinia 182
undatus, Micromelo 180
undosus, Cantharus 128
undulata, Emarginula 30
undulata, Paphia 273
unedo, Unedogemmula 169
unguis, Scutus 30
UNGULINIDAE 237
unifasciale, Pusia 156
urceus urceus, Strombus (Canarium) 61
uropigimelana, Anadara 210

vaillanti, Solamen 216
vaillantii, Acanthopleura 190
valentiana, Eburna 126
valtonis, Tellina 252
Vanikoro sp. 68
VANIKORIDAE 67
vanninii, Callochiton 189 (not illus.)
varia, Stomatella 37
variabilis, Brachidontes 214
variabilis, Funa 167
variabilis, Pagodatrochus 37
variabilis, Scintillula 239
variabilis, Serpulorbis 58
variabilis, Tricolia 43
varians, Euplica 129
varicosum, Cirsotrema 107
variegata, Cardita 241
variegata, Ficus 92
variegatus, Heliacus (Heliacus) 174
VASIDAE 143
vaticinator, Mitra 149
velum, Hydatina 180

VENERIDAE 265
veneriformis, Donax 258
VENERINAE 265
VENEROIDEA 264
venosa, Niso 112
venosa, Rapana 124
ventricosa, Arca 206
ventricosa, Harpa 143
venusta, Eulimella 177
verconis [cf], Seilarex 105 (not illus.)
verecundum, Argyropeza 51 (not illus.)
VERMETIDAE 58
vermiculata, Monodonta 33
vernalis, Tellina 252
vertebrata, Tomopleura 168
vespaceum, Cymatium (Turritriton) 100
vestiarium, Umbonium 36
vexillum, Atrina 224
vexillum sumatrensis, Conus 165
vexillum, Vexilla 124
viali, Limatulella 225
victorialis, "Lucina" 235
violascens, Asaphis 259
virgata, Alaba 55
virgo, Conus 165
vitellus, Cypraea 81
vitellus, Natica 87
vitrea [cf], Haminoea 180
vittulata, Turritella 58 (not illus.)
VITRINELLIDAE 38
VOLUTIDAE 141
voorhoevei, Eomiltha 236
vulgaris, Lima 225
vulsella, Vulsella 222

wagneri, Cantharus 128
wallaceae, Pharaonella 254
walshi, Crepidula 69
watsoni, Amygdalum 216
weaveri pseudogracilis, Phenacovolva (Pellasimnia) 82
weinkauffi, Gari 260
wilkinsi, Strioterebrum (Partecosta) 171
winckworthii, Ischnochiton 189 (not illus.)
woodwardi, Acanthochitona 191 (not illus.)

xanthias [cf], Pelycidion 49
xenicima, Amaea 107
XENOPHORIDAE 70
XENOPHOROIDEA 70
xuthedra, Ocinebrina 121

yemenensis, Pygmaepterys 120
yerburyi, Ischnochiton 189
yerburyi, Pitar 270
yerburyi, Strombus 61
YOLDIIDAE 204

zalosa, Semele 261
zanclaeus meridionalis, Pseudomalaxis 174 (not illus.)
zea, Engina 128
zebra, Pterelectroma 220
zeylanicus, Conus 166
ziczac, Cypraea 81
zmitampis, Mathilda 175
zonalis, Engina 128 (not illus.)
zonata, Hydatina 180
zonata, Tonna 91
ZONULISPIRINAE 169

Acknowledgements

Our debt, as the length of the following list makes abundantly clear, is a large one, extending to persons and institutions from all over the globe. They lent us shells for study and illustration, gave specialist advice upon many puzzling shells, helped us collect and sort shells, gave hospitality, often accommodated us, and showed encouragement, kindness and understanding. Without their help, often given at short notice and always unstintingly, this book could never have been completed. We hope they consider it all justified by the result and apologise for the brevity of our acknowledgement to them; they deserve a book to themselves. To all we extend our heartfelt thanks.

We are particularly indebted to the following people (alphabetically listed) for their outstanding contributions to this book:

Mr John Baxter, for providing virtually the entire text of the section on chitons.

Commander Steve Bennett, Hydrographer with the Royal Navy of Oman, for helping to organise an expedition to the Kuria Muria Islands.

Mr J Christiaens, for identifications of patelliform shells.

Mrs Una Dance, for photography and moral support.

Mr Martyn Day, collector and diver who has spent long periods of time at Musandam and Masirah Island, for supplying us with rare and unusual specimens.

Mr Henk Dekker, knowledgeable student of the Mollusca, for much help with identifications, especially of patelliform shells, and for extensive criticism of proof sheets.

Mr R Michael Dixon, for identifications of naticid species.

Ms Mathilde Duffy, for original drawings showing shell features.

Mr Michael Gallagher, Senior Adviser to the Oman Natural History Museum, whose help and cooperation has made it possible to utilise the material in the Museum.

Mr Steve Green, of Bahrain, an enthusiastic collector who has helped us to establish the range of many species and supplied specimens for study.

Mrs Christina Hagström, a long-term resident of Oman who accompanied us on some expeditions and has shared her magnificent collection with us.

Mr Harry Henseler, a dedicated collector who lived for many years on Masirah Island and has shared with us many species from that island.

Mr Horst Kauch, of Dubai, a diver and collector, who very generously made his extensive collection available to us. Thanks to him we have been able to include numerous species from the United Arab Emirates.

Mr and Mrs Bill Larkworthy, providers of gracious hospitality and shell specimens from the United Arab Emirates.

Mr Phil Palmer, for identifying all our scaphopods.

Mr Charles W Pettitt, Keeper of Invertebrates at the Manchester Museum, England, for the Index.

Mr Khamees Salih, an Omani fisherman whose collecting skills have made it possible to include species which would otherwise not have been found.

Mr Hans Turner, for identifying many mitrid species.

All of the team at Motivate Publishing who were involved with the design and production of this book.

We are also indebted to the following (alphabetically listed) for their various contributions:

Georgina Armour; U Aubry (Terebridae); Eveline Baker; Rüdiger Bieler (Architectonicidae); David and Paul Bosch and families (shells from Oman and Saudi Arabia and notes on eastern Arabian geology); Eloise Bosch (encouragement and companionship); Elizabeth Bostock; British Eastern Relay Station Staff on Masirah Island (hospitality); Bill and Judy Charter; Bunnie Cook; Robert Dance (computer assistance); J de Visser (microshells, Arabian Gulf); Peter Dow (shells from Khor Fakkan); Kathie and Hilary Fry; Neil Hart; J Hoenselaar (sorting sediment samples for microshells, Arabian Gulf); Roland Houart (Muricoidea); Bill Howard; Lynne Hubers; Lorraine Johansen (shells from Saudi Arabia); Barrie Knight; H Kool (Nassariidae); R Kurkjian; Carolyn Lehmann (shells from the Emirates and hospitality); Alma Macdonald; Eli Morrison; Mubarek Nasser; Roy and Mavis Owen; Laura Romans; Richard Salisbury; A Samale (University of Teheran); Richard Skinner; Jan and Peter Smythe (access to Kathie Smythe's material and hospitality); E Staal (Epitoniidae); H Strack; Martti and Pirko Tuhkanen; T Keukelaar-van den Berge (sorting sediment samples for microshells, Arabian Gulf); WM van der Hijden; Peter L van Pel (Cypraeidae and collecting expertise); A Verhecken (Cancellariidae); Ann Waring.

Museum participation, alphabetically listed.

The American Museum of Natural History, New York: Walter Sage. The Manchester Museum, England: Charles W. Pettitt. The Natural History Museum, London: Joan Mulrooney, David Reid, John D Taylor and Kathie Way. The Natal Museum, South Africa: R Kilburn, D Herbert. The National Museum of Natural History, Washington DC: Alan R Kabat (Naticidae). The National Museum of Wales, Cardiff: Alison Trew. The New Zealand Geological Survey: Alan Beu. The Royal Museums of Scotland, Edinburgh: David Heppell (zoological nomenclature). The Zoological Museum of Amsterdam: AN van der Bijl, R de Bruyne.